Tourism and Socio-Economic Transformation of Rural Areas

W0113616

This book aims to fill a gap in the current literature by tracing the rural transformation process and the development of rural tourism functions in Poland over the last 30 years.

It examines the socioeconomic transformation between 1989 and 2019 that resulted in the formation and development of Polish rural tourism and the various practices associated with it. This timely topic is addressed in a central and eastern European context and sparks interest in further in-depth analysis due the diversity and magnitude of the transformation processes undertaken by the Polish rural areas. Since Polish rural areas constitute as much as 30% of the total rural areas in all new European Union member states, this book adds value through an in-depth statistical analysis of the pace of socioeconomic changes in Polish rural areas. It delves into the creation and consumption of tourism services locally, as well as the impact of global trends on the development of rural tourism in Poland.

This book will be of interest to economists, sociologists, political scientists and postgraduate students across eastern and central Europe who deal with rural tourism issues.

Joanna Kosmaczewska is an Associate Professor at the Poznań University of Life Sciences, Poland. Her basic interests in research encompass tourism economics on rural areas and tourism behaviour. She is author of about 70 publications and a number of expert assessments and projects in the field of tourism.

Walenty Poczta is Full Professor, Dean of the Faculty of Economics at the Poznań University of Life Sciences. He is an author of over 400 publications, a member of numerous scientific councils and an expert of government institutions.

Perspectives on Rural Policy and Planning

Series Editors: Andrew Gilg and Mark Lapping

This well-established series offers a forum for the discussion and debate of the often-conflicting needs of rural communities and how best they might be served. Offering a range of high-quality research monographs and edited volumes, the titles in the series explore topics directly related to planning strategy and the implementation of policy in the countryside. Global in scope, contributions include theoretical treatments as well as empirical studies from around the world and tackle issues such as rural development, agriculture, governance, age and gender.

For more information about this series, please visit: www.routledge.com/Perspectives-on-Rural-Policy-and-Planning/book-series/ASHSER-1035

Tourism and Socio-Economic Transformation of Rural Areas

Evidence from Poland

Edited by
Joanna Kosmaczewska and Walenty Poczta

 Routledge
Taylor & Francis Group

LONDON AND NEW YORK

First published 2021
by Routledge
2 Park Square, Milton Park, Abingdon, Oxon OX14 4RN

and by Routledge
52 Vanderbilt Avenue, New York, NY 10017

Routledge is an imprint of the Taylor & Francis Group, an informa business

British Library Cataloguing-in-Publication Data
A catalogue record for this book is available from the British Library

Library of Congress Cataloging-in-Publication Data
A catalog record has been requested for this book

ISBN: 978-0-367-63887-0 (hbk)
ISBN: 978-1-003-12117-6 (ebk)

Typeset in Times New Roman
by Taylor & Francis Books

Contents

Figures

Maps

Tables

Contributors

Agnieszka Baer-Nawrocka, Associate Professor in the Department of Economics and Economic Policy in Agribusiness, Faculty of Economics, Poznań University of Life Sciences (Poland). Author of about 80 publications and a number of expert assessments and projects in the fields of agriculture and food economics, European integration economics, agricultural policy, European and world agriculture, and regional development.

Jakub Hadyński, Assistant Professor in the Department of Economics and Economic Policy in Agribusiness, Faculty of Economics, Poznań University of Life Sciences (Poland). He deals with topics related to local and regional development and EU cohesion policy.

Joanna Kosmaczewska, Associate Professor in the Department of Economics and Economic Policy in Agribusiness, Faculty of Economics, Poznań University of Life Sciences (Poland). The basic interests of the researcher encompass tourism economics on rural areas and tourism behaviour.

Walenty Poczta, Full Professor in the Department of Economics and Economic Policy in Agribusiness, Faculty of Economics, Poznań University of Life Sciences (Poland). Author of about 450 publications in the field of agriculture and food economics, European integration economics, economic and policy, agricultural policy, international agricultural trade, and financing the development of the agri-food sector and rural areas.

Lucyna Przezbórska-Skobiej, Associate Professor in the Department of Economics and Economic Policy in Agribusiness, Faculty of Economics, Poznań University of Life Sciences (Poland). She deals with topics related to agritourism, rural tourism, and multifunctional development of rural areas.

Monika Wojcieszak-Zbierska, Assistant Professor in the Department of Economics and Economic Policy in Agribusiness, Faculty of Economics, Poznań University of Life Sciences (Poland). She deals with topics related to agritourism, rural tourism, nature tourism, and social economics (with particular emphasis on the role and importance of Polish care farms and female entrepreneurship in rural areas).

1 Rural areas as a place for non-agricultural economic activity in a Central and Eastern European context

Jakub Hadyński

This chapter presents the most important definitions of rural areas developed based on various systems in Poland and in other EU countries. Diversification of rural areas has been presented in relation to the administrative structure of a given country, including the historical background. This part of the book covers deruralization as well as urbanization, migration and multifunctional development in rural areas, as phenomena referred to the contemporary concepts for development of rural and suburban areas, as well as cities and urbanized areas.

1.1 Definition, characteristics, and diversity of rural areas

Rural areas play a crucial role in the food production system while also being of importance to other parts of the economy, social development, and preservation of the natural environment. As a consequence, rural issues are of interest to the representatives of many scientific disciplines, such as economics, sociology, agricultural sciences, anthropology, ethnography, geography, and more. The relevant scientists differ in how they define the subject of their research. Also, rural areas are defined in a variety of ways by rural institutions active at local, regional, national, and international level. The broad scope of activity of rural development institutions and of representatives of different scientific disciplines is the reason why no homogeneous definition of rural areas has yet been proposed despite numerous discussions.

Moreover, no single definition exists for rural areas that would be applicable across European Union member states. This is because rural areas strongly differ between countries in their social, economic, and environmental aspects. At the same time, they undergo rapid processes involved in the transformation of both economic and social structures, and experience the related human migration which often results in administrative changes driven, without limitation, by urbanization, counterurbanization, and suburbanization. Some areas are no longer considered rural, while others adopt a rural status. The above is exacerbated by the fact that the rural category is formed by different territories in each European Union country. Therefore, rural areas covered by research can considerably differ in size or population.

It can be therefore assumed that the definition of rural areas is context-sensitive and depends on the type of analysis, policy or management practice. As a consequence, rural areas have a meaning that varies depending on the delimitation criteria (Czapiewski, 2010). Note also that rural areas are viewed in a dual light; sometimes, they are defined "as a settlement unit (village, rural estate, town) or space (territory, area)" (Wieliczko, 2006; Czarnecki, 2005).

Attempts to define rural areas at an international level include those made by the Organization for Economic Co-operation and Development (OECD). In 1994, they developed a terminology which identified rural areas at a local level depending on population density.

According to the OECD, rural areas have a population density below 150 people per km^2 at the basic administrative unit level (NUTS 5). In turn, at regional and sub-regional levels (larger administrative units; district groups, in the Polish context), the OECD determines how rural an area is based on a criterion related to the share of the rural population (Poczta & Wysocki, 2000). This method allows us to identify three types of regions (Rural Developments, 1997):

- predominantly (definitely) rural areas, where the rural population accounts for more than 50%;
- intermediate areas (with a mostly rural character), where the share of the rural population varies in the range of 15% to 50%;
- predominantly (definitely) urban areas, where the rural population accounts for less than 15%.

Moreover, in OECD countries, three categories of rural areas were identified based on how much they are integrated with the national economy (cf., e.g. Kłodziński & Wilkin, 1998):

- economically integrated areas which experience demographic and economic development and are located next to urban centers (they account for ca. 20% of rural areas);
- intermediate rural areas which experience sluggish economic development and demographic stagnation, and are dominated by an agricultural economy (ca. 60%);
- remote rural areas which are sparsely populated and located away from urban centers (they account for up to 20% of rural areas).

Currently, the OECD uses a broader typology of rural areas (as shown in Table 1.1). The delimitation criteria used are the population density at local level, the share of the rural population in the region concerned, and the population of neighboring cities.

In addition to the existing typology of rural areas, the European Commission's Directorate-General for Regional Policy also takes account of a classification based on the accessibility of the nearest city. This involves calculating the time needed for no less than half of an area's population to travel to the nearest city

Table 1.1 Classification of rural areas as per the original OECD typology

Delimitation criterion	Regional division	
Local level, by population density	Rural area	Population density < 150 persons/km^2
	Urban area	Population density > 150 persons/km^2
Regional level: share of the rural population in a territory	Predominantly rural	> 50% of the region's population live in rural areas
	Intermediate	15%–50% of the region's population live in rural areas
	Predominantly urban	< 15% of the region's population live in rural areas
Regional level; additional division by city population	A predominantly rural area which becomes an intermediate area	Includes a city with a population over 200,000 which accounts for no less than 25% of the region's population
	An intermediate area which becomes a predominantly urban area	Includes a city with a population over 500,000 which accounts for no less than 25% of the region's population

Source: Own elaboration based on Eurostat, https://ec.europa.eu/eurostat/statistics-explained/index.php?title=Glossary:Rural_area (2020); NSPLiM (2015).

with a population of no less than 50,000. This allowed us to identify five types of regions (Table 1.2), i.e. predominantly urban areas; intermediate areas (located close to or away from a city); and predominantly rural areas (located close to or away from a city).

The OECD methodology enables simple comparisons between different types of areas. However, the outcomes often fail to reflect the nature of rural areas, especially in densely populated regions (cf. Zwoliński, 2009).

Table 1.2 Extended typology of regions of the Directorate-General for Regional Policy

Region type	Description of the region
Predominantly urban	Less than 15% of the region's population live in rural areas
Intermediate area located close to a city	An intermediate area with a traveling time of up to 60 minutes
Intermediate area located away from a city	An intermediate area with a traveling time over 60 minutes
Predominantly rural area located close to a city	A predominantly rural area with a traveling time of up to 60 minutes
Predominantly rural area located away from a city	A predominantly rural area with a traveling time over 60 minutes

Source: Own study based on NSPLiM (2015); Eurostat, https://ec.europa.eu/eurostat/statistics-explained/index.php?title=Glossary:Rural_area (2020).

Rural areas were also defined in documents of the Parliamentary Assembly of the Council of Europe on April 23, 1996 (in the European Charter for Rural Areas).[1] Accordingly, rural areas mean areas located inside an administrative unit, or coastal areas, which can also include villages and small towns. As defined in the Charter, rural areas include territories where most land is used for (Heller, 2000):

- agriculture, forestry, pisciculture, and fisheries;
- economic and cultural activity of the local population;
- leisure outside urban areas;
- other, such as rural residential housing.

In Central and Eastern European (CEE) countries, rural areas are delimited based on diverse criteria, mainly including population density indicators or administrative decisions which attribute an area to a specific category. For instance, the Polish Central Statistical Office delimits rural areas based on an official national register which includes the country's territorial division, and defines them as "areas located outside the administrative limits of cities." This means rural areas are the territories of rural and urban-rural communes (without cities) because both of them have some characteristics of remote areas. That definition is of a very general nature, does not take account of rural heterogeneity, and therefore provides only an outline of what it refers to, presenting it as a mere opposition to urban areas (cf. Kapusta, 2005). In Poland, communes were established under the Act of March 8, 1990. In 1999, a three-level territorial division was put in place, with the local government model spanning over communes, districts and voivodeships. Communes were divided into rural, urban–rural and urban units).

A definition of rural areas can also be found in the provisions of different national-level strategic documents, e.g. in the 2020 Polish Medium-term National Development Strategy (ŚSRK, 2020) and the 2010–2020 Regional Development Strategy. Regions, cities, rural areas (KSRR 2010–2020, 2010) and rural territories are defined depending on how far they are located from an urban center, which results in the identification of three types:

- urbanized rural areas;
- rural areas within the sphere of influence of the biggest cities;
- remote urban areas.

The Concept of National Land Development 2030 (KPZK 2030, 2011) includes references to:

- functional rural areas involved in development processes;
- rural areas which require support for their development processes.

The "Strategy for the sustainable development of rural areas, agriculture and fisheries by 2020" (SZRWRiR, 2020) defines rural areas based on the sphere of influence of the biggest cities and on the links between cities and countryside. This allowed us to identify the categories of rural areas strictly related to a big city, rural areas located within the sphere of influence of a big city, and remote rural areas.

Rural areas are an important part of socioeconomic diversity of CEE countries. Making the right use of the development potential of rural areas and their residents, and meeting the challenges they face are currently the basic goals of the rural development policy. However, it is difficult to attain them because of the diverse definitions of rural areas which result in the emergence of different, often inconsistent, rural development measures. The relevant literature (including Clapson, 2003, Miller & Luloff, 1981, Wilkin, 2005, 2007, Zawalińska, 2009) provides many interesting terms and definitions which see rural areas from different perspectives. In summary, the most frequently used definitions present rural areas as territories characterized by (Bański, 2012):

- a relatively low population density;
- a specific, open landscape;
- an extensive use of land (mainly in agriculture and forestry);
- scattered buildings and a low-density settlement pattern;
- a dominant share of people related to agriculture and forestry;
- a traditional lifestyle of the population and respect for customs and traditions;
- most people being convinced they live in rural areas.

The difficulties in defining rural areas are related to fact that they have for centuries been viewed as a pool of resources used in agriculture. Hence, rural issues were initially considered to be agricultural issues, and that was the perspective adopted in describing the socioeconomic situation of these territories. Social issues were seen in a context of how the rural population live and carry out their agricultural production activities. However, as the society and economy progressed, rural areas started to be viewed as a place for doing business not only for farmers but also for the part of the society who previously had nothing to do with agriculture. Because of modernization and technological progress, a part of the labor force was released from the agricultural sector; these people either stayed in rural areas or found urban employment. The development of services brought diversity to rural areas and resulted in the emergence of a society whose lives, jobs, and human activities were not closely related to agricultural production. Rural mobilization efforts were based on the creation of non-agricultural jobs (Kobiałka, 2003). As a consequence, several concepts of local development were created (Siemiński, 1994), multipurpose development being among the most important ones. A multipurpose nature was identified and defined as a development of non-agricultural activities and seeking new rural jobs. Thus, rural areas became a place for doing business for a large part of the society. These people were increasingly aware of being a separate community while having a need to maintain a high quality of life comparable to what is enjoyed by urban dwellers.

As the integration into the EU progressed, and as the Community policy changed the perception of rural areas, CEE countries increasingly focused their efforts on addressing the problems of the rural population. Attention started to be paid to their particularities, including socio-cultural aspects as well as moral and ethical issues which considerably affect the nature and functioning of rural areas (Rosner & Stanny, 2014). Also, the rural community was becoming increasingly larger due to suburbanization and migration processes making people move out of cities to nearby or remote locations. This, in turn, was driven by factors which include the society becoming wealthier, the availability of transport facilities, the declining quality of urban life, and the growing environmental awareness related to a commitment to create conditions for living in a sound natural environment.

In Western European countries, these processes started in the mid-1900s and were reflected by attempts to standardize the legal regulations. For instance, in Germany, the amendment to the 1953 Act on agricultural land reparcelling[2] set out new development targets for rural areas (by reinforcing a practice that existed in Baden-Württemberg). In the late 1970s, federal authorities launched a program to support development investments whose assumptions included reviving the rural economy. In the 1980s, European Union member countries started to change the way they looked at rural areas; in addition to agriculture, other important drivers of rural development were identified (Wilczyński, 2003).

According to Zioło (2014), the description of rural areas gradually shifted from an approach focused on agricultural activities to one based on population density. Efforts were made to analyze the situation of rural areas otherwise than in relation to the agri-food economy, by focusing on non-farming activities, industrialization, location of labor and production resources or raw material flows.

Based on experience gained in Germany and, afterwards, in Austria, dedicated funds[3] started to be established in the mid-1980s to support rural areas and enhance international cooperation[4]. This resulted in the concept of rural change becoming widespread across Europe, and provided a stimulus for designing an international cooperation formula which promotes rural development (Wilczyński, 2003). Ultimately, the rural development concept was included in the European Union's programming and support system.

The socioeconomic system of CEE countries was formed in the period from the end of World War II to the late 1980s under the political influence of the Soviet Union. After the economic shift which started in 1989, that part of Europe gradually moved to a free market economy. Seen from this perspective, the problems of rural areas can be considered through the prism of different objectives, targets, and scopes of changes. Four fundamental phases can be identified (cf. e.g. Pouliquen, 2011):

I   from the centrally planned economy to the economic shift in 1989;
II   transition, system changes: 1990–2000;
III   integration into and accession to the European Union: 2001–2004/2006;
IV   common EU policy: from 2004/2006 to this day.

During phase I (with a centrally planned economy), CEE countries paid little attention to rural development. Research on rural development was mainly equated with the development of agricultural production. However, note that in Poland, for instance, the Institute of Agricultural and Food Economics – the National Research Institute was established in 1971 to investigate the situation of rural areas. In the late 1980s, the emergence of the rural development concept and the shift away from agriculture as the sole dominant function of rural areas was a response to the difficulties of what the economic model was at that time. In most countries covered by this analysis, the political change in 1989–90 resulted in a shift from a centrally planned economy to a market economy system. Measures were taken to ensure convertibility and stabilization of the national currency and to liberalize trade. The subsidy system for companies was phased out, and the privatization process started. The reforms began in 1991 in the Baltics, but only in 1996 and 1998 in Bulgaria and Romania; this had an adverse impact on the functioning and structural transformation of rural areas and affected the whole CEE territory.

Although phase II was marked by attempts made to address the rural problems, relevant research efforts were still rare (cf. Table 4.3). Nevertheless, this resulted in certain rural issues being taken into account in government documents (e.g., in Poland, the Strategy for Poland and the Assumptions for the socioeconomic policy for rural areas, agriculture and food economy until 2000). Some other documents were also drawn up in the 1990s which laid down the goals and assumptions of the policy for rural and agricultural development in the years to come. CEE countries were mostly focused on problems facing the food economy because the economic transformation resulted (to a varying degree) in high inflation rates and in a reduction in incomes, consumption levels and GDP. Combined with the discontinuation of subsidies for operators active in the food sector, the decline in prices of agricultural products caused a real deterioration in the socioeconomic conditions of rural areas. At the same time, intensive socioeconomic processes caused a strong heterogeneity in their functional structure.

In phase III, rural development was driven by European Union integration processes and the availability of pre-accession funds. Agriculture experienced an upturn fueled by persistent economic growth and by the preparations for becoming Community members. The first hands-on experience gained at that time (including in using PHARE and other pre-accession schemes, in particular the Special Accession Program for Agriculture and Rural Development, SAPARD) allowed CEE countries, once they had joined the European Union, to be covered by the instruments of the Common Agricultural Policy (CAP) and of the structural policy which included measures designed to promote rural development. During 2000–2006, a total of EUR 3.7 billion was allocated to

SAPARD implementation. The development of the rural infrastructure (roads, pipelines, and sewerage systems) and economic diversification were of particular importance among the proposed scope of rural development measures. Also, rural development was supported through the renewal of villages and their cultural heritage.

Phase IV was the inclusion of CEE countries in certain areas of the European Union's common policy system. Following the accession to the EU, the annual amount of EU support allocated to rural development was ten times that offered under SAPARD. That period saw rapid economic growth (until 2008), including in Romania and Bulgaria. In a context of growing emigration, some countries (i.e. Poland and Romania) experienced a significant decline in unemployment rates (until 2007). Indeed, in the ten new member states, unemployment rates recorded in 2007 were close to those seen in EU-15 countries. Furthermore, when CEE countries joined the Community, measures were taken for the shaping of rural areas.

In CEE countries that are EU members, the differences in the situation of rural areas are driven by the following factors (in addition to the spatial diversity of economic structures which can be explained by historical events): location relative to industrial hubs, to markets for agricultural produce, and to non-agricultural labor markets; and the quality of access routes which facilitate the development of non-agricultural economic functions in a regional and international context. Note that while CEE rural areas hold considerable labor resources, their quality (measured as education levels, for instance) is lower than in big cities. Moreover, rural areas in CEE countries also differ in their settlement patterns.

It needs to be emphasized that CEE countries are heterogeneous in socio-economic and geographic terms; as a consequence, their rural areas differ in how fast they embrace non-agricultural economic activities. In most of these countries, the share of the rural population is above the EU-27 average level (Table 1.3). In CEE countries, 37.5% of the population live in rural areas, compared to 29.2% in the European Union as a whole. In 2018, Lithuania had the highest share of rural residents (54.3%) of all CEE countries. High shares could also be observed in Slovenia and Romania (each with 45.7% of rural population) and in Slovakia (42.7%) and Poland (41.1%) (Eurostat, 2020a).

Based on changes in the number of rural residents in CEE countries between 2010 and 2018, it can be concluded that the greatest rural exodus (a decline by more than ten percentage points) was experienced in Estonia, Bulgaria, Romania, Hungary, and Latvia. Only two countries, Slovenia and Slovakia, recorded growth in the rural population, which was consistent with the trend that prevailed in the European Union (UE-27 as at 2020) in the study period. A drop by over ten percentage points was witnessed in Poland.

When looking at rural areas from a regional perspective, in 2016, the largest pool of predominantly rural areas (as per the OECD classification) was recorded in the largest countries, i.e. Poland (166,883 km^2) and Romania (161,667 km^2). In turn, of all CEE countries, the highest rural population density was reported in Estonia, Lithuania, Romania, and Slovenia (Eurostat, 2020a).

Table 1.3 Share of areas grouped by urbanization degree in the EU and in CEE countries, 2010–2018

Area	Rural areas					Intermediate areas	Urban areas
	2010	2012	2014	2016	2018	2018	2018
EU-27 (2020)	27.7	31.0	29.7	30.2	29.2	31.6	39.2
Bulgaria	49.6	34.6	34.2	33.0	31.9	22.8	45.3
Croatia	47.4	51.2	46.9	42.8	38.4	32.3	29.2
Czech Republic	39.9	38.8	37.3	37.4	36.0	34.0	30.0
Estonia	51.0	41.1	41.2	40.7	31.8	50.2	18.0
Hungary	48.2	37.1	39.6	46.9	33.0	34.2	32.8
Latvia	51.1	46.8	46.9	37.6	37.2	19.4	43.4
Lithuania	58.2	47.3	46.7	54.8	54.3	2.3	43.5
Poland	45.3	41.6	42.3	42.6	41.1	24.4	34.4
Romania	62.5	43.5	49.5	45.2	45.7	25.4	28.9
Slovakia	41.1	44.4	41.3	41.5	42.7	36.5	20.8
Slovenia	44.1	43.9	44.8	48.1	45.7	35.1	19.3

Source: Eurostat, last update on February 24, 2020.

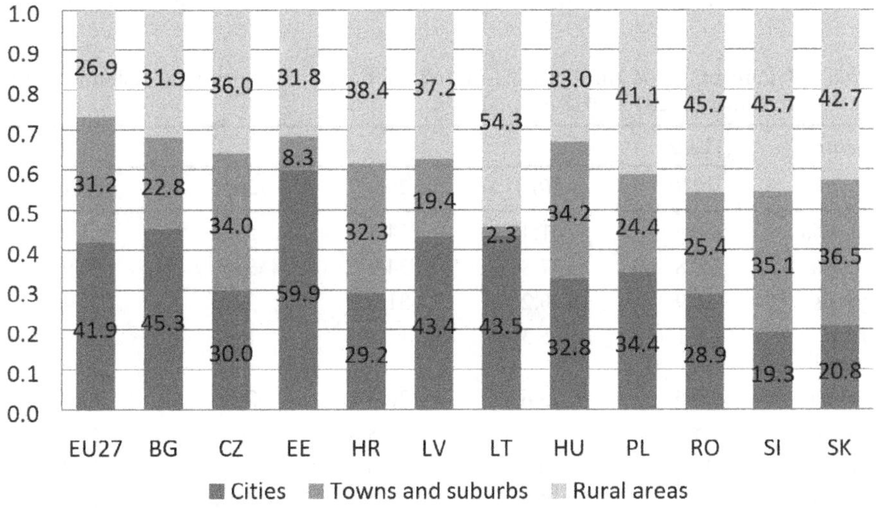

Figure 1.1 CEE population grouped by place of residence.
Source: Eurostat, February 24, 2020.

In CEE countries, the concept of rural development was formed under the influence of systemic, social, and economic changes, while also being the consequence of noticeable migration processes which, in addition to creating new jobs and alleviating unemployment, also had some negative impacts (uneven industrial and urban development; an unfavorable demographic structure; and socioeconomic degradation of rural areas). Most of these areas face disorders which are typical of remote areas located away from economic centers. Usually, rural development in the largest CEE countries (mostly in Poland and Romania) struggles with the following problems:

- small capital resources and the inability to procure investment capital;
- having a single function related to an excessive labor force;
- registered and hidden unemployment;
- an agrarian structure which requires large labor inputs;
- an obsolete socio-occupational structure with farmers and non-agricultural blue-collar workers as the largest group;
- an underdeveloped business and institutional infrastructure;
- lack of traditions and skills related to starting and running an own business;
- an unfavorable demographic situation (primarily including low education levels) which affects rural areas and poses a serious barrier to development.

This situation is reflected by the rate of people at risk of poverty and social exclusion which is covered by the European Union's Europe 2020 strategy (Table 1.4). In 2018, of all CEE countries covered by this study, the largest number of rural people at risk of poverty and social exclusion was recorded in Poland (over 3.8 million), followed by Bulgaria (over 1 million). In Poland

Table 1.4 Percentage of rural people at risk of poverty or social exclusion in CEE countries, 2010–2018

Country	Years				
	2010	2012	2014	2016	2018
Bulgaria	58.5	61.4	54.4	53.8	47.4
Croatia	38.8	37.8	34.9	33.5	30.9
Cyprus	26.9	30.2	31.0	32.2	26.4
Czech Republic	16.1	15.3	15.2	11.4	11.6
Estonia	24.5	25.6	26.5	26.1	29.5
Hungary	34.7	39.2	37.8	31.3	25.8
Latvia	40.5	40.1	38.6	35.0	32.3
Lithuania	36.9	38.2	32.4	37.6	35.4
Malta	–	–	34.0	2.9	–
Poland	33.9	33.2	31.2	27.9	25.3
Romania	48.8	56.6	51.9	51.7	45.5
Slovakia	24.8	24.8	20.8	21.8	19.1
Slovenia	19.8	20.2	21.6	–	16.6

Source: https://appsso.eurostat.ec.europa.eu/nui/show.do (June 17, 2020).

and Bulgaria, these groups accounted for ca. 30% and over 55% of the rural population, respectively. Conversely, when it comes to Polish cities, that ratio was among the lowest in the European Union. The fastest changes in this respect were recorded in Bulgaria where the number of rural people at risk of poverty and social exclusion dropped by more than 50% between 2010 and 2018. In Poland, the decline was much slower (ca. 33 percentage points over the same period). In the group of countries considered, the slowest changes were witnessed in Lithuania and Estonia. When looking at the percentages of rural population at risk of poverty and social exclusion, it can be concluded that—despite a large drop—the biggest problems were faced in Bulgaria (47.4%) and in the Baltics: Lithuania, Latvia, and Estonia (35.4%, 32.3%, and 29.5%, respectively).

In CEE countries, rural areas continue to face the challenge of diversification of economic functions; this means not only allocating the available rural resources to non-agricultural activities but also developing new businesses, including services, which are the key indicator of development. The functional structure of rural areas of CEE countries is strongly heterogeneous in this aspect, too. The basic economic functions fulfilled by these areas are still primarily related to agriculture, forestry, tourism, and leisure, while residential housing, industry, and services play a less significant role. This is illustrated in Figure 1.2. which presents data on different uses of land in the countries considered.

The above suggests that in most CEE countries, the area of land allocated to agricultural uses is above the EU-27 average level. Moreover, the countries strongly differ one from another; for instance, Estonia has nearly 50% less land allocated for agricultural uses than Hungary; at the same time, the share of forests in total landmass in Estonia and Hungary is nearly 32% and ca. 17%, respectively. In turn, the greatest share of forests in total landmass is recorded in Slovakia (41%). Wetlands and water bodies have the largest share in Estonia (nearly 6%) and Latvia (5%).

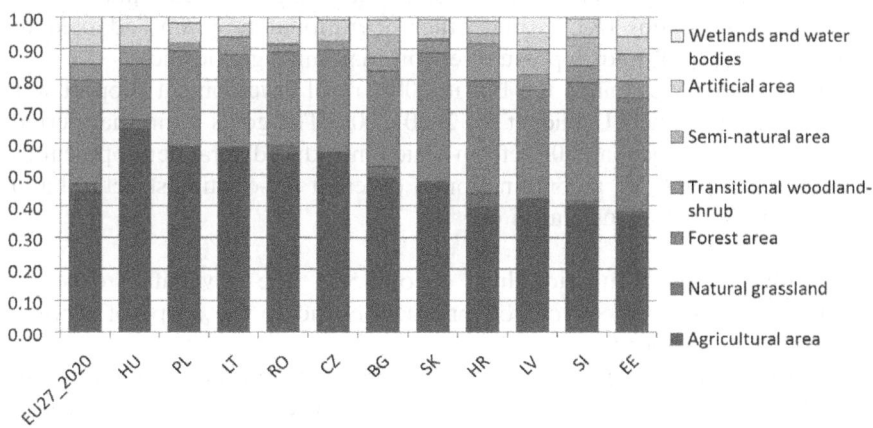

Figure 1.2 Uses of land in Central and Eastern European countries, 2018.
Source: Eurostat, February 24, 2020.

The differences in land use can be explained by differences in functional structures between the countries covered by this study. Each land use listed above can be assigned a function fulfilled by rural areas; note that the basic distribution of functions of rural areas is a continuum that runs between agriculture, industry, and services. This provides a basis for the following classification: areas where agricultural and forestry functions dominate; areas where non-agricultural (industrial, residential, service) functions dominate; and areas where tourism and leisure functions (including tourism based on forestry and agriculture; and tourism and leisure based on non-agricultural functions) dominate (cf. Bański & Stola, 2002). When looking at the function of rural areas from different perspectives, it can be concluded that modern rural areas are those which fulfill other tasks in addition to agriculture. Based on that criterion, considerable differences exist both between and within CEE countries. This can be partly explained by natural and climate conditions and by socioeconomic development at regional level. Other aspects of importance are large urban centers and the development of transport routes inside rural areas. Although Romania is the country with the largest area dominated by agricultural functions, it also offers above-average resources which could be used in developing the service sector. On the other hand, there is growing importance of rural tourism which has considerably affected rural development over the last decade. Hence, rural areas in CEE countries have the potential they need to address the growing demand for tourism services.

1.2 Rural areas in selected strategic documents

Rural development is an issue addressed in a number of strategic documents adopted at different levels of public authority in CEE countries. This includes documents issued by the Community and member states as well as internal ones applicable at regional and local level.

At EU-level, rural development priorities are implemented under Rural Development Programs (RDP) and other operational programs adapted to specific conditions of, and opportunities for, maximizing efficiency.

After CEE countries joined the EU in 2004, rural development support was implemented under the EU budget for 2000–2006. The goals set in that period were related to the Agenda 2000 reform which introduced rural development as the 2nd pillar of the CAP, a combination of different pre-existing structural and territorial funds, with focus placed on:

- agricultural multi-functionality; recognizing the diversified role of agriculture with respect to food production and to the scope of services provided by farmers;
- adopting an integrated multi-sectoral approach to rural economy in order to diversify the activity, create new sources of employment and protect the rural heritage;

- flexible support (which is based on subsidiarity and promotes decentralization) and consultancy at regional, local and partner level;
- transparent program development and management based on a simplified and more accessible legislation.

The 2007–2013 perspective was marked by the evolution of rural development targets. According to initial statements, it was important to implement structural changes in the agricultural sector in an effort to shift to a model where agriculture has a multifaceted impact on rural areas. This was followed by rural empowerment, with emphasis being placed on the need for implementing a common rural development strategy across the entire European Union. Initiated in Europe in the early 2000s, the development of the cohesion concept resulted in the preparation of rural development principles for 2007–2013 and of instruments available to member states and regions (which actually continue to be applicable today).

Pursuant to Council Regulation (EC) No. 1698/2005, the Union's policy for rural development during 2007–2013 was focused on three topics:

- improving the competitiveness of agricultural and forestry sectors;
- enhancing the environment and rural areas;
- improving the quality of rural living and encouraging the diversification of the rural economy.

The goals listed above were in consonance with the objectives of Europe 2020, a follow-up of the development vision outlined in the 2000 Lisbon Strategy. It laid down the following priorities (Europe, 2020a):

- smart growth, i.e. development of an economy based on knowledge and innovation;
- sustainable growth which involves a transformation towards a competitive, low-carbon, resource-efficient economy;
- inclusive growth, i.e. supporting an economy with a high employment rate which ensures economic, social and territorial cohesion.

The challenges and key changes which must be implemented in rural areas are issues that can be considered from the perspective of objectives and indicators set out in the Europe 2020 strategy. The strategy defined the goals followed by decision makers as regards rural development.

In the 2014–2020 budget, the scope of support for rural development measures was set under the interinstitutional agreement between the Council of the European Union, the European Parliament and the European Commission. The common strategic framework for all funds used in financing the sustainability of regional development also addressed the issue of rural development in CEE countries. The integrated relevant measures were outlined in the 3rd (from 2004), 4th (from 2007), 5th (from 2010), and 6th Cohesion Report (from 2014), in the

Community Strategic Guidelines on Cohesion (2006) and in the Green Paper on Territorial Cohesion. The deepest insight into spatial cohesion was presented in the Green Paper which, following a summary of progress in moving towards economic and social cohesion, defined the goal of territorial cohesion: to support harmonious, sustainable development which would achieve a more even and sustainable use of assets, "bringing economic gains from less congestion and reduced pressure on costs, with benefits for both the environment and the quality of life" (Green Paper on Territorial Cohesion, 2008, p. 6). The Green Paper also placed emphasis on building relationships with other regions with a view to ensure a coordinated and sustainable use of shared potentials, and on the role the infrastructure plays in that process.

During the 2014–2020 programming period, the Common Agricultural Policy promoted rural development processes while also stimulating the market and supporting rural incomes. During 2014–2020, three long-term rural development goals were set, including:

- supporting the competitiveness of agriculture;
- ensuring sustainable management of natural resources and climate action;
- achieving sustainable territorial development of rural economies and communities, including creating and maintaining jobs.

The European Agricultural Fund for Rural Development (EAFRD) was the CAP instrument used to support rural development strategies and projects, and a component of European Structural and Investment Funds. Fixed at ca. EUR 100 billion, the EAFRD budget for 2014–2020 was disbursed through the implementation of rural development programs valid until the end of 2023.

Aligned with national or regional conditions, rural development programs are focused on at least four of the six EAFRD priorities (EC, 2020a):

- supporting the transfer of knowledge and innovation in agriculture, forestry, and rural areas;
- enhancing viability and competitiveness of all types of agriculture and promoting innovative agricultural technologies and sustainable forest management;
- promoting food chain organization, animal welfare, and risk management in agriculture;
- promoting resource efficiency and supporting the transition to a low-carbon and climate change resilient economy in agriculture, food, and forestry sectors;
- restoring, preserving, and enhancing ecosystems related to agriculture and forestry;
- promoting social inclusion, poverty reduction, and economic development in rural areas.

As an additional criterion, European Union countries must allocate no less than 30% of EAFRD's financial resources to actions centered around the environment and climate change (this includes subsidies and payments for farmers who use environmentally sound practices). Another assumption is that no less than 5% of funds are allocated to local action groups operating under the LEADER approach. The concept of this activity consists in that local communities manage funds in order to implement local development strategies which address the needs of the local community. Rural development programs may also support smart villages (EC, 2020b).

In the future, the European Union will continue to follow that axis of rural development because in June 2018, the European Commission proposed future rural development targets in relation to the upcoming 2021–27 programming period. The rural development policy was assumed to provide an effective response to changes in the socioeconomic and environmental situation, i.e. climate change and generational renewal. At the same time, importance continues to be attached to assistance for farmers in measures taken to ensure sustainable development and improve the competitiveness of the rural sector.

Rural development support is the 2nd pillar of the Common Agricultural Policy which provides member states with a pool of EU funds allocated to national or regional-level management activities under multiannual co-financed programs. A total of 118 programs were planned for all 28 member states. Figure 1.3 presents the number of rural development plans in the European Union. It suggests that each CEE country implements one RDP, while some Western European countries have several plans in place. For instance, in Germany, rural development issues are governed by 15 RDPs, whereas France has as many as 30 of them (EC, 2020c).

In an effort to pursue the socioeconomic and environmental priorities, rural development support measures for 2014–2020 were based on precisely defined goals. This included setting the levels of specific indicators to be attained. These measures were coordinated and consistent with the commitment to maximize synergies with other European structural and investment funds and with each member state (EC, 2020c).

Rural development programs for 2014–2020 were accepted for implementation in most CEE countries in 2015 (in Poland, acceptance was granted already in December 2014) (Table 1.5). The largest amounts of funds were planned to be allocated to Poland (EUR 13.6 billion), Romania (9.5), and Hungary (4.2) (Table 1.5).

When analyzing rural development programs launched in CEE countries under the 2014–2020 budget, it can be assumed that improving the competitiveness of the agri-food sector was the key rural development priority for most of them (EC, 2020d). This means implementing priority 2 which includes enhancing farm viability, enhancing competitiveness of all types of agriculture, and promoting innovative technologies. It follows from the above that support for CEE rural areas was supposed to be focused on increasing the marketability of agricultural produce while another priority was to enhance the competitiveness of non-agricultural activities. Strong

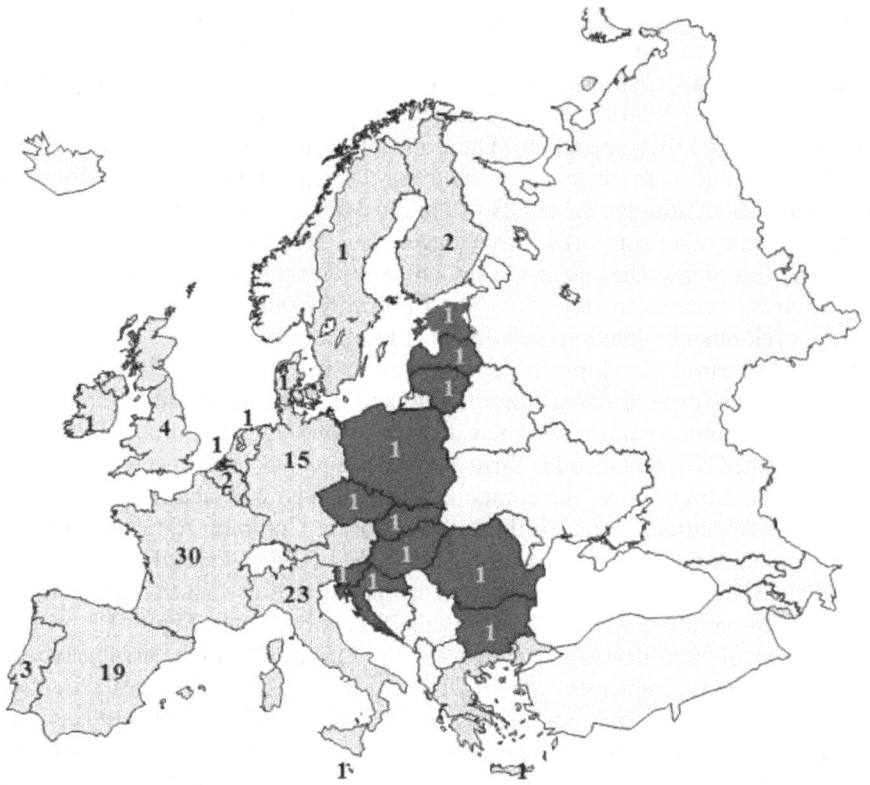

Figure 1.3 Number of rural development programs in European Union countries, 2014–2020.
Source: own study based on https://ec.europa.eu/info/sites/info/files/fod-farming-fish eries/key_policies/documents/number-of-rdp-per-country-2014-20_en.pdf.

emphasis was also placed on developing new technologies designed to improve the competitiveness of local economic operators. Priority 4, i.e. restoring, preserving, and enhancing ecosystems related to agriculture and forestry, was a matter of importance, too. In most CEE countries, these two axes were the development priorities for rural areas. At the same time, measures were taken to preserve and protect the natural environment. This was a crucial line of action, all the more since rural areas in the entire European Union (and especially in CEE countries) were, and continue to be, at risk of degradation caused by adverse natural and environmental impacts driven by historical events, mostly related to infrastructural underdevelopment. When considering this issue, it can be noticed that the biggest problems related to pollution, waste, and other consequences of environmental degradation were experienced in Lithuania, Latvia, Slovenia, and Bulgaria. In this context, note that in Bulgaria, these issues have gained in

Table 1.5 2014–2020 RDPs in CEE countries (by acceptance date and by budget)

Country	2014–2020 RDP			
	Date accepted	Budget (EUR billion)		
		Total	EU funds	National funds
Poland	December 12, 2014	13.6	8.70	4.90
Estonia	February 13, 2015	0.9	0.82	0.17
Slovenia	February 13, 2015	1.1	0.84	0.26
Latvia	February 13, 2015	1.5	1.07	0.48
Lithuania	February 13, 2015	2.1	1.61	0.49
Slovakia	February 13, 2015	2.1	1.56	0.54
Croatia	May 22, 2015	2.3	2.03	0.30
Bulgaria	May 26, 2015	2.9	2.37	0.53
Czech Republic	May 26, 2015	3.1	2.31	0.77
Romania	May 26, 2015	9.5	8.13	1.34
Hungary	August 10, 2015	4.2	3.43	0.74

Source: Own study based on https://ec.europa.eu/info/fod-farming-fisheries/key-policies/common-agricultural-policy/financing-cap/cap-funds_en#eafrd

importance over the last decade, which suggests an increase in pollution levels (Eurostat, 2020a). A similar situation exists in Poland, which has seen its environmental problems exacerbate over the last ten years. Conversely, in the group of CEE countries, Czech Republic and Croatia have demonstrated the best performance in eliminating the adverse consequences of pollution and environmental issues, and have achieved considerable progress in this respect (Eurostat, 2020a). In addition to environmental challenges, the form of RDPs can differ, mainly depending on internal resources and problems affecting CEE countries.

Below is a list of key priorities pursued in CEE countries under the European Union's budget implemented until the end of 2020. In Hungary, measure 4 (investments in physical assets) was the priority of the RDP in terms of financing amounts (total public financing); a total of EUR 1.38 billion was allocated to it. EUR 757 million was allocated to measure 10 (agri-environmental and climate measure). In terms of budget amounts (total public financing), the Slovakian RDP was focused on measure 4: production investments (EUR 569 million). In turn, EUR 459 million was allocated to measure 13: payments for areas with natural handicaps and under other natural constraints. Bulgaria allocated the largest amounts from the RDP budget to measure 4: investments in physical assets (EUR 841 million) and measure 7: basic services and village renewal in rural areas (EUR 626 million). In Croatia, priority 2 (enhancing farm viability, enhancing competitiveness of all types of agriculture in all regions, and promoting innovative agricultural technologies and sustainable forest management) was of key importance.

Another crucial measure consisted in supporting the transfer of knowledge and innovation in agriculture, forestry, and rural areas. In Latvia, important measures taken under the RDP were those focused on social integration and local development as well as on promoting the development of small and medium enterprises. The most important part of RDP funds (over EUR 540 billion) was allocated to investments in physical assets. In Romania, the RDP was focused on the following priority axes: promoting competitiveness and restructuring of the large Romanian agricultural sector; environmental protection and climate change; stimulating economic development, creation of jobs, and improvements in the quality of life in Romanian rural areas. In the Czech Republic, the largest parts of the RDP budget were allocated to measure 10 (related to climate change and agri-environmental activities) (EUR 905 million) and measure 13 (related to payments for areas with natural handicaps and under other natural constraints) (EUR 786 million). In Estonia, 29% of RDP funds (EUR 292 million) were allocated to investments in improving the performance of farms; 22% (EUR 220 million) were allocated to support for environmentally sound management practices. As regards total public financing in the Lithuanian RDP, EUR 630 million were invested in measure 4: productive and non-productive investments, and EUR 380 million were allocated to measure 13: agriculture in areas with natural constraints. In Poland, the RDP was focused on all six rural development priorities, with priority 2 (farm viability and competitiveness) being of key importance. The largest amounts of funds under the RDP are allocated to investments in physical assets (measure 4, EUR 2.31 billion), development of farms and economic activity (measure 6, EUR 1.47 billion), payments to areas facing natural constraints (measure 13, EUR 1.22 billion), agri-environmental programs (measure 10, EUR 870 million) and other activities. Support under the RDP was also provided to community-led local development projects. Slovenia allocated the largest part of the RDP budget (EUR 266 million) to measure 13: areas facing natural constraints. An amount of EUR 206 million was dedicated to measure 4: investments; and EUR 204 million were allocated to measure 10: agri-environmental and climate measures (EC, 2020d).

The diversity of the potential behind natural resources and productive inputs sets the development targets for each geographic area, as laid down in strategic documents adjusted to specific conditions of each CEE country.

In addition to the 2nd pillar of the Common Agricultural Policy, the European Union's support measures for CEE rural areas also include the Cohesion Policy.

It is essentially focused on closing the development gaps between regions in order to achieve socioeconomic and territorial cohesion, and is financed with structural and investment funds. The European Fund for Regional Development (EFRR) is a basic fund used in implementing the Cohesion Policy. It mostly supports regions whose development is lagging behind, considering GDP per capita at NUTS 2 level. The EFRR finances the European Territorial Cooperation and is primarily focused on tangible investments. The European Social

Fund (ESF) is mostly used in financing intangible measures related to education, employment, social exclusion and poverty. The Cohesion Fund (CF) invests in projects related to transports and environmental protection.

During 2007–2013, the principles and objectives laid down in the National Strategic Reference Framework provided a basis for the implementation of national policies in function of socioeconomic and environmental conditions. Based on a diagnosis of these conditions, national development strategies were implemented which provided for the use of European funds in investment measures taken in the pursuit of defined objectives, including:

- convergence, i.e. measures taken to close the socioeconomic development gap between regions;
- regional competitiveness and employment in regions not covered by objective 1;
- European Territorial Cooperation focused on implementing partnership projects.

During 2014–2020, Partnership Contracts became the basic document used in the implementation of the European Union's Cohesion Policy, and were intended to pursue two of its objectives, namely:

- investments in economic growth and employment;
- European Territorial Cooperation.

The fact that the Cohesion Policy combined three previously implemented policies, i.e. the Cohesion Policy, the Common Agricultural Policy, and the Common Fisheries Policy, was a novelty in that context. Eleven thematic objectives were set in this area, including:

- strengthening research, technological development, and innovation;
- enhancing access to and use and quality of information and communication technologies;
- enhancing the competitiveness of small and medium-sized enterprises, the agricultural sector (for the EAFRD), and fisheries and aquaculture sector (for the EMFF);
- supporting the shift towards a low-carbon economy in all sectors;
- promoting climate change adaptation, risk prevention, and management;
- protecting the environment and promoting resource efficiency;
- promoting sustainable transport and removing bottlenecks in key network infrastructures;
- promoting employment and supporting labor mobility;
- promoting social inclusion and combating poverty and any discrimination;
- investing in education, training, and lifelong learning;
- enhancing institutional capacity of public institutions and stakeholders and an efficient public administration.

The management of European Union funds was largely decentralized in order to ensure a better compatibility of problems and needs of local communities. This was the basis for defining the areas of strategic intervention under two axes which support:

- rural areas developing based on endogenous potential—support is allocated to enable their sustainable multipurpose development with the use of their endogenous potentials;
- areas at risk of permanent marginalization, mostly including small towns and related rural areas as well as remote areas within voivodeships, usually located in north and east Poland; support was allocated to stimulate entrepreneurship and investments in public services.

In the budget for 2014–20, focus was also placed on community-led local development (pursued based on the LEADER approach), considering the fact that local action groups can access all funds allocated under the Cohesion Policy. At national level, the Cohesion Policy was implemented based on operational programs guided by specialized national or regional-level management institutions.

For instance, in Poland, 12 operational programs were defined which differed in how much they supported rural development issues. Table 1.6. presents the operational programs and their impact on rural development.

Table 1.6 presents a selection of operational programs implemented in Poland during 2007–2013 and 2014–2020, taking into account the measures designed to support rural areas. According to data shown in the table, although rural areas were addressed in most documents, the scope of support differed between them and was aligned with the country's administrative division.

The priorities and policy principles (as provided for in documents which set rural development targets in European Union countries) address the particularities of each area; this requires different priorities and objectives to be defined. Socioeconomic and environmental characteristics provide a basis for determining the type of the area concerned and for defining the budgets. Political cohesion is ensured by the European Commission which approves and monitors the pursuit of goals under aid schemes. In turn, the selection of projects and payments is the responsibility of national or regional-level authorities.

Similar activities are carried out at national level in European Union member states when developing program documents. For instance, in Poland, the allocation of European Union funds to the development of rural areas, agriculture, and fisheries is governed under the Strategy for the sustainable development of rural areas, agriculture, and fisheries for 2012–2020 (SZRWRiR, 2012). It was designed primarily to specify the key development targets for rural areas, agriculture, and fisheries in the 2020 perspective. This was supposed to enable the right targeting of public interventions financed with national and Community funds. The key long-term objectives of rural development measures, as provided for in the Strategy, include enhancing the quality of rural life and making an efficient use of their resources and

Table 1.6 Operational programs in Poland which included rural development, 2007–2013 and 2014–2020

Operational program	References to rural development	Funds allocated to rural areas
Infrastructure and Environment 2007–2013	The program was demonstrated to be complementary only to the 2007–2013 RDP in accordance with the demarcation between the Cohesion Policy and the CAP	Not specified
Infrastructure and Environment 2014–2020	Rural areas identified as the beneficiary of direct and indirect support, especially when it comes to environmental and nature protection	No less than 1.5% directly under "Strategic Intervention in Rural Areas"
Innovative Economy 2007–2013	A reference was made to rural problems involved in: an unfavorable employment structure (a high share of agriculture); the need for supporting agri-food exports and digitization	Not specified
Smart Development 2014–2020	No measures intended exclusively for rural areas. The program assumed that urban development would have a diffusion effect on rural areas	Not specified
Human Capital 2007–2013	Rural areas explicitly indicated as the beneficiary of support under three objectives: economic activation, education, and territorial cohesion	Ca. 19%
Knowledge, Education Development 2014–2020	Specific support was intended for rural areas as part of higher education and youth employment schemes	Ca. 10% (a territorial approach)
Development of Eastern Poland 2007–2013	A reference was made to the direct impact of some interventions on rural areas, especially as regards road, cycling, and information infrastructure (territorial cohesion)	Not specified
Eastern Poland 2014–2020	A mostly indirect impact on rural areas is planned (transport accessibility, support for the agri-food sector)	Ca. 8% (mostly indirectly)
Digital Poland 2014–2020	Interventions in rural areas regarding investments in broadband Internet access and ICT	Ca. 37%
European Territorial Cooperation 2007–2013 (10 programs)	Interventions in rural areas regarding transformation of the economic structure and environmental protection. It was indicated that rural areas often account for most of the territory covered by support	Not specified
European Territorial Cooperation 2014–2020 (10 programs)	Intervention measures in rural areas taken to increase transport accessibility and protect common natural and cultural heritage	14%

(Continued)

Table 1.6 (Cont.)

Operational program	References to rural development	Funds allo-cated to rural areas
RDP of the War-mińsko–Mazurskie voivodeship 2007–2013	Certain rural interventions were planned, including reinforcement of transport links and support for the information society	44%
RDP of the War-mińsko–Mazurskie voivodeship 2014–2020	Interventions in rural areas regarding infra-structural investments and mobilization of the rural population	No less than 11% (esti-mated to be much more)
RDP of the Wielk-opolskie voivodeship 2007–2013	The program was demonstrated to be com-plementary to the 2007–2013 RDP in accor-dance with the demarcation between the Cohesion Policy and the CAP; some output and result indicators were measured at rural level	32%
RDP of the Wielk-opolskie voivodeship 2014–2020	Intervention measures focused on stimulat-ing entrepreneurship, growth of human capi-tal, and infrastructural projects	Ca. 27%
RDP of the Mał-opolskie voivodeship 2007–2013	In rural areas, the territorial dimension of the Cohesion Policy was implemented espe-cially under two priority axes: tourism / cul-tural industry and intra-regional cohesion	35%; plus 10% allocated to mountain areas
RDP of the Mał-opolskie voivodeship 2014–2020	Most measures are implemented in both urban and rural areas; non-urban areas are addition-ally supported under village renewal projects	26%
RDP of the Zachodniopo-morskie voivodeship 2007–2013	Interventions focused on rural areas, mostly related to energy, telecommunications, social and healthcare infrastructure	29%
RDP of the Zachodniopo-morskie voivodeship 2014–2020	The three Strategic Intervention Areas cov-ered by intervention include the Special Inclusion Zone, which mostly consists of rural areas facing social problems and poor transport accessibility	No less than 11%

Source: IRWIR (2019).

potentials, including agriculture and fisheries, in the sustainable development of the country (Ministry of Agriculture and Rural Development, 2020). Cur-rently, the updated Strategy for the sustainable development of rural areas, agriculture, and fisheries 2020 (2030) sets the development pillars and lines of action for Polish public institutions active in promoting rural and agricultural development by 2030.

The implementation of the Strategy was supported with other docu-ments, including the Rural Pact 2020 (Pakt [Pact], 2018), one of the first documents used a basis for the Polish rural policy. It was based on three interlinked pillars:

1 supporting the agriculture and its environment;
2 supporting entrepreneurship and the creation of non-agricultural jobs;
3 supporting a comprehensive social policy for rural areas and agriculture, and a human development environment in rural areas.

The Rural Pact 2020 (2030), a document which programmed rural development in the 2030 perspective, was adopted in 2018. The Pact was assumed to be a document supporting the implementation of the Strategy for the sustainable development of rural areas, agriculture, and fisheries 2020 (2030), and therefore the structure of both documents is based on four development pillars (Pakt [Pact], 2018):

Pillar 1: profitability of agricultural production;
Pillar 2: quality of rural living and environment;
Pillar 3: non-agricultural jobs and an active society;
Pillar 4: an efficient agricultural administration.

In Poland, regional and local development strategies were created at a below-national level (mostly at commune level) and set the development targets for specific administrative units. These measures were initiated by the reforms of the territorial system in 1990 and 1998. The first reform was the adoption of the Commune-level Local Government Act of March 8, 1990 (Journal of Laws [Dz. U.] of 1990, No. 16) which triggered local development at commune level. In turn, the District-level Local Government Act (Journal of Laws [Dz. U.] of 1998, No. 91, Item 578) and the Voivodeship-level Local Government Act of June 5, 1998 were adopted in 1998. With the reforms, Poland finally introduced a three-level local government structure, one which is common in Western European countries. Local government units became self-reliant. No hierarchical subordination exists between local government levels; their competencies are defined in accordance with the principle of subsidiarity, under the assumption that they undertake complementary action in addressing the needs of local and regional communities (Domagała, 2010). This administrative division enables rural development to be driven based on regional and local-level development concepts and plans.

1.3 Key changes affecting rural regional and national transformation

Over the last three decades, CEE countries have witnessed a number of socioeconomic transformations related to the political shift, their aspirations of becoming a member of the European Union, and their participation in its structures. The economic changes have had a positive effect on the standards of living for a vast majority of the CEE population. At the same time, they have also resulted in increasing the differences between territories and social groups in their ability to seize the opportunities provided by new development conditions. During that period, rural areas were becoming less and less dependent on the agricultural sector, the consequences of which include the growing number of landless families. Multipurpose rural development became a matter of importance; rural areas experienced rapid changes in their demographic situation, resulting in a permanent restructuring of the population. This was accompanied by urbanization and suburbanization processes related to migrations (including emigration and immigration), and by a decline in mortality and birth rates and an extension of life expectancy.

The consequences of migrations were of particular importance to CEE countries. Generally, due to their particularities and diversity, migration processes can be divided by different criteria, including into external and internal emigration. The former means people moving between countries. The Polish Central Statistical Office defines external emigration as traveling abroad or coming to Poland either in order to become a permanent resident or for a temporary stay (GUS, 2019a). People who decided to leave their country are referred to as emigrants. In turn, immigration means an influx of people who stay for some time in a territory (Kawczyńska-Butrym, 2009).

Large-scale migration affects both the social and economic structure of the population. This can have a temporary destabilizing effect on local labor markets. Furthermore, if intense and accompanied by negative natural growth, emigration leads to a considerable reduction in the country's population. Table 1.7 presents the population of CEE countries in 2019 and population change between 2000 and 2019 and between 2013 and 2019.

Table 1.7 Population and its changes as of January 1 by citizenship [migr_pop2ctz]

Country	Population		
	Total in 2019	*Change between 2000 and 2019*	*Change between 2013 and 2019*
Bulgaria	7,000,039	–	−284,513
Croatia	4,076,246	–	−185,894
Czech Republic	10,649,800	201,747	133,675
Estonia	1,324,820	−45,232	4,646
Hungary	9,772,756	−270,468	−136,042
Latvia	1,919,968	−461,747	−103,857
Lithuania	2,794,184	−717,890	−177,721
Poland	37,972,812	−290,491	−89,723
Romania	19,414,458	–	−605,616
Slovakia	5,450,421	51,764	39,585
Slovenia	2,080,908	93,153	22,087

Source: Own elaboration based on Eurostat, https://ec.europa.eu/eurostat/en/ web/products-data sets/-/MIGR_POP2CTZ (2020).

Data presented in Table 1.7 suggests that during 2000–2019, most CEE countries witnessed a decrease in their population, although it was the contrary for the Czech Republic, Slovenia, and Slovakia. Also, Estonia reported population growth between 2013 and 2019. The greatest depopulation between 2000 and 2019 was recorded in Lithuania where the number of residents declined by nearly 718,000, i.e. by approximately one-third of its 2019 population (Eurostat, 2020a). This situation clearly has a socioeconomic impact on all countries. The shortages of workforce can be partly remedied by immigrants from other countries, often from outside the European Union, as is the case in CEE countries. For instance, Poland recorded 153,200 people who came for a temporary stay in 2017 (51,100 more than in 2016). The largest part of that group comprised Ukrainian immigrants: 131,900 persons in 2017 and 209,000 persons in 2019 (migracje.gov.pl, accessed in May 2020).

The rapid demographic change in CEE countries was driven by migration processes which took place in the 1990s as a consequence of the political shift which provided the CEE population with the freedom to move. In the 2000s, the integration into the European Union and the partial opening of labor markets was the second important stimulus for increased migration (especially among the young generation). In the past, the CEE population willingly migrated both to European countries and to other continents to find lucrative employment (Holzer, 1999). However, after 2004, that process gained unprecedented momentum (Daszkowska, 2014). In the 1990s, the first phase of migration was a movement of young people. A typical emigrant was a poorly educated male aged 24–30, originating from areas hit by high unemployment rates, who looked for employment as a manual worker, mainly on German construction sites. In subsequent phases, the emigrants were older people (aged 30–39) in a stable professional and family situation. They were well educated and highly qualified, and expected social security (Daszkowska, 2014). Hence, many rural areas experienced a demographic imbalance caused by the migration of the young population at early stages of life when people usually start a family. In the countries of origin, this resulted in accelerated demographic aging, depletion of human capital (including what is referred to as brain drain), deterioration in business competitiveness and accumulation of public debt. Upon the enlargement of the EU in 2004, the UK, Ireland, and Sweden opened their labor markets for employees from new member states. In the following years, more countries decided to do so.

When considering emigration issues, one can clearly see that 2008 marks a shift in the number of people migrating from CEE countries. Indeed, that year, the number of CEE emigrants grew by 300%, on average, compared to what was recorded earlier. This can be explained by further measures taken to liberalize the European Union's labor markets. Finland, Spain, Portugal, and Greece completely removed the restrictions two years following the accession of EU-10 countries, in May 2006. Italy and the Netherlands did so in June 2006 and May 2007, respectively. The Luxembourgian and French labor markets have been open to the Polish population since November 2007 and

July 2008, respectively. As at 2018, the average annual number of emigrants from a CEE country was nearly 59,000, translating into a total of almost 9 million people during 1990–2000 (see Figure 1.4). During that period, Poland saw a considerable increase in the number of emigrants between 1990–2008 and 2009–2018 (growth in average annual numbers from 24,514 to 243,716 people) (Eurostat, 2020a, Emigration by age and sex). Restrictions applicable to the employment of Bulgarians and Romanians were removed in 2014; this triggered yet another upsurge in emigration (Figure 1.4).

Adding to this were major changes in external immigration which affected the situation of rural areas. When analyzing the immigration to CEE countries during 1998–2000, it can be noticed that Poland received the greatest number of immigrants (over 2 million). High levels were also recorded in the Czech Republic which became home to nearly a million incomers over the study period. Considering the average annual growth in the number of immigrants to CEE countries, it can be noted that immigration surged right after they joined the European Union. Until 2004, the annual number of immigrants did not exceed 15,000. Afterwards, there was an increase in the number of migrants to CEE countries. For instance, a major increment was recorded in Poland as the annual number of migrants went up from ca. 15,000 in 2008 to over 150,000 in 2009 (Eurostat, 2020a, Immigration by age group, sex, and citizenship [migr_imm1ctz]).

Internal migration means people moving from rural areas to cities, or in the opposite direction, within a single country. This consists in individuals changing their place of (permanent or temporary) residence in the national territory and involves crossing the administrative borders between communes or relocating from rural to urban areas, or in the opposite direction, within a single commune (GUS, 2019b). The flow of people from rural areas to cities is mostly caused by the opportunities offered by urban centers. In addition to a broader choice of jobs in the labor market and the ability to earn a higher

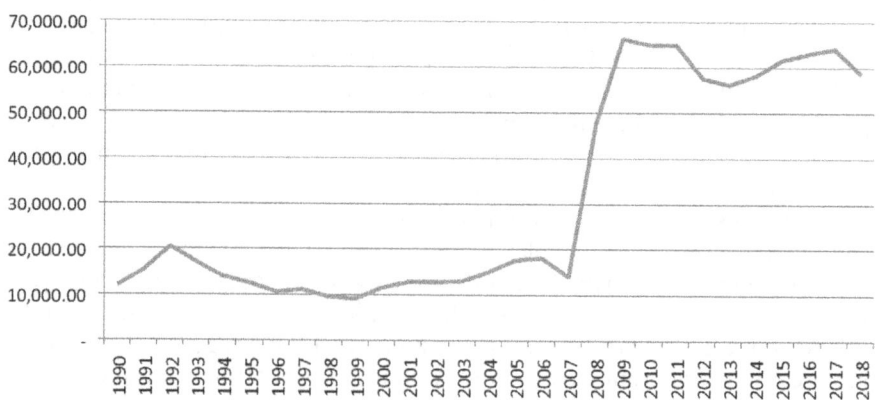

Figure 1.4 CEE emigrants, 1990–2018.
Source: Eurostat 2020: Emigration by age and sex.

salary, important aspects also include access to culture and education. On the other hand, moving from cities to rural areas is a popular migration direction, too, and is particularly noticeable in big cities (Kawczyńska-Butrym, 2009). This involves urbanization, an urban restructuring process which consists in the emergence of new agglomerations, reconstruction and expansion of urban areas, and population growth caused by rural immigration (Wrona, 2012). A complex and multifaceted socioeconomic process, urbanization means scientific and technological progress, concentration of production potentials, a specific form of social relationships (related to the widespread adoption of an urban lifestyle and a change in social relations and ties), and modernization of the whole settlement network (Szymańska, 2007). Similar processes take place in rural areas as the rural landscape can be observed to become increasingly consistent with what exists in cities. This is the consequence of the development of residential housing and of changes in the rural employment structure. Indeed, rural residents more often choose a job in the industrial, service or tourism sectors. The above is referred to as rural urbanization or semi-urbanization (Wrona, 2012). Four basic stages can be identified in the territorial, economic, and social development of a city. During the initial urbanization, rural people migrate to cities, which is conducive to the creation of new towns and to the increase in the area and population of existing ones. The secondary urbanization stage, also referred to as suburbanization, consists in a continued development of the agglomeration and an increase in the number of people living in its outer zones. This is followed by counterurbanization, which means a reduction in the number of residents in both the central and remote zones of the urban agglomeration. Afterwards, the population moves to territories located farther away from it. The next stage, after the population stabilizes, is re-urbanization: a process where population outflow slows down while the number of urban residents grows due to urban revitalization efforts (focused especially on the central part of the city) (Szymańska, 2007).

In CEE countries, including in Poland, local migrations were clearly visible after World War II; at that time, people mostly moved from rural areas to cities. This was caused by the accelerated development of urban centers due to growing importance of the industrial sector (Lewandowska-Gwarda & Antczak, 2015). Today, most big CEE cities experience suburbanization processes which mean people moving from the centers to outer (suburban) zones. As a consequence of this form of migration, the number of agglomeration residents grows although the population of the central city is on a decline (Gołata, 2012). This results in growth of suburban areas which witness rapid development of infrastructure, commerce, and services. These processes become increasingly important in big CEE cities. Big city depopulation, a term often used in this context, means the decrease in the number of urban residents due to urbanization processes. The consequence is the deconcentration of activities previously taking place in the center of the territory concerned (Szukalski, 2014). This can be illustrated by statistical data from

Eurostat (2020a, Population structure—functional urban areas, Population change over 1 year) which suggests that most big CEE cities experienced constant population decline between 2000 and 2019 (at an average annual rate of –3%). However, on the other hand, note that some big cities (especially the capitals of CEE countries) witness opposite processes, i.e. an increase in the number of residents. In that group, the greatest increments were recorded in Prague and Warsaw (Eurostat, 2020a) which can be explained by re-urbanization processes and other factors.

That situation can also be analyzed using the housing cost overburden rate by degree of urbanization. In this respect, Bulgaria and Romania stand apart from other CEE countries as they report a rate considerably below the average level for European Union countries (27). In CEE countries other than Romania, Bulgaria, and Croatia, the housing cost overburden is an issue which mostly affects urban areas. Lithuania is the only country where this problem is faced by the rural population. The relevant data is as shown in Table 1.8.

Another important phenomenon which has an effect on changes in CEE countries and impacts the regional and national situation of rural areas is the extension of life expectancy and a decline in birth numbers. In the early 2000s, CEE countries witnessed accelerated demographic aging; Poland was the fastest-aging country of all European Union members (Stańczyk & Szałtys, 2016). At the same time, compared to other European Union countries, the Polish population is relatively young. According to Eurostat, the median age for the Polish population in 2018 was 40 (vs. 43 in European Union

Table 1.8 Housing cost overburden rate by degree of urbanization, 2015 (%)

Area	Cities	Towns and suburbs	Rural areas
EU-28[a]	13.3	10.6	9.1
Romania	13.8	13.9	18.6
Bulgaria	13.9	14.6	16.2
Croatia	5.7	6.1	8.6
Lithuania	10.1	12.5	8.3
Poland	9.8	8.8	7.9
Slovakia	12.1	8.6	7.8
Hungary	9.2	9.5	7.6
Latvia	9.0	7.1	7.5
Czech Republic	15.6	10.6	5.9
Estonia	8.5	6.1	5.2
Slovenia	9.2	6.2	4.9

Source: Eurostat (online data code: ilc_peps13). *Eurostat Regional Yearbook 2017*, Chapter 14—Focus on rural areas.

Note: ranked on rural areas.
a Rural areas: estimate.

countries) (Eurostat, 2020b). The falling birth rate is also a factor in demographic change. From the end of World War II until the mid-1950s, Poland followed a policy of rebuilding the demographic potential lost as a consequence of wartime hostilities. Afterwards, the socialist authorities discontinued their family support programs and started to promote a 2+1 family model instead. This was supposed to result in a reduction of population growth which would guarantee a higher quality of life. As a consequence of the abortion act adopted in 1956, birth numbers fell by more than 260,000 between 1955 and 1968 (Dzienio & Latuch, 1983). The reduction in fertility and the change in social attitudes resulted in the nation's inability to ensure generation replacement. The drop in fertility accelerated in the 1990s, mostly because of the socioeconomic and political situation related to the political shift which entailed the systemic transformation. The period 1980–2018 saw a major decline in the young population aged up to 14 and 15–29. The number of people aged 0 to 14 dropped from 8.6 million to 5.8 million between 1980 and 2018. A similar situation occurred in the 15–29 age bracket, with a decline from 12 million to 6.5 million (nearly 50%). At the same time, the population aged over 60 kept growing, with 3.6 million in 1980 and over 9.5 million in 2018. The changes which occurred in the demographic condition during 1980–2000 resulted in an increase in the old-age dependency ratio. In 1980, there were more than 21 persons aged 60 and over per 100 people aged 15–59; in 2018, that value nearly doubled, reaching 41.3 (Kucharska, 2020).

According to estimations by the Central Statistical Office (GUS, 2014), in the years to come, demographic changes in Poland will continue to follow the same trend at the same pace. It is forecasted that in 2050, the elderly (aged over 60) will have the greatest share in the Polish population, and 38% of them will live in rural areas. In turn, people aged below 14 will have the smallest share (ca. 12%). In the next 30 years, the share of people aged 15–59 in the total population is forecasted to follow a downward trend. Until 2050, the old-age dependency ratio will reach nearly 85 persons aged 60 and over per 100 people aged 15–59 (Figure 1.5).

In 2018, the average life expectancy for people aged below 1 year in CEE countries was 77.3 years (vs. 81 years for EU-27 countries), i.e. 6.2 years more than in 1980. The longest average life expectancy is recorded in Slovenia (80.5) and the shortest in Bulgaria (75) which also demonstrates the slowest improvements in that respect. Also, an extension of the last phase of human life (over 60 years) can be observed in the recent years. In 1980, a statistical Polish sexagenarian had over 17 years (17.3) of life ahead of him/her; in 2018, 60-year-olds could expect to live 21.6 years more (GUS, 2019b).

Over the years, the declining share of people in youngest age groups in the total population (combined with the extension in life expectancy) has resulted in increasing the median age. In 2018, the average median age for CEE countries was 42.6 years, i.e. 9.2 years more than in 1980 (33.4 years) (Eurostat, 2020b). A median falling within the interval of 30 to 34 years means a demographically old population; CEE countries entered that phase already in the 1960s. In the mid-1990s, they crossed the threshold of a very old

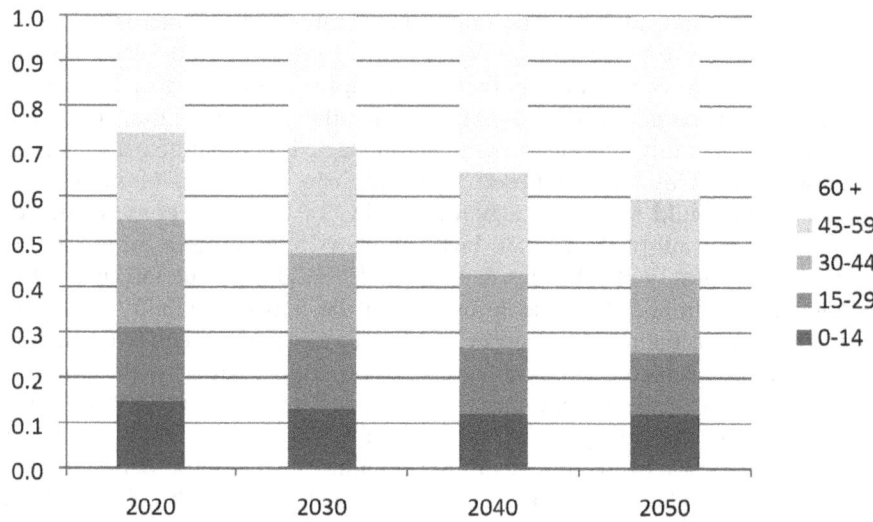

Figure 1.5 Forecasted age structure (%) of the Polish population, 2020–2050.
Source: own study based on Central Statistical Office data: Population forecasts for 2014–2050, 2014.

population (a median of 35 years and more). The 5-year increase in median age (from 30 to 35 years) took place between 1961 and 1994, i.e. within a 34-year period. However, another 5-year increment took them only 22 years, from 1995 to 2009 (Eurostat, 2020b), and the next leap can be expected to be even quicker. Having the above in mind, it can be concluded that the rate of demographic aging of CEE countries is extremely fast.

The demographic processes which have been observed since the early 1990s indicate that no change can be expected that would guarantee a sustainable demographic development. Also, CEE countries report low fertility rates (1.6 in 2019) which will have an adverse impact on the future number of births (Eurostat, 2020b). In 2019, the average fertility rate was 1.55 for EU-27 countries and 1.6 for CEE countries, with the lowest level being recorded in Poland (1.46). In most countries, this situation is exacerbated by the intensity of emigration processes. If the fertility rates and birth numbers are low while life expectancy increases (which is a favorable development), the number and share of the oldest people in total population grows increasingly faster. However, it can be assumed that the fertility rate in CEE countries grows quicker (by 0.15) than in all EU-27 countries (0.03) (Table 1.9).

Adding to the changes in the demographic structure are social problems. Some rural regions of CEE countries are at risk of poverty and social exclusion. In 2015, these problems were faced by at least half of the Bulgarian and Romanian population (50% and more). In Latvia, Croatia, Lithuania, Cyprus, Hungary, Poland, and Estonia, that rate went beyond the European Union average level (25.5%). Conversely, poverty and social exclusion pose a relatively minor

Table 1.9 Aggregate fertility rate for CEE countries, 2007–2019

Years	EU-27	Bulgaria	Croatia	Czech Republic	Estonia	Hungary	Latvia	Lithuania	Poland	Romania	Slovakia	Slovenia
2007	1.52	1.49	1.48	1.45	1.69	1.32	1.54	1.36	1.31	1.45	1.27	1.38
2008	1.57	1.56	1.55	1.51	1.72	1.35	1.58	1.45	1.39	1.6	1.34	1.53
2009	1.56	1.66	1.58	1.51	1.7	1.32	1.46	1.5	1.4	1.66	1.44	1.53
2010	1.57	1.57	1.55	1.51	1.72	1.25	1.36	1.5	1.41	1.59	1.43	1.57
2011	1.54	1.51	1.48	1.43	1.61	1.23	1.33	1.55	1.33	1.47	1.45	1.56
2012	1.54	1.5	1.51	1.45	1.56	1.34	1.44	1.6	1.33	1.52	1.34	1.58
2013	1.51	1.48	1.46	1.46	1.52	1.35	1.52	1.59	1.29	1.46	1.34	1.55
2014	1.54	1.53	1.46	1.53	1.54	1.44	1.65	1.63	1.32	1.56	1.37	1.58
2015	1.54	1.53	1.4	1.57	1.58	1.45	1.7	1.7	1.32	1.62	1.4	1.57
2016	1.57	1.54	1.42	1.63	1.6	1.53	1.74	1.69	1.39	1.69	1.48	1.58
2017	1.56	1.56	1.42	1.69	1.59	1.54	1.69	1.63	1.48	1.71	1.52	1.62
2018	1.55	1.56	1.47	1.71	1.67	1.55	1.6	1.63	1.46	1.76	1.54	1.6

Source: Own elaboration based on Eurostat, https://ec.europa.eu/eurostat/databrowser/view/TPS00199__custom_1434/default/table?lang=en (2020).

problem to Slovakian, Slovenian, and Czech rural areas (the corresponding rates are 20.6%, 19.4%, and 12.8%). These issues are the consequence of historical events and of the socioeconomic development model which has been implemented over the last 30 years (Eurostat, *Eurostat Regional Yearbook 2017*, Chapter 14—Focus on rural areas). At the same time, note that in Western and Northern European Union countries, the risk of poverty or social exclusion is often higher for the urban population, whereas in Eastern, Baltic, and Southern EU members it is higher for rural dwellers.

When considering the key changes affecting the regional and national transformation of rural areas, emphasis also needs to be placed on problems related to their multi-functionality, a concept which dates back to the 1990s. At that time, alternative forms of development were sought for rural areas hit by a deep crisis. It was caused by changes in the economic policy and by the decision made early that decade to discontinue agricultural support. This is the context in which the first attempts were made to comprehensively analyze the problems of multipurpose development (Stasiak & Kulikowski, 1996). The concept provided for an economic mobilization of rural areas based on the creation of non-agricultural jobs (Kobiałka, 2003). Multi-functionality was identified and defined as developing non-agricultural activities and seeking new rural jobs. It was viewed as being directly related to local development and to entrepreneurship, strategic planning, agricultural diversification, and infrastructural development (Kłodziński, 1995). As a consequence of research efforts, several concepts of territorial development were created (Siemiński, 1994), multipurpose development being among the most important ones. The development of CEE countries followed the multipurpose development doctrine established in the EU-15.

Multipurpose rural development is a complex process which differs over time and depends on diverse factors, such as the environment, infrastructure, education and skills, marketing activity at commune level, development of entrepreneurship, and internal and external investments (Kłodziński & Rosner, 1995). The key objective of multipurpose rural development is to provide better living and working conditions for the local population. This can be done by enhancing the diversity of jobs and, later on, by increasing the rural population's incomes and making rural areas a more attractive place to live and work. The fact that rural areas take over the functions defined as above drives a reduction in unemployment while also making people more active. In addition to agricultural and non-agricultural economic functions, multipurpose rural development also takes the fulfillment of environmental and social functions into account (Zarębski, 2001; Duczkowska-Małysz, 1997).

The process of political and economic changes in CEE countries, together with a shift in mentality, formed a new way of thinking of multi-functionality. This made multi-functionality an empowered concept and resulted in it being implemented in the national policy, especially in a context of the European integration. The implementation of the assumptions behind multipurpose development was manifested by an increase in the levels of rural activity and entrepreneurship.

1.4 Rural infrastructure development and institutions

The conclusion from the analysis of rural development in CEE countries is that each geographic area has its unique territorial potential which can be used in promoting development based on a dedicated, adequately structured development policy. In addition to natural conditions, important development drivers include the endogenous potential of rural areas which can be derived from land use patterns and depends on whether the local infrastructure is or is not developed.

The economic importance of infrastructure was mentioned already in the 1700s by Adam Smith (2020) who justified the responsibility of the government to invest in and maintain facilities of public use and public institutions. According to Buhr (2003), an economic approach to infrastructure was discussed in works by List (1841). In turn, Hirschman (1958) defined infrastructure as "capital which ensures the provision of public services" (after: Fourie, 2006).

Infrastructure is the total of tangible institutional and personal facilities and information. It is the whole set of assets, equipment, and working capital (which generate income to the economy and enable the supply of energy and of transport and telecommunication services) together with structures used to protect natural resources and transport roads (in their broadest meaning). It includes buildings and installations used in public administration, education, research, healthcare, and social assistance (Jochimsen, 1966). The analysis of infrastructure takes account of relevant sectors, such as transport, energy, electricity, and telecommunications; at the same time, infrastructure is defined as social facilities related to such services as hospitals, schools, and government institutions. That perspective provides a basis for identifying the economic infrastructure: one which addresses the needs related to the pursuit of an economic activity, i.e. water pipes, sewerage systems, roads, highways, railways, airports, seaports, electricity, and telecommunications. Conversely, social infrastructure is defined as one which promotes health, education and cultural standards for the population, i.e. measures with both a direct and indirect impact on the quality of life. Hence, defined in broad terms, social infrastructure can span different institutions, such as schools, libraries, universities, clinics, courts, museums, theaters, playgrounds, parks, fountains, and monuments. All of them include capital goods used by the public.

Infrastructure is discussed as a combination of equipment and institutions which fulfill auxiliary tasks relative to other territorial systems. Also, infrastructure is presented in a context of settlement patterns (referring to how the population is distributed across a territory). Note that infrastructural equipment is intrinsically linked to a specific location. Furthermore, infrastructure exhibits complexity (e.g. the construction of a water supply network should involve the construction of a sewerage network which, in turn, should be related to a sewage treatment plant etc.) and complementarity (in order for equipment to be used, supplementary facilities, e.g. roads, must be constructed).

CEE countries saw major changes in the development pace of their rural infrastructure in the late 1990s as they accessed the first European Union funds allocated to rural development. Of the pre-accession programs prepared by the European Union, the Special Accession Program for Agriculture and Rural Development (SAPARD) was of major importance to rural development. It was focused on both agricultural and rural development, and derived its legal basis from the following EU-level legal acts: Council Regulation (EC) No. 1268/99 of May 21, 1999 and Commission Regulation (EC) No. 2222/2000 of June 7, 2000.

Just like other pre-accession funds, SAPARD was designed for the ten aspiring members of the European Union. It was supposed to promote rural structural transformation in candidate countries and to assist in preparing local institutions and beneficiaries to access the instruments of the Common Agricultural Policy after they join the European Union. Hence, the rapid development of rural infrastructure was the consequence of CEE countries joining the European Union, as well as the outcome of implementing a coordinated policy and of becoming eligible for rural development funds. Since that time, measures have been taken to enhance the rural structure by seeking new forms of land use while also focusing on the local community (Durydiwka, 2012).

The analysis of the infrastructural condition of CEE countries faces certain difficulties in accessing consistent data. This is due to differences between systems for the collection and definition of infrastructure components. Nevertheless, under the assumption that infrastructure means socially useful technical solutions which contribute to improving the quality of life for local residents, rural infrastructure can be analyzed in a context of the quality of life and of how satisfied are the residents with how they live.

Table 1.10 shows the distribution of the population aged 18–64 by health status and by how satisfied they are with how they live, based on the results of a survey contracted by the European Union. The findings illustrate the changes in how people viewed their living conditions between 2010 and 2018 (based on the number of the highest ratings, *very good*) and the distribution of scores recorded in 2018 between the following categories of living conditions: *very good, good, fair, bad, very bad*. The results show that as regards CEE countries, the highest indicators of how people view their quality of life were reported in Romania (the respondents found their living conditions to be *very good* and *good*). Note that in CEE countries, the indicators of how the population view their living conditions were above the aggregate score (74.7) for the entire European Union (EU-27). The lowest indicators of the quality of life and the smallest number of people viewing their lives as being *good* or *very good* were recorded in Lithuania (50.9), Latvia (57.2) and Estonia (58.4). The lowest indicators of how rural people view their living conditions (*bad* and *very bad*) were recorded in Croatia, Lithuania and Hungary. While the results of this survey do not prejudge the quality and maturity of infrastructure, they provide an insight into the quality of rural living.

Pollution and environmental problems might reflect the condition of the infrastructure related, for instance, to waste management (Table 1.11). When

Table 1.10 Distribution of the population aged 18–64 by health status, age group, gender and degree of urbanization; an EU-SILC survey

Area	Rural areas								
	Very good					Good	Fair	Bad	Very bad
	2010	2012	2014	2016	2018	2018	2018	2018	2018
EU-27 (2020)	24.1	24.2	23.0	21.3	22.9	51.8	19.3	1.0	1.0
Bulgaria	18.5	20.7	20.9	22.1	21.4	51.5	19.0	1.4	1.4
Croatia	16.8	16.4	25.6	27.8	31.5	37.9	19.5	2.0	2.0
Czech Republic	23.3	22.6	21.8	20.7	24.1	47.4	21.1	1.2	1.2
Estonia	12.6	10.7	13.1	11.5	9.2	49.2	30.8	1.5	1.5
Hungary	15.3	16.8	17.8	17.8	16.8	50.5	24.6	1.7	1.7
Latvia	4.5	4.6	4.9	6.1	6.6	50.6	34.8	0.9	0.9
Lithuania	7.7	8.2	8.5	5.9	7.4	43.5	39.7	2.0	2.0
Poland	19.7	19.2	18.6	17.1	15.8	51.9	24.0	1.3	1.3
Romania	31.7	30.5	33.0	29.6	28.9	52.2	15.7	0.7	0.7
Slovakia	22.1	22.9	23.9	23.6	22.2	51.5	18.8	1.3	1.3
Slovenia	19.3	21.3	24.2	22.2	22.1	50.4	20.8	1.3	1.3

Source: Eurostat, last update on February 24, 2020, retrieved on July 15, 2020.

Table 1.11 Pollution and other environmental problems facing rural areas (by degree of urbanization) in CEE countries and in the EU (%) (pollution, grime or other environmental problems by degree of urbanization)

Area	Years				
	2010	2012	2014	2016	2018
EU-27 (from 2020)	8.8	8.4	8.4	8.2	8.5
Bulgaria	10.4	8.3	9.7	10.4	10.8
Croatia	7.1	5.0	4.4	4.2	2.8
Czech Republic	13.1	12.9	8.6	10.0	7.7
Estonia	9.1	11.3	8.9	7.7	7.6
Latvia	23.9	19.2	17.6	11.7	13.6
Lithuania	8.0	7.4	8.8	10.5	12.7
Poland	6.2	6.3	6.6	7.1	9.4
Romania	13.9	9.4	10.1	8.0	8.9
Slovakia	16.3	11.8	9.1	7.5	7.8
Slovenia	14.2	12.6	12.0	12.6	13.5

Source: Eurostat, last update on February 24, 2020, retrieved on July 15, 2020.

looking at CEE countries from this perspective, it can be noticed that the biggest problems related to pollution and other environmental aspects are faced in rural Latvia, Slovenia, Lithuania, and Bulgaria. In these countries, the above problems were reported in over 10% of the total rural area, which is above the EU-27 average level of 8.5%. The corresponding rates for Poland and Romania (9.4% and 8.9%, respectively) exceed the EU average, too. Croatia is the least affected by these issues as only 2.8% of its rural territories struggle with environmental quality and pollution (Table 1.11).

This phenomenon can also be looked at in the context of measures taken by CEE countries between 2010 and 2018. It can be concluded that Croatia demonstrated the best performance in eliminating environmental problems and pollution: in 2018, only 2.8% of rural areas were affected by these issues (vs. 7.1% in 2010). A large drop could also be observed in the Czech Republic (from 13.1% to 7.7% of the total rural area). Conversely, Poland and Bulgaria see their pollution and environmental problems grow; in Poland, that rate went up from 6.2% in 2010 to 9.4% in 2018, whereas Bulgaria reported an increase by 0.4 percentage points (from 10.4% in 2010 to 10.8% in 2018) (Table 1.11).

The issue of infrastructural development in CEE countries in a social context can also be tackled from the perspective of availability of different formal modes of childcare. As regards this issue, Eurostat has carried out surveys with the population of children aged below 3 and below 12 to determine the percentage of children receiving cost-free or paid childcare services and of children not covered by care services at all. Generally, in European Union (EU-27) countries, the ratio of children aged up to 3 years receiving childcare services (Table 1.12) was nearly 25% in 2016 (as per the most recent available data). For children aged below 12, that rate was almost 32%. When analyzing how CEE countries deal with this issue, it can be concluded that childcare infrastructure for children aged under 3 years is relatively poorly developed. Some of them, for instance the Czech Republic and Lithuania, have failed to establish a childcare system for children aged up to 3 years. In other countries, the ratio of children aged up to 3 years receiving childcare services is not in excess of 10% (except for Poland where it exceeds 10%). As regards formal care services for children aged up to 12 years, all countries have already put in place dedicated systems, though to a varying degree. The average rate for the European Union is 31.8%; Poland is the only CEE country with a higher score (54.6%). The rates are lower in other CEE countries; note that in Croatia, only 3.8% of children below 12 years of age can access formal modes of care. In Estonia and Hungary, that rate does not exceed 10%, whereas the levels recorded in other countries vary in the range of 10% to 30% (Table 1.12).

As regards the situation of childcare services for children aged below 3 and 12 years of age in CEE countries, almost all indicators—whether for rural or urban areas—are below the average values recorded in EU-27 countries (Table 1.13). The analysis also reveals that rural areas are in a much worse position to provide the essential infrastructure for the functioning of a childcare system. Poland is a positive outlier in this respect as its childcare system

Table 1.12 Children receiving childcare services by age in CEE countries in 2016 total (%)

Area	Less than 3 years			12 years or less		
	Full or reduced price	*Cost-free*	*No costs as not used*	*Full or reduced price*	*Cost-free*	*No costs as not used*
EU-27 (2020)	23.0	1.9	75.1	24.6	7.2	68.2
Bulgaria	0.5	0.0	99.5	8.1	13.4	78.6
Croatia	1.7	0.6	97.7	2.8	1.0	96.2
Czech Republic	0.0	0.0	100.0	18.2	5.0	76.8
Estonia	1.2	0.7	98.1	0.5	8.4	91.0
Hungary	6.7	8.0	85.3	3.2	6.6	90.3
Latvia	0.9	0.0	99.1	1.8	10.7	87.6
Lithuania	0.0	0.0	100.0	5.0	6.9	88.1
Poland	10.1	0.6	89.3	38.8	15.8	45.4
Romania	1.1	9.3	89.7	0.9	13.5	85.6
Slovakia	0.5	0.0	99.5	15.6	0.0	84.4
Slovenia	0.5	1.3	98.2	16.8	13.7	69.5

Source: Eurostat, last update on February 24, 2020, retrieved on July 15, 2020.

for children aged 12 years or less has a coverage ratio of 45.6% (which is definitely above the European Union average level of 30%). A similar situation can also be observed in urban areas where childcare services are received by 64.5% of children, nearly twice the average level recorded in European Union countries. However, in other countries, that ratio is not in excess of the Union average figure. The most disadvantageous situation exists in Estonia, with only 6% of children aged below 12 accessing formal modes of childcare. In some CEE systems, childcare facilities are more accessible in rural areas than in urban areas. Examples include the Czech Republic where 21% of urban children aged below 12 receive formal childcare services (vs. over 28% for their rural peers) and Estonia (6% vs. over 12%) (Table 1.13).

In the childcare system for the whole population of children aged below 12 years, free services are more common in rural areas, and the percentage of children receiving childcare services is higher in CEE countries than in the European Union as a whole.

The development of modern infrastructure can be illustrated based on the number of people who use Internet on a daily basis, grouped by degree of urbanization, as shown in Table 1.14. Data suggests that the average ratio for the rural population of the European Union (28) was 62% in 2016. With 74% of residents having a personal Internet access, Estonia had the most developed Internet access infrastructure of all CEE countries. Together with Latvia, Estonia exceeded the average ratios for the European Union as a

Table 1.13 Children receiving care services by age and degree of urbanization: rural and urban areas in CEE countries in 2016 (%)

Area	Less than 3 years						12 years or less					
	Cities			*Rural areas*			*Cities*			*Rural areas*		
	Full or reduced price	Cost-free	No costs as not used	Full or reduced price	Cost-free	No costs as not used	Full or reduced price	Cost-free	No costs as not used	Full or reduced price	Cost-free	No costs as not used
EU-27 (from 2020)	24.7	1.6	73.7	22.4	1.5	76.1	26.2	6.4	67.4	21.9	8.1	70.0
Bulgaria	1.1	0.0	98.9	0.0	0.0	100.0	10.4	10.3	79.3	5.3	14.7	80.0
Croatia	6.1	2.2	91.7	0.3	0.0	99.7	8.9	2.2	88.8	0.8	0.8	98.4
Czech Republic	0.0	0.0	100.0	0.0	0.0	100.0	18.6	2.4	79.0	20.0	8.1	71.9
Estonia	1.2	0.7	98.1	0.0	1.0	99.0	0.6	5.4	94.1	0.3	12.1	87.7
Hungary	7.0	8.1	84.9	4.1	4.0	91.9	3.2	8.7	88.1	2.4	5.4	92.2
Latvia	0.4	0.0	99.6	1.3	0.0	98.7	2.3	10.7	87.0	1.0	11.5	87.5
Lithuania	0.0	0.0	100.0	0.0	0.0	100.0	8.2	4.2	87.6	2.8	9.1	88.1
Poland	15.4	0.2	84.4	6.1	0.3	93.6	50.1	14.4	35.5	27.7	17.9	54.4
Romania	2.0	5.9	92.1	0.0	0.6	99.4	2.1	10.8	87.1	0.4	11.0	88.6
Slovakia	2.5	0.0	97.5	0.0	0.0	100.0	19.2	0.0	80.8	15.0	0.0	85.0
Slovenia	0.0	0.0	100.0	0.8	1.8	97.4	20.5	8.8	70.6	15.9	16.2	67.9

Source: Eurostat, last update on February 24, 2020, retrieved on July 15, 2020.

Table 1.14 Individuals accessing the Internet on a daily basis, by degree of urbanization, 2016 (% of all individuals)

Area	Cities	Towns and suburbs	Rural areas
EU-28	75	72	62
Bulgaria	60	47	33
Croatia	70	63	59
Czech Republic	73	64	61
Estonia	81	74	74
Hungary	80	72	61
Latvia	74	64	64
Lithuania	72	61	50
Poland	69	56	49
Romania	57	42	32
Slovakia	74	70	62
Slovenia	77	64	60

Source: Eurostat (online data code: ilc_peps13), *Eurostat Regional Yearbook 2017*, Chapter 14—Focus on rural areas.

whole. A relatively favorable situation also exists in Slovakia, Czech Republic and Hungary where the respective ratios are only slightly below the EU average figure. Conversely, Internet access continues to be a problem in Romania and Bulgaria where the ratios do not exceed 33%. In all CEE countries, Internet access is an issue faced mostly in rural areas. This can be explained by the technological capabilities related to the development of the relevant infrastructure.

The analysis of infrastructure allows us to determine its condition in relation to the quality of rural life. Table 1.15 presents some selected components of the economic and social infrastructure which had an effect on the quality of life in Polish rural areas in 2018. Data shown in the table suggests that in rural areas, the average values of indicators of household equipment were lower than in intermediate areas and in predominantly urban areas. As a consequence, the same is true for indicators of access to the economic infrastructure. In rural areas, most households (94.4%) have access to water-supply networks; a high percentage of dwellings are equipped with a bathroom and a lavatory (87.2% and 90.2%, respectively). Also, a relatively large percentage (76.9%) of dwellings are equipped with central heating (only 8.6 percentage points less than in predominantly urban areas). A large gap exists between predominantly rural and other (predominantly urban and intermediate) areas when it comes to the share of dwellings with access to gas networks. Only 31.8% of residents of predominantly rural areas have access to this kind of infrastructure, which can be explained by the technical barriers to the development of the gas network. Nevertheless, most households use gas thanks to the bottled-gas distribution system. The analysis of the social infrastructure

Table 1.15 Selected components of the economic and social infrastructure in Poland by degree of urbanization according to OECD (2018)

Type of sub-region	Dwellings (in % of total dwellings) fitted with					Pupils per	Population per		
	Water supply system	Lavatory	Bathroom	Gas supply system	Central heating	School	Library	Cinema	Hospital bed
PU	99.0	96.7	95.0	68.6	85.5	289.7	8185.6	150.0	175.0
IN	97.3	94.6	92.3	56.5	83.6	226.6	4588.7	221.3	275.9
PR	94.4	90.2	87.2	36.7	76.9	174.9	3537.3	291.6	247.9

Source: Central Statistical Office. https://stat.gov.pl/obszary-tematyczne/infrastruktura-komunalna-nieruchomosci/nieruchomosci-budynki-infrastruktura-kom unalna/infrastruktura-komunalna-w-2019-roku,10,3.html

Note: PR: predominantly rural; IN: intermediate; PU: predominantly urban.

reveals considerable gaps when it comes to the number of pupils per school and the number of residents per library, per cinema, and per hospital bed. In rural areas, there are nearly 175 pupils per school, which is almost 115 less than in predominantly urban areas (289.7). The average population per library ratio was 3,537.3 in rural areas, i.e. less than half that recorded in predominantly urban areas (8,185.6).

Also, the population per cinema in rural areas was nearly one and a half that recorded in urban areas. A similar pattern was revealed by the analysis of the number of rural residents per hospital bed (almost 73 persons more than in urban areas). At the same time, the ratios reported in rural areas were more favorable than in intermediate areas (275.9). When considering the ratio of pupils per school, it can be even assumed that rural areas find themselves in a more favorable situation than other territories covered by this analysis. Nevertheless, due to rural particularities, some components of both the economic and social infrastructure cannot attain the levels recorded in other areas. This can be explained by factors such as the technical specificity of infrastructural systems discussed.

In addition to the technical infrastructure (which mostly involves transport and energy supply, cf. Małkowska, 2011), the components of the economic infrastructure presented above can have a decisive impact on socioeconomic development. The development level of the technical and social infrastructure is a crucial part of the rural potential which is decisive for the capacity to attract potential investors. It also is essential for a high quality of life and, as such, often becomes the basis for migration decisions. CEE countries are strongly heterogeneous in this respect; some of them face a mix of severe problems related to infrastructural underdevelopment which affect their capacity to develop in the long run.

In a context of rural tourism development, infrastructure is a major driving force for the tourism market. Four infrastructural categories directly related to rural tourism development can be identified (Raina, 2005): physical, cultural, service-related, and governance-related aspects (Figure 1.6).

Improvements to the quality of infrastructure in these domains foster the development of the tourism market in rural areas and beyond. Currently, the development of the rural tourism infrastructure is among the priorities for many CEE countries.

Table 1.16 presents selected components of the tourism infrastructure; account was taken of the number of beds offered in CEE countries by degree of urbanization. The data suggests that the largest and smallest number of beds is found in Croatia (11 million) and Latvia (54,000), respectively. The European Union (27) has a total of over 28 million beds, including slightly over 4 million (15%) offered in CEE countries. Although their hospitality sector has a large development potential, CEE countries lag behind the EU-15 in this respect.

Both in the European Union (27) and in countries covered by this study, the greatest number of beds are offered in rural areas (45.5%). This is

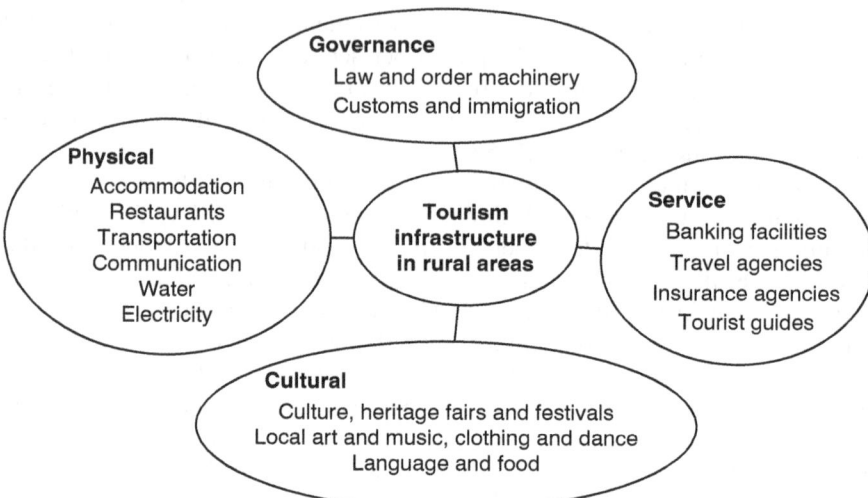

Figure 1.6 Components of tourism infrastructure in rural areas.
Source: own compilation based on Raina, A. (2005) Ecology, Wildlife and Tourism Development: Principles, Practices and Strategies, Sarup and Sons, Delhi.

especially true for Croatia and Czech Republic where rural areas account for more than half of the total number of beds (67.9% and 57.9%, respectively) (Table 1.16). A higher development level of the tourism infrastructure can contribute to production efficiency and to enhancements in tourism service distribution while also accelerating economic development.

Table 1.16 Tourism infrastructure by urbanization degree in CEE countries

Area	Total	Rural areas		Towns and suburbs	Cities
	No. of beds	No. of beds	% of total	% of total	% of total
EU-27 (2020)	28 293 307	12 857 729	45.4	33.6	20.9
Bulgaria	335 597	115 711	34.5	37.7	27.8
Croatia	1 115 659	757 069	67.9	26.4	5.7
Czech Republic	741 235	429 215	57.9	19.5	22.6
Estonia	61 193	29 256	47.8	17.9	34.3
Hungary	419 199	153 530	36.6	38.0	25.4
Latvia	53 948	21 616	40.1	23.8	36.2
Lithuania	89 813	41 565	46.3	26.6	27.1
Poland	798 723	332 225	41.6	33.1	25.3
Romania	348 592	109 681	31.5	36.0	32.6
Slovakia	190 773	95 505	50.1	32.2	17.7

Source: Eurostat (2020a).

Note: no data is available for Slovenia.

Infrastructural investments are the responsibility of public authorities, and are implemented depending on an adequate diagnosis of needs and available funds. As it generates high costs, the development of infrastructure depends on the economic development of the territory concerned. The availability and development level of technical and social infrastructure are factors that substantially affect the attractiveness of rural areas to investors and residents. To become attractive places to invest and live in, rural areas must embrace new functions (reflected in urbanization processes and in infrastructural development) which change the way they are viewed. It can be assumed that rural development trends include a transformation embedded in an urban context, i.e. urbanization processes accompanied by infrastructural modernization. As a consequence, rural living conditions become similar to what is enjoyed by urban dwellers, while the quality of rural life follows an upward trend.

1.5 Transition to private property and rural business development

The reforms started by CEE countries in the 1990s clearly highlighted weaknesses in their economies. However, they also had a stimulating effect on social activity and initiatives, and provided momentum for the development of rural entrepreneurship. The situation was tough, having in mind that for the first time ever, so many countries simultaneously embarked on the political transformation path. Due to the absence of reference models, Poland somewhat pioneered the reforms of the post-socialist economy. Moreover, because of its size, it became a CEE leader which set new goals for the transformation process (Siemiątkowski, 2011). In these countries, the lack of a rural development policy perpetuated the adverse phenomena related, for instance, to low levels of economic and social activity, high unemployment rates, deterioration in the quality of the technical infrastructure, and the outmigration of young educated people.

The economic changes became a major problem to be faced by CEE governments during the transformation period. High inflation rates, supply shortfalls, corruption, deteriorating condition of national budgets, as well as underdeveloped financial and banking institutions were the reasons behind the deficiencies in capital required for the reforms. Economic activation measures were taken, including in rural areas, by opening the economies for foreign capital in the form of direct foreign investment and foreign portfolio investment. During 1990–2000, direct foreign investment considerably contributed to the transformation and modernization of the Polish economy. The capital was partly invested in the privatization of state-owned companies, including in rural areas. In these conditions, large state-owned enterprises were privatized, which had a mobilizing effect on the population and played a crucial role in economic modernization. New owners quickly switched to modern management styles, implemented new technologies and restructured their plants, making them fit to face market competition (cf. Siemiątkowski, 2011). These measures also provided momentum for the economic activation of a considerable part of the rural population, and could contribute to changing the socioeconomic situation of rural areas and boost their development (cf. Bański & Czapiewski, 2008).

Later on, the activation of the rural population was based on measures arising from the European Union membership perspective. In this context, CEE countries initiated socioeconomic reforms focused on developing economic and regional policy tools which stimulate activity and entrepreneurship while being consistent with what is in place in the EU. The territories differed in their resilience to transformation risks and their ability to easily adapt to new conditions. As a consequence, economic and social activities tended to concentrate in specific geographic areas, resulting in the impoverishment of other localities and in a widening development gap. In these circumstances, the cohesion policy became a line of action for measures taken to reduce development disparities.

Some candidate countries, such as Lithuania and Latvia or Romania and Bulgaria, initially achieved little progress in their economic reforms, although Bulgaria witnessed an acceleration in the late 1990s. Note that—similarly to what was in place in Western European countries—measures related to increasing rural activity and competitiveness were often restricted by the demands of the farming lobby, primarily including protection for the agricultural sector and the request that funds be transferred to the farming population (World Bank, 2000) instead of being allocated to support for rural activity and development. This was actually the case as the goals set under the CAP remained unchanged for many years. Only the reforms initiated in the early 1990s resulted in a transformation of policy provisions; the membership perspective for CEE countries boosted these efforts. A separate budget was allocated to supporting rural diversity. At the same time, people were becoming increasingly aware of the need to have a balanced diet and to protect and preserve the natural environment as well as food security and quality. Emphasis was placed on: plant and animal health and welfare; preservation and protection of rural landscape; biodiversity; and climate change. Initiatives were taken in this respect to stimulate the activity of rural communities. An assumption was made that entrepreneurial and active attitudes of the rural population who access European Union funds can be the most efficient way to contribute to rural change. These challenges had an effect on realigning the priorities for the development of agricultural areas so as to take account of the rural population and support their activity.

Currently, rural development increasingly depends on the activity of residents and local entrepreneurs. In this context, note the social and economic activity of the rural population.

Social activity, an aspect of crucial importance to socioeconomic development, often requires qualitative research, and its measurement is highly problematic. In most cases, it is measured as the residents' participation in local community undertakings, initiatives, and organizations (Stanny et al., 2018). It can be considered in a context of being a member of and holding a function in organizations, taking measures for the local community, engaging in voluntary service, and participating in elections (Czapiński, 2013). Another component of activity is the ability to establish organizational structures of common interest, self-organizational capacity, and willingness to engage in causes considered to be important. Social activity also includes measures that affect local institutions, e.g. structures attached to commune authorities, farmer's wives' associations, self-help organizations or folk bands (cf. Kamiński, 2008). According to Stanny et al. (2018), the basic indicators of social activity in rural areas include: turnout in local (commune head or town mayor) elections or in presidential elections; number of non-government organizations; share of personal income tax (PIT) payers who allocate 1% of their tax to public benefit purpose foundations; number of applications for European Union funds used in financing projects implemented under the LEADER approach.

The analysis of social activity in terms of turnout in the first round of the 2015 presidential election in Poland shows that the Polish population is heterogeneous in this respect. The basic conclusion is that the turnout was higher (above 50%) in big cities. The second finding is the geographic distribution of voter turnout, with higher levels being recorded in the southeast of the country (State Election Commission, 2015). A higher local election turnout was recorded in central and eastern Poland, especially in communes located in outer zones of their voivodeships, in relatively poorer areas where agriculture plays a dominant role. In turn, the lowest levels were reported in areas where the largest numbers of economic migrants originate from. In areas which were formerly home to large state-owned agricultural holdings, the election turnout was relatively low, too. This suggests low involvement in local issues among people previously related to the state-owned agricultural sector. Often, these are poorly educated individuals who fail to show much initiative not only in solving local issues but also in addressing their own and their loved ones' problems. Relatively low levels of social activity are characteristic of areas located away from larger cities and from urban networks, and of those where agriculture (traditional and family farming) continues to be the dominant economic function (Stanny et al., 2018).

One of the forms of rural mobilization used by CEE countries after joining the European Union was LEADER, the support program for multipurpose rural development. It was planned to create conditions that promote rational management of the development potential and endogenous resource efficiency at local level. It consisted in a local development strategy being conceived and implemented by the local community based on a bottom-up approach. Local action groups established under the program grouped together representatives of local environments. Due to their social nature, they provided an opportunity to stimulate the social capital which—supported with financial resources—became a tool in the mobilization of rural dwellers (cf. Hadyński & Borucka, 2015).

In Poland, between 2004 and 2006, LEADER was implemented under the LEADER+ Pilot Program as measure 2.7 of the Sectoral Operational Program "Restructuring and modernization of the food sector and rural development 2004–2006." The program was focused on supporting potential partnerships in the creation and registration of local action groups (as foundations, associations or association unions) which represented the population of a specific territory spanning rural areas. The mission of these social groups was to develop Integrated Rural Development Strategies (Scheme 1). The activity of LAGs was mainly focused on tourism development, protecting and promoting the natural environment and historical assets, and promoting the wide adoption and development of regional products. The number of applications filed (249 under Scheme 1; 187 under Scheme 2) and agreements signed (167 and 149, respectively) reflected the high interest in the program (cf. Janiak, 2008; Kościelecki et al., 2010, Hadyński, 2006).

In the next programming period (2007–2013), support under LEADER was allocated to measures focused on the implementation of local development strategies, the functioning of Local Action Groups (LAGs), acquisition of skills, and activation of and cooperation between LAGs. The implementation of local development strategies involved the financing of operations including diversification into non-agricultural activities, establishment and development of micro-enterprises, village renewal and development, and small projects which contributed to improving the quality of life, the economic diversity in the LAG's activity area, and more (Łukasiewicz, 2009).

As regards measure "diversification into non-agricultural activities," agreements were entered into with farmers who intended to engage in an economic activity, and mostly focused on: services for farms or forestry (48.5%); tourism and recreation services (17.6%); and services provided to the public (12.5%). The operations were mostly carried out by farm owners (84%) (MRiRW, 2012).

When it comes to the "establishment and development of micro-enterprises," most applications addressed services provided to the public (39.1% of operations), and tourism services and services related to sports (18.6%). Other measures covered construction works and services, handicraft, and small-scale processing. Aid granted under this measure was primarily allocated to existing enterprises (which accounted for 75% of beneficiaries) (cf. Hadyński & Borucka, 2015).

Rural activation measures were also implemented under the "village renewal and development" program to create conditions for socioeconomic development of rural areas. Efforts made to mobilize the rural population were based on investment support granted for the implementation of projects related to maintenance, restoration, and upgrading of the cultural and natural heritage of rural areas and to improving the attractiveness of rural areas to tourists. The applications concerned: construction works related to the public infrastructure (nearly 66%); addressing tourism needs in society (29%) and preserving cultural heritage (5%). Financing was mostly accessed by communes, followed by cultural and ecclesiastical institutions. Some projects were also implemented by non-government organizations (MRiRW, 2012).

Small projects included support for minor measures taken to promote sustainable rural development. The directory of available projects included an ample category of support which contributes to improving the quality of life for the population by enabling them access to technological innovation or by organizing cultural, recreational, and sports events. Measures taken in this area included: the organization of cultural, recreational, and sports events (over 40%), repair and equipment of rural community centers (28%), and investments in tourism infrastructure (ca. 20%). The largest number of projects was implemented by local government units and non-government organizations (nearly 60%) (MRiRW, 2012).

In addition to the activation targets, LEADER supported domestic and international partners in their cooperative efforts to implement common projects. Collaborative projects contributed to promoting the LAGs and the region, to a better use of the local potential, to the development of entrepreneurship and more. In the implementation period, rural residents progressively increased their activity levels; it was reflected by aspects which include an increase in the number of aid applications. This was due to LAGs gaining experience and strengthening their position in the territory concerned (MRiRW, 2012). An important mission of LAGs was to increase the awareness of the local community by organizing trainings and other mobilization projects targeted at local leaders, managers and contributors to the implementation of local development strategies (Ministry of Agriculture and Rural Development, 2014).

The economic activity of the rural population can be viewed through the prism of education, employment, unemployment, and activity in non-agricultural sectors. Note, however, that in CEE countries, agriculture continues to be an important aspect in analyzing the social activity of rural dwellers (Wrzochalska, 2014).

The indicators of activity of the rural population include education levels. On the one hand, a high proportion of educated people shows how many residents took the effort to complete an education program and acquire new skills. On the other hand, these people form the social potential which can help boosting rural development. When analyzing the share of people aged 30–34 with a tertiary education in CEE countries, it can be concluded that in most of them, the proportion of rural population having a tertiary level of educational attainment was above the EU-28 average level (Table 1.17, Figure 1.7). The highest ratios were recorded in Lithuania, Slovenia, Estonia, Latvia, and Poland, each being above the European Union average of 27.9% in 2015.

Figure 1.7 shows the percentage of people aged 30–34 with tertiary education (ISCED levels 5–8) attainment, by degree of urbanization. With a ratio of 7.8% and 9.4%, respectively, Romania and Bulgaria presented the most unfavorable situation in this respect. As regards other countries, it should be noted that the Czech Republic, Croatia, Slovakia and Hungary should intensify their efforts to at least attain the EU-wide average level.

The situation of young people (aged 18–24) not in education, employment or training could pose a problem to the socioeconomic activity of the CEE population. The issue is specifically acute for rural areas in the countries

Table 1.17 Share of people aged 30–34 with tertiary education (ISCED levels 5–8) attainment, by degree of urbanization, 2015 (%)

Area	Cities	Towns and suburbs	Rural areas
EU-28	48.1	33.4	27.9
Lithuania[a]	72.4	44.6	39.3
Slovenia	53.9	43.5	38.4
Estonia	54.2	33.0	35.1
Latvia	50.4	30.7	34.6
Poland	59.7	40.5	28.6
Czech Republic	42.8	26.7	20.7
Croatia	49.4	25.9	19.3
Slovakia	50.1	26.6	18.5
Hungary	51.1	32.3	17.6
Bulgaria	46.6	26.5	9.4
Romania	46.4	19.8	7.8

Source: Eurostat (online data code: ilc_peps13) *Eurostat Regional Yearbook 2017*, Chapter 14—Focus on rural areas.

Note: ranked on rural areas.
a Towns and suburbs: low reliability.

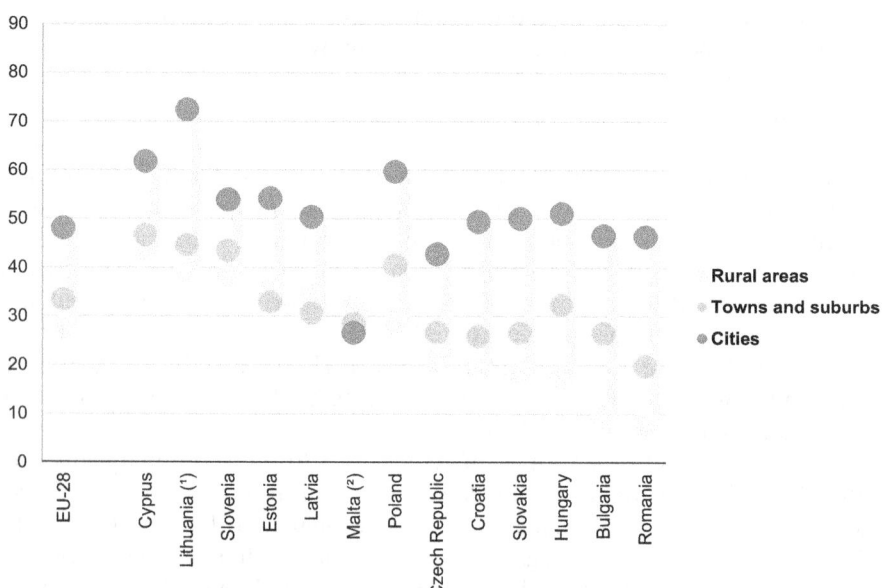

Figure 1.7 Share of people aged 30–34 with tertiary education (ISCED levels 5–8) attainment, by degree of urbanization, 2015 (%).
Source: Eurostat (online data code: ilc_peps13) Eurostat Regional Yearbook 2017, Chapter 14—Focus on rural areas.

covered by this study. The Union (EU-28) average ratio for rural areas was 17.9% in 2015, with only Lithuania, Latvia, Estonia, Slovenia, and Czech Republic being at a lower level. The biggest problems are faced in Bulgaria, with over 40% of the population aged 18–24 being inactive. This is double the average level for European Union countries. In most CEE countries, the problems discussed above affect rural areas, although in the Czech Republic, Slovenia, and Romania, they are mostly faced in intermediate areas.

Rural activation can be considered in the context of employment. The basic indicator, employment ratio for people aged 20–64, is used in the Europe 2020 strategy which assumes that, on average, 75% of that group will be employed by 2020 (while also setting target levels for each country). When analyzing employment in that age group in rural areas of CEE countries, it can be concluded that all of them failed to achieve the desired level by 2015 (Table 1.19). In the EU-28, the average attainment rate of the goal set under the Europe 2020 strategy was 69.2%; only the Czech Republic, Estonia, Lithuania, and Slovenia went above the average level for the European Union as a whole. The least favorable situation in rural areas was found in Bulgaria and Croatia, with the share of employed people aged 20–64 being below 60% (56.7% and 57.2%, respectively) (Table 1.19). In subsequent years, the situation changed so that most countries, i.e. the Czech Republic, Estonia, Croatia, Latvia, Lithuania, Slovenia, and Slovakia attained the desired levels by 2019,

Table 1.18 Share of young people aged 18–24 neither in employment nor in education or training (NEETs), by degree of urbanization, 2015 (%)

Area	Cities	Towns and suburbs	Rural areas
EU-28	14.2	16.5	17.9
Bulgaria	11.2	28.9	40.9
Croatia	17.7	22.5	30.2
Romania	13.9	26.7	26.3
Slovakia	14.2	14.2	20.3
Hungary	10.2	13.7	18.7
Poland	9.7	15.4	18.0
Latvia	10.2	15.4	16.1
Estonia[a]	10.4	13.6	15.9
Lithuania[b]	8.9	–	15.9
Slovenia[c]	9.5	14.1	12.7
Czech Republic	8.8	10.2	10.0

Source: Eurostat (online data code: ilc_peps13) *Eurostat Regional Yearbook 2017*, Chapter 14—Focus on rural areas.

Note: ranked on rural areas.
a Towns and suburbs: low reliability.
b Towns and suburbs: not available.
c Cities: low reliability.

Table 1.19 Employment indicators for the CEE population aged 20–64 by degree of urbanization (%)

Country	Areas											Total	Europe 2020 target
	Rural												
	1999	2001	2003	2005	2007	2009	2011	2013	2015	2017	2019	2019	
EU-27	–	–	68.5	69.0	68.8	67.0	66.5	67.5	69.2	71.6	73.3	73.1	75.0
Bulgaria	–	–	–	–	63.9	64.5	58.2	54.7	56.7	61.9	66.1	75.0	76.0
Croatia	–	–	52.6	66.8	63.5	65.5	59.5	55.0	57.2	59.2	63.3	66.7	62.9
Czech Republic	–	–	71.6	71.8	72.1	70.4	70.4	71.9	74.9	78.1	79.8	80.3	75.0
Estonia	68.8	66.8	69.1	69.3	74.2	68.0	69.4	72.3	73.4	76.1	76.9	80.2	76.0
Hungary	–	57.4	59.1	57.7	58.3	55.7	56.9	58.7	65.8	71.0	73.3	75.3	75.0
Latvia	–	64.5	64.9	67.5	72.4	66.0	65.3	67.2	69.9	71.2	73.6	77.4	73.0
Lithuania	67.7	64.5	70.9	68.6	70.3	62.9	62.2	63.7	68.2	68.3	73.5	78.2	72.8
Poland	–	–	–	–	62.8	63.9	63.5	64.3	66.5	69.1	70.7	73.0	71.0
Romania	–	–	–	–	–	63.6	63.2	66.7	65.5	67.7	69.4	70.9	70.0
Slovakia	–	–	–	–	63.9	62.9	61.8	61.4	66.1	68.9	72.1	73.4	72.0
Slovenia	–	–	–	72.5	73.6	73.0	69.2	68.2	69.6	74.0	76.2	76.4	75.0

Source: Eurostat, https://appsso.eurostat.ec.europa.eu/nui/setupDownloads.do (2020).

both in rural areas and generally. Bulgaria, Hungary, Poland, and Romania failed to attain the target levels of Europe 2020 indicators in rural areas. However, at national territory level, only Bulgaria failed to do so by 2019, as illustrated in Table 1.19.

In analyzing the pace of changes in employment of the CEE population aged 20–64, it can be noted that in some of these countries, the situation changes faster than the EU-27 average level. At the same time, this data reflects the way labor markets responded to the 2007–2009 financial crisis (Table 1.19). This provides a basis for analyzing the quality of the entrepreneurial potential. Generally, there was a decline in overall employment figures. In 2009, the reduction in the number of employed people in CEE countries was above the European Union average level. However, in the next two years, CEE countries saw an improvement in employment numbers. It was the contrary for the total EU-27 group which started to record steady growth in the share of employed people aged 20–64 only in 2014. The above can be assumed to reflect the flexibility of the economy and of the labor market in how they counteracted the economic crisis, which differed from one country to another. Croatia, Slovenia, and Romania took the most time to emerge from the crisis (as they followed a declining trend in employment until 2013). In Bulgaria, the drop was experienced from 2009 to 2011, whereas Poland recorded a decline only during 2009–2010. In Romania, Lithuania, and Hungary, the reduction in employment started to be noticeable already during 2007–2008. When it comes to rural areas, it can be concluded that the labor market's inflexibility in responding to the economic crisis in the late 2000s had the strongest effect on the Croatian labor market, which experienced the greatest decline in employment for a prolonged period (from 2009 to 2013). A similar situation was witnessed in Slovenia. Generally, CEE rural areas were more severely affected by the crisis than European Union countries on average. In Poland, employment declined during 2009–2010 but then embarked on an upward trend. Lithuania and Hungary saw the emergence of problems affecting their labor markets during 2007–2008. However, in Hungary this situation ended in 2009, which marked an upturn in employment (Table 1.19). The data illustrates the activity of economic operators in overcoming the impacts of the labor market crisis. In CEE countries, employment of people aged 20–64 is a problem which mostly affects rural areas. Only in Poland and Czech Republic, the lowest employment rates for people aged 20–64 are recorded in intermediate areas.

Education and employment are indicators of the potential for rural mobilization. The more educated the population of a territory, and the larger the group of educated residents, the higher the potential for entrepreneurship and mobilization. Conversely, the scale of unemployment is a barrier.

When considering the unemployment rate for people aged 15–74 by degree of urbanization in 2015, it can be concluded that the biggest problems were faced by the rural population of Croatia, with 18% (nearly double the average figure for EU-28 which was 9.1%) (Table 1.20). Major problems were also facing the Bulgarian, Slovakian, Lithuanian, and Latvian population. In

Table 1.20 Unemployment rate, persons aged 15–74, by degree of urbanization, 2015 (%)

Area	Cities	Towns and suburbs	Rural areas
EU-28	10.0	9.0	9.1
Croatia	13.9	16.1	18.3
Bulgaria	6.2	9.2	14.7
Slovakia	8.3	10.6	13.5
Lithuania	5.9	8.6	12.4
Latvia	8.9	9.5	11.3
Slovenia	9.3	9.3	8.7
Poland	6.5	8.3	8.1
Hungary	5.7	6.8	7.8
Romania	6.0	8.1	6.8
Estonia	5.8	6.1	6.7
Czech Republic	4.7	5.4	5.1

Source: Eurostat (online data code: ilc_peps13). *Eurostat Regional Yearbook 2017*, Chapter 14—Focus on rural areas.

Note: ranked on rural areas.

these countries, the unemployment rate exceeded 10% and was above the EU-28 average level. Conversely, the Czech Republic, Estonia, and Romania were the least affected by unemployment (5.1%, 6.7%, and 6.8%, respectively).

In summary, the political and economic transformation of CEE countries from a centrally planned economy model to a market system resulted in the need to address a number of major development issues, including rural activity and entrepreneurship. The socioeconomic development gap between regions and areas (also observed between rural and urban areas) necessitated the measures to change the existing situation. As an important part, the transformation included institutional reforms which accelerated in CEE countries in 1995, driven by requirements involved in accessing the European Union. Measures taken by public authorities were initially supported by European Union's pre-accession programs (SAPARD, PHARE, ISPA) which promoted activity and entrepreneurship among those interested in accessing the funds. The rural support system, in force until 2006, was extended to cover CEE countries as they joined the European Union. Subsequent budgets (2007–2013, 2014–2020) set a direction for rural change and gave fresh momentum for rural activity and entrepreneurship among operators and individuals.

Notes

1 The European Charter for Rural Areas was created on the initiative of the European Commission's Committee on Agriculture and Rural Development in order to support the policy for the development and protection of rural areas. It specified the economic, environmental, cultural, and social characteristics and functions of rural areas and

confirmed the need for the governments to promote a rural policy for which some guidance is provided in the Charter (for a broader description, see: Anttila, Report, European Charter for Rural Areas, Doc. 7507, March 20, 1996, https://assembly.coe. int/nw/xml/XRef/X2H-Xref-ViewHTML.asp?FileID=7441&lang=en.

2 Called *Flurbereinigungsgesetz* in German (BGBl. I, p. 591, 1953, No. 37), the Act of July 14, 1953 entered into force on January 1, 1954.

3 For instance, in 1984, rural development in Germany was supported with the Fund for Improvements in the Agrarian Structure and Coastline Protection.

4 Held in November 1987, the first European Congress for Village Renewal, organized by Austrian associations for the agricultural and forestry policy, was attended by representatives of nine countries.

References

Bański, J. (2012). Problematyka definicji i zasięgu przestrzennego obszarów wiejskich i sfer podmiejskich, Acta Scientiarum Polonorum, *Administratio Locorum* 11(3), 5–15.

Bański, J., & Czapiewski, K. (2008). *Identyfikacja i ocena czynników sukcesu społeczno-gospodarczego na obszarach wiejskich*, Instytut Geografii i Przestrzennego Zagospodarowania PAN, Warszawa.

Bański, J., & Stola, W. (2002). *Przemiany struktury przestrzennej i funkcjonalnej obszarów wiejskich w Polsce*. Studia Obszarów Wiejskich, 3, Warszawa.

Buhr, W. (2003). *What is Infrastructure?*Department of Economics, School of Economic Disciplines, University of Siegen. Siegen Discussion Paper No. 107–103.

Clapson, M. (2003). *Suburban Century: Social Change and Urban Growth in England and the USA*, Berg, Oxford.

Community Strategic Guidelines on Cohesion (2006). *DECYZJA RADY z dnia 6 października 2006 r.*, w sprawie strategicznych wytycznych Wspólnoty dla spójności (2006/702/WE).

Czapiewski, K. (2010). *Koncepcja wiejskich obszarów sukcesu społeczno-gospodarczego i ich rozpoznanie w województwie mazowieckim*, Studia Obszarów Wiejskich, 22, IGiPZ PAN, PTG, Warszawa.

Czapiński, J., (2013). *Diagnoza Społeczna 2013 Warunki i jakość życia Polaków*. Cały Raport (PL), Współczesna ekonomia, t. 7, wydanie 3.1.

Czarnecki, A. (2005). Obszary wiejskie, urbanizacja wsi, rozwój wielofunkcyjny, rolnictwo wielofunkcyjne-przegląd pojęć, w: *Uwarunkowania i kierunki przemian społeczno-gospodarczych na obszarach wiejskich*, A. Rosner (red.). IRWiR PAN, Warszawa.

Daszkowska, E. (2014). *Emigracja Polaków po 2004 roku i jej skutki*. Biuletyn Opinie FAE9.

Domagała, M. (2010). Polski model decentralizacji, w: *Rola samorządu terytorialnego w modernizacji Polski*, M. Barański, A. Czyż, & S. Kubas (red.). Wydawnictwo Naukowe "Śląsk", pp. 27–37.

Duczkowska-Małysz, K. (1997). Strategie rozwoju obszarów wiejskich w Polsce, w: K. Duczkowska-Małysz, *Strategia doradztwa w realizacji rządowego, regionalnych i lokalnych programów rozwoju obszarów wiejskich w Polsce*. Olsztyn: Centrum Rozwoju Obszarów Wiejskich (35).

Durydiwka, M. (2012). *Czynniki rozwoju i zróżnicowanie funkcji turystycznej na obszarach wiejskich w Polsce*, Wydział Geografii i Studiów Regionalnych UW, Warszawa.

Dz. U. [Journal of Laws] (1990). Nr 16, poz. 95, Ustawa z dnia 8 marca 1990 r. o samorządzie gminnym.

Dz. U. [Journal of Laws] (1998). Nr 91 poz. 578 ustawa o samorządzie powiatowym.

Dzienio, K., & Latuch, M. (1983). *Polityka ludnościowa europejskich krajów socjalistycznych*. Warszawa.

EC (2020a). https://ec.europa.eu/regional_policy/sources/docgener/studies/pdf/erdf_supp ort_inclusive_growth_en.pdf

EC (2020b). https://ec.europa.eu/info/sites/info/files/food-farming-fisheries/key_policies/ documents/rur-dev-small-villages_en.pdf

EC (2020c). https://ec.europa.eu/info/sites/info/files/fod-farming-fisheries/key_ policies/ documents/ number-of-rdp-per-country-2014–2020_en.pdf

EC (2020d). https://ec.europa.eu/info/food-farming-fisheries/key-policies/common-agri cultural-policy_en

Europe (2020a). *Strategia na rzecz inteligentnego i zrównoważonego rozwoju sprzyjającego włączeniu społecznemu*, Komunikat Komisji, Bruksela, 3.3.2010 KOM(2010) 2020 wersja ostateczna.

Europe (2020b). *15 Krajowy Program Reform – Europa 2020: Strategia na rzecz inteligentnego rozwoju sprzyjającego włączeniu społecznemu*, Ministerstwo Gospodarki, Warszawa, s. 3.

Eurostat (2020a). https://data.europa.eu/euodp/pl/data/dataset/VrbbG2kP6U8YElh5oRUw.

Eurostat (2020b). Struktura ludności i starzenie się społeczeństwa. Pobrano 04 27, 2020 z lokalizacji. https://ec.europa.eu/eurostat/statistics-explained/index.php?title= Population_structure_and_ageing/pl#Mediana_wieku_jest_najwy.C5.BCsza_we_W. C5.82oszech

Fourie, J. (2006). Economic infrastructure: a review of definitions, theory and empirics, *South African Journal of Economics*, 74 (3), September.

Gołata, E. (red.) (2012). *Migracje mieszkańców dużych miast*, Wydawnictwo Uniwersytetu Ekonomicznego w Poznaniu.

Green Paper on Territorial Cohesion (2008). *Zielona księga w sprawie spójności terytorialnej, Przekształcenie różnorodności terytorialnej w siłę Komunikat Komisji do Rady, Parlamentu Europejskiego, Komitetu Regionów i Komitetu Ekonomiczno-Społecznego* {SEC(2008) 2550}, Bruksela, dnia 6.10.2008 COM(2008) 616 wersja ostateczna.

GUS (2014). *Prognoza ludności na lata 2014–2050*. Warszawa: Główny Urząd statystyczny.

GUS (2019a). Zeszyt metodologiczny pt. Migracje ludności, https://stat.gov.pl/obsza ry-tematyczne/ludnosc/migracje-zagraniczne-ludnosci/zeszyt-metodologiczny-migra cje-ludnosci,15,1.html?pdf=1.

GUS (2019b). *Informacja o rozmiarach i kierunkach czasowej emigracji z Polski w latach 2004–2018*. Warszawa: Główny Urząd statystyczny.

Hadyński, J. (2006). Program Leader+ jako przykład partnerstwa, w: *Strategie zarządzania współpracą*. Poznań: Wyd. AR w Poznaniu.

Hadyński, J., & Borucka, K. (2015). Efekty polityki rozwoju obszarów wiejskich UE w kontekście programu Leader w Wielkopolsce. *Journal of Agribusiness and Rural Development*.

Heller, J. (2000). *Regionalizacja obszarów wiejskich w Polsce*. Studia i Monografie, z. 99. Warszawa: IERiGŻ.

Hirschman, A. O. (1958). *Strategia rozwoju gospodarczego*, Yale Studies in Economics, New Haven, CT: Yale University Press.

Holzer, J. (1999). *Demografia*. Warszawa: Polskie Wydawnictwo Ekonomiczne S.A.

IRWIR (2019). *Wpływ polityk spójności na rozwój obszarów wiejskich, Raport Końcowy* (*Impact of cohesion policies on rural development, Final Report*),October,

Institute of Rural and Agriculture Development of the Polish Academy of Sciences, Wolański sp. z o.o.,Warsaw, 51–54.

Janiak, K. (2008). *Leader szansą dla polskiej wsi*. Warszawa: FAPA.

Jochimsen, R., ed. (1966). *Theorie der Infrastruktur: Grundlagen der marktwirtschaftlichen Entwicklung*. Tübingen, J.C.B. Mohr.

Kamiński, R. (2008). *Aktywność społeczności wiejskich. Lokalne inicjatywy organizacji pozarządowych*, Instytut Rozwoju Wsi i Rolnictwa PAN, Warszawa.

Kapusta, R. (2005). Uwarunkowania zrównoważonego rozwoju wsi i rolnictwa w Polsce, w: *Zrównoważony rozwój – doświadczenia polskie i europejskie*, Wydawnictwo I-Bis, Wrocław.

Kawczyńska-Butrym, Z. (2009). *Migracje*. Wybrane zagadnienia, Wydawca UMCS.

Kłodziński, M. (1995). Uwarunkowania wielofunkcyjnego rozwoju gminy, w: L. Klank (red.), *Wieś i Rolnictwo w okresie transformacji systemowej*. Warszawa: IRWiR PAN.

Kłodziński, M., & Rosner, A. (1995). Wielofunkcyjny rozwój terenów wiejskich a polityka regionalna, w: M. Kłodziński & A. Rosner (red.), *Polityka regionalna w rozwoju obszarów iejskich*. Warszawa: SGGW.

Kłodziński, M., & Wilkin, J. (1998). Rozwój obszarów wiejskich w Polsce, w: *Identyfikacja priorytetów w modernizacji sektora rolno-spożywczego w Polsce*. Warszawa: FAPA.

Kobiałka, A. (2003). Rola gminy w rozwoju obszarów wiejskich, w: M. Adamowicz (red.), *Strategie rozwoju lokalnego (Tom t.1)*. Warszawa: SGGW, 65–70.

Kościelecki, P., Bloch, E., Śpiewak, R., & Zalewska, K. (2010). *Podręcznik tworzenia systemów i ewaluacji wskaźników w lokalnych strategiach rozwoju*. Opracowanie EGO s.c. dla MRiRW.

KPZK 2030 (2011). *Koncepcja przestrzennego zagospodarowania kraju 2030*. Ministerstwo Rozwoju Regionalnego, Warszawa, s. 193.

KSRR 2010–2020 (2010). *Krajowa strategia rozwoju regionalnego 2010–2020. Regiony, miasta, obszary wiejskie, Ministerstwo Rozwoju Regionalnego*, Warszawa, s. 150.

Kucharska, J. (2020). *Proces starzenia się społeczeństwa a rozwój gospodarczy (na podstawie województwa wielkopolskiego)*. Uniwersytet Przyrodniczy w Poznaniu, Wydział Ekonomiczno-Społeczny, Poznań.

Lewandowska-Gwarda, K., & Antczak, E. (2015). *Migracje wewnętrzne w polskich miastach- analiza z wykorzystaniem przestrzennej dynamicznej metody przesunięć udziałów*, Roczniki Kolegium Analiz Ekonomicznych Szkoła Główna Handlowa w Warszawie.

List, F., ed. (1841). *Das nationale System der politischen Ökonomie*. Stuttgart: Cotta Verlag.

Łukasiewicz, K. (2009). *Wdrażanie lokalnych strategii rozwoju w ramach Osi IV Leader Program Rozwoju Obszarów Wiejskich na lata 2007–2013 (s. 3)*. Warszawa: MRiRW.

Małkowska, A. (2011). Rola infrastruktury ekonomicznej w rozwoju społeczno-gospodarczym. Zeszyty Naukowe Uniwersytetu Ekonomicznego w Krakowie, Nr 850 2019, s. 65–76. https://r.uek.krakow.pl/bitstream/123456789/938/1/171189551.pdf.

Miller, M. K., & Luloff, A. E. (1981). Who is rural? A typological approach to the examination of rurality. *Rural Sociology* 46, 608–625.

Ministry of Agriculture and Rural Development (2014). https://www.gov.pl/web/rol nictwo/leader1

Ministry of Agriculture and Rural Development (2020). https://www.gov.pl/web/rolnictwo/strategia-zrownowazonego-rozwoju-wsi-rolnictwa-i-rybactwa-na-lata-2012-2020

MRiRW (2012). *Sprawozdanie z realizacji Programu Rozwoju Obszarów Wiejskich na lata 2007-2013.* Sprawozda nie za 2011 rok, Numer sprawozdania 5/2011. Warszawa: MRiRW.

NSPLiM (2015). *Narodowy Spis Powszechny Ludności i Mieszkań 2011*, Obszary wiejskie, GUS, Warszawa–Olsztyn.

OECD (2018). http://www.oecd.org/regional/regional-policy/42392313.pdf

Pakt (Pact) (2018). Pakt dla obszarów wiejskich, projekt z dnia 14 czerwca 2018 r. Dokument uwzględniający uwagi zgłoszone przez członków Komitetu Społecznego Rady Ministrów. http://www.krir.pl/files/Pakt_dla_obszar%C3%B3w_wiejskich.pdf.

Poczta, W., & Wysocki, F. (2000). *Strategia Rozwoju Rolnictwa i Obszarów Wiejskich w Wielkopolsce.* PWRiL, Poznań, s. 118.

Pouliquen, A. (2011). Integracja krajów Europy Wschodniej z unią europejską: od ożywienia do kryzysu w rolnictwie, *Zagadnienia Ekonomiki Rolnej / Problems of Agricultural Economics*, 327(2), 3–40.

Raina, A. (2005). *Ecology, Wildlife and Tourism Development: Principles, Practices and Strategies.* Delhi: Sarup & Sons.

Rosner, A., & Stanny, M. (2014). *Monitoring rozwoju obszarów wiejskich. Etap I. Przestrzenne zróżnicowanie poziomu rozwoju społeczno-gospodarczego obszarów wiejskich w 2010 roku (wersja pełna)*, Fundacja Europejski Fundusz Rozwoju Wsi Polskiej, IRWiR PAN, Warszawa.

Rural Developments (1997). *CAP 2000. Working Document.* European Commission. Directorate General for Agriculture, DG VI.

Siemiątkowski, P. (2011). Rola kapitału zagranicznego w transformacji polskiej gospodarki, *Toruńskie Studia Międzynarodowe*, 1(4).

Siemiński, J. L. (1994). Główne koncepcje rozwoju lokalnego (zarys problematyki). *Człowiek i środowisko*, 18(3).

Smith, A. (2020). *Badania nad naturą i przyczynami bogactwa narodów*, Tom 1i2, Warszawa, Wydawca: Wydawnictwo Naukowe PWN.

ŚSRK (2020). *Średniookresowa strategia rozwoju kraju 2020*, Ministerstwo Rozwoju Regionalnego, Warszawa, wrzesień 2012, s. 165.

Stańczyk, J. i Szałtys, D. (2016). *Sytuacja demograficzna Polski na tle Europy. Perspektywy demograficzne jako wyzwanie dla polityki ludnościowej*, J. Hryniewicz & A. Potrykowska (red.). Warszawa: Rządowa Rada Ludnościowa.

Stanny, M., Rosner, A., & Komorowski, Ł. (2018). *Monitoring rozwoju obszarów wiejskich. Etap III, Struktury społeczno-gospodarcze, ich przestrzenne zróżnicowanie i dynamika*, Fundacja Europejski Fundusz Rozwoju Wsi Polskiej, Instytut Rozwoju Wsi i Rolnictwa PAN, Warszawa.

Stasiak, A., & Kulikowski, R. (1996). Gminy wiejskie w Polsce, *Wieś i Państw*, 2–3.

State Election Commission (2015). https://parlament2015.pkw.gov.pl/347_Wyniki.html

Strategia lizbońska (2000). 12 Strategia Lizbońska na rzecz wzrostu i zatrudnienia, przyjęta przez Radę Europejską na posiedzeniu w Lizbonie 2000 r. www.strategia lizbonska.pl.

SZRWRiR (2012). *Strategia zrównoważonego rozwoju wsi, rolnictwa i rybactwa na lata 2012-2020.*

SZRWRR (2020). *Strategia zrównoważonego rozwoju wsi, rolnictwa i rybactwa do 2020 r.*, Ministerstwo Rozwoju Regionalnego, Warszawa, kwiecień 2012, s. 140–141.

Szukalski, P. (2014). *Demografia i gerontologia społeczna*, Biuletyn Informacyjny nr 7.

Szymańska, D. (2007). *Urbanizacja na świecie.* Warszawa: Wydawnictwo Naukowe PWN.

Wieliczko, B. (2006). *Polityka Unii Europejskiej wobec obszarów wiejskich*, Studia i Monografie IERiGŻ, nr 134, Warszawa.

Wilczyński, R. (2003). *Odnowa wsi perspektywą rozwoju obszarów wiejskich w Polsce.* Poznań, Fundacja Fundusz Współpracy – Program Agro-Info, Krajowe Centrum Doradztwa Rozwoju Rolnictwa i Obszarów Wiejskich.

Wilkin, J. (2005). O potrzebie i założeniach długookresowej wizji rozwoju wsi w Polsce, w: *Rozwój obszarów wiejskich, doświadczenia krajów europejskich*, Instytut Rozwoju Wsi I Rolnictwa Polskiej Akademii Nauk, Warszawa.

Wilkin, J. (2007). Obszary wiejskie w warunkach dynamizacji zmian strukturalnych, w: *Ekspertyzy do Strategii Rozwoju Społeczno-Gospodarczego Polski Wschodniej do 2020*, Ministerstwo Infrastruktury I Rozwoju, Warszawa.

World Bank (2000). *Rural Development Strategy Eastern Europe and Central Asia.* World Bank Technical Paper No. 404. Europe and Central Asia Environmentally and Socially Sustainable Rural Development Series. Work in progress WTP484 for public discussion, September.

Wrona, J. (2012). *Słownik geografii społeczno-ekonomicznej*, Universitas, Kraków.

Wrzochalska, A. Ed. (2014). *Rural Economies in Central Eastern European Countries after EU Enlargement*, Instytut Ekonomiki Rolnictwa i Gospodarki Żywnościowej – Państwowy Instytut Badawczy, Warszawa.

Zarębski, M. (2001). Koncepcja wielofunkcyjnego rozwoju terenów wiejskich, w: P. Jaworowski, C. Sobków, M. Zarębski, *Agrobiznes – problemy negocjacji z Unią Europejską*. Toruń: UMK (110).

Zawalińska, K. (2009). *Instrumenty i efekty wsparcia Unii Europejskiej dla regionalnego rozwoju obszarów wiejskich w Polsce*, Instytut Rozwoju Wsi i Rolnictwa Polskiej Akademii Nauk, Warszawa.

Zioło, Z. (2014). Modele funkcjonowania i rozwoju obszarów wiejskich w Polsce po 1945 r, w: W. Kamińska & K. Heffner (red.), *Polityka spójności UE a rozwój obszarów wiejskich stare problemy i nowe wyzwania*. Warszawa: Polska Akademia Nauk Komitet Przestrzennego Zagospodarowania Kraju.

Zwoliński, (2009). *Zmiany społeczno-demograficzne na terenach wiejskich w państwach Unii Europejskiej*, A. Sikorska (red.). Warszawa: Instytutu Ekonomiki Rolnictwa i Gospodarki Żywnościowej – Państwowego Instytutu Badawczego.

2 Thirty years of transformation in Poland as a basis for changes in the socioeconomic structure of rural areas

Agnieszka Baer-Nawrocka and Walenty Poczta

This chapter presents the changes undergone by the Polish agricultural sector in the last three decades. These transformations were co-defined by two key processes, i.e. the economic shift and the integration with the European Union. Topics addressed in this chapter are related to multipurpose rural development defined as reducing the dominant role of agriculture in the rural economy while developing non-agricultural activities, including tourism. The analysis covers the evolution of the role of agriculture in the national economy and of socioeconomic structures in rural areas. The authors present the changes in productive inputs and in relationships between them. Productive inputs remain strictly related to production structures, and therefore this chapter makes a reference to changes in farm structures (in terms of size and ownership). These are important topics from the perspective of the potential outflow of capital, land, and labor resources (especially the latter) to non-agricultural activities. The authors emphasize that the Polish agricultural sector continues to be affected by an excessive labor potential which has adverse economic effects on both the agricultural economy and on farming families. This suggests there is need for continuing the shift to non-agricultural activities and to a non-agricultural use of labor resources engaged in farming. Also, the authors present the role of funds derived from the European Union budget in agricultural and rural support. The above topics are presented in a context of the political and economic restructuring in Poland compared to the experience of other Central and Eastern European countries.

2.1 Transformation of political and economic structures: a comparison with Central and Eastern European countries

The history of political systems and economics has never seen anything that could match the Central and Eastern European (CEE) transformation which started 30 years ago. As Orłowski (2019a) noted, there are three reasons which make it a period of crucial importance to all CEE countries under transformation (Blanchard, 1997):

- Firstly, that process was known from the very beginning to be unique and of historical importance.
- Secondly, the successes and failures of the economic transformation involved political change of an importance rarely seen in history, and had a profound impact on the formation of the whole socio-political system in CEE countries.
- Thirdly, from the economic point of view, it was an extremely difficult process which rapidly turned out to be much more complex, painful, and risky than initially expected (Orłowski, 2019a).

In that part of the world, the transformation involved the fall of a political and economic system which for several decades had determined the functioning of tens of countries from many continents with a population of hundreds of millions. Actually, from the end of World War 2 to 1989–90, that system completely dominated the Central and Eastern European countries.

The fall of real socialism as a political system and of the planned economy—which was an intrinsic part of it and *de facto* turned to a command-and-quota economy—resulted in the formation of a particular kind of economy. According to Samecki (2012) that condition of the economy was referred to in English as misdevelopment: an economy affected by structural errors and distorted by the lack of market forces. That term is something different than underdevelopment where insufficient amounts of capital, combined with weak institutions and poor human capital, restrict a country's sustainable development capacity. In practice, the command-and-quota economy referred to above means that prices were managed by public authorities, the central planner. This consisted in either setting or controlling the prices. Also, the government's price policy was combined with the system of allowances.[1] In a country affected by misdevelopment, structural differences can also be found in the banking sector and in financial services. Moreover, the government controlled the flows of physical goods, i.e. the deliveries of machinery, equipment and raw and other materials, while also playing a certain role in setting the production volumes of consumer goods and services. The economy of real socialism countries was largely a closed economy; international trade was controlled by the government and handled by state-owned companies. The government's supervision over the entire economy and the absence of market price signals resulted in a faulty allocation of resources. As a consequence, Central and Eastern European countries suffered from a distorted economic structure. According to Matyska et al. (2019) this resulted in these countries becoming over-industrialized, with the industrial sector contributing 45–60% to value added. Things looked similar in agriculture, which could be referred to as an "excessively agrarian" sector. This was true not only due to the agricultural sector's share in the economies of these countries but also because of its physical dimension. Indeed, in CEE, most rural resources were used for agricultural purposes. As a consequence, during the transformation, CEE countries faced the problem of how to shift to non-agricultural activities in these territories. The reallocation of some resources from agriculture to tourism was found to be among the important transformation measures taken to enable that shift.

Central and Eastern European countries were also severely affected by international imbalances, manifested by extreme amounts of external debt in some of them. In Hungary, Poland, and Bulgaria, it was 68%, 62%, and 158% of GDP, respectively, in 1991 (Åslund, 2002). This was largely the consequence of errors in governmental investment policies.

Another important problem faced by countries under transformation was the lack of institutions characteristic of a democratic and market system, or the way they differed in legal and institutional aspects. According to Winiecki (2012), all communist economies entered the transformation period with a considerable, if not enormous, imbalance at the macro level (and elsewhere, too). The internal and external imbalance affecting most centrally managed countries required them to promptly restore macroeconomic stability. This was a priority because without stabilizing the economy, they would be unable to initiate a successful transformation in the area of liberalization and institutional reconstruction. Hence, stabilization was a *sine qua non* condition for further action (Csaba, 1995, Åslund et al., 1996).

After 1990, in addition to their economies being in a deplorable condition, challenges caused by internal and external imbalances, and the absence or low quality of institutions, CEE governments lacked relevant knowledge and faced the uncertainty of the transformation process which was supposed to convert the existing economies into market economies. They needed to make parallel transformations in state structures: build a democratic system and a market economy. The journey from communism to both a pluralist democracy and a market economy was an unprecedented process. Therefore, no patterns or models existed to guide such a profound transformation facing Central Europe (Matyska et al., 2019). Jarmołowicz and Piątek (2013) take a slightly different look at the initiation of the transformation process. They find that the political change experienced in socialist countries in the late 1980s enabled a systemic transformation, including in the economy, which means they emphasize that the economic transformation was preceded by political change. Kornai (2006) takes the middle ground and claims that in Central and Eastern European countries, the transformation was a complete process that took place in all spheres: in the economy and political structure, in the world of political ideology, in the legal system, and in social stratification. It was non-violent, took place under peaceful circumstances, and was not preceded by war. The society was not forced to accept the changes as a result of foreign military occupation. Whether the political change preceded or was parallel to the economic transformation, it is a fact that the latter (which started in the 1990s in Central and Eastern Europe) was a transition from a centrally managed to a market economy. It included a comprehensive, radical shift in general goals and conditions of doing business by all operators. Thus, it was not yet another attempt to reform the existing system but a qualitative change which involved breaking with the previously applicable economic logic. However, there was agreement that the economic transformation was a must (Jarmołowicz & Piątek, 2013).

A major problem, if not a dispute, was the speed and depth of the transformation process, i.e. the transformation strategy. As Winiecki (2012) wrote, initially, "where do we go?" was not the only question the reformers needed to answer. Another important aspect was the path towards capitalism. Should it be steep or gently sloping? Should a big-bang or a gradual approach be followed? These were two opposite strategies. Big bang is a shock transformation, also referred to as a radical, comprehensive economic program. Conversely, a gradual transformation is called a progressive or evolutionary approach (Jarmołowicz & Piątek, 2013). According to Åslund (2002), Poland, Czech Republic, and the Baltics implemented a shock treatment, whereas Hungary, Slovakia, and other post-communist countries embarked on a gradual path (Matyska et al., 2019).

As Balcerowicz (1997), the author of the Polish reforms, noted the great political breakthrough which took place in Poland and Central Europe in early 1990s was followed by a period of an "extraordinary" policy which, however, was gradually replaced by a "normal" one. One characteristic aspect of the extraordinary policy period is that in a context where the former elites lost trust and support, and new political structures were yet to be formed, both new leaders and ordinary citizens were more willing to focus their minds and acts on reforms and be committed to common goals. This made it possible to implement economic changes even at a high cost (already at the early stage) because right after the political shift, the society was ready to accept it (Winiecki, 1999). Moreover, Balcerowicz (1997) emphasized that the implementation of this transformation strategy allowed to break the inertia of economic operators and take advantage of the "extraordinary" policy period and attain synergies. Finally, it made the Polish transformation a successful project.

The duration of the transformation process is an important though debatable issue. Usually, the transformation of CEE economies is believed to have started during 1990–1991. In Poland, there are no doubts that the transformation started at the beginning of 1990, with the implementation of the "Balcerowicz plan,"[2] a reform bundle named after the then deputy prime minister. According to Orłowski (2019a), the plan liberalized the economy by abolishing most of the remaining components of the command-and-quota system, deregulating the prices, making the zloty a convertible currency in the domestic market and enabling free trade with the external world. On the other hand, it initiated the process of economic stabilization by pegging the zloty to the U.S. dollar, restricting the allowances, generating a budget surplus and reducing (with special fiscal instruments) the existing wage indexation mechanisms (through a high taxation rate applicable above the defined wage growth level). The third key element of reforms, i.e. privatization and development of market institutions (a process spread over many years), also started in that period (Åslund & Orłowski, 2014).

Orłowski (2019a) indicates that it could be more problematic to specify the moment where the transformation process came to an end. He specifies five possible dates for Poland which could be viewed as the completion of the transformation (the corresponding dates for other CEE countries can be determined in a similar way):

- 1995, when GDP went above the level recorded in 1989 and the economy embarked on a rapid growth path;
- 1994–2006, according to the European Bank for Reconstruction and Development (EBRD), i.e. 1994 as the year when the liberalization of the economy reached a level characteristic of highly developed market economies, and 2006 as the year where a similar level of institutional development was attained;
- 2004, when Poland joined the EU (Orłowski, 2019b);
- 2009, when the World Bank classed Poland among high-income economies;
- 2015, when Poland outperformed Greece (as the first European country which was part of western market economies before 1990) in terms of GDP per capita.

Whether one believes that one of the abovementioned dates marks the end of the transformation process or that Poland and other CEE counties continue to undergo transformation, it has to be admitted that these countries are still at a lower level of economic development than western market economies. This means they imperatively need to continue implementing processes aimed at economic modernization and strengthening of institutions that are characteristic of developed market democracies.

Similarly to other CEE countries, the Polish economy was initially hit by a deep transformation recession (Table 2.1). During 1990–1991, these countries saw their economies decline by 10–20%. In Poland, the economic downturn was accompanied by high inflation and a high and growing unemployment rate. However, the bounce back in started in 1992. According to Åslund and Orłowski (2014), the main reason behind the success is that the difficult early-1990s reforms bore fruit. The tough macroeconomic conditions resulted in the failure of many enterprises and in the rise of unemployment, on the one

Table 2.1 Average annual GDP growth rate (%)

Specification	1990–1991	1992–1999	2000–2003	2004–2008	2009–2013	2014–2018
Poland	−7.1	5.2	2.7	5.4	2.6	4.0
Other CEE countries (arithmetic mean)	−8.4	0.0	4.4	5.8	−0.2	3.7
UE-15	2.4	2.0	2.1	2.1	−0.4	1.9

Source: International Monetary Fund data, Orłowski (2019a).

hand, but drove an unprecedented explosion of entrepreneurship, on the other. Although inflation declined, it was not entirely suppressed (a two-digit inflation rate persisted). The Polish economy somehow slowed down in the pre-accession period; this was related to economic restructuring processes and macroeconomic consolidation that preceded the accession to the EU.

In the initial years that followed the accession, Poland (just like other CEE countries) recorded high economic growth rates and embarked on an economic convergence path with EU-15 members. The Polish economy proved to be highly robust to the global financial crisis; unlike other CEE and EU-15 countries, Poland reported a positive growth rate. This provided grounds for rapid economic growth in Poland which lasted from 2014 until the crisis caused by the COVID-19 pandemic.

From the extremely deep 1989 recession to 2019, the Polish economy grew by 156% (Table 2.2), which is the highest rate of all CEE countries. As a consequence, in 2018, per capita GDP in purchasing power parities went above 70% of the average GDP for EU-27. In 1991 and in 2004 (when Poland accessed the EU) that proportion was 32.0% and 51.5%, respectively.

Table 2.2 GDP growth and levels in Central and Eastern European countries, 1990–2019

Country	Average annual growth rate (%), 1990–2019	GDP in 2019				GDP per capita in PPS in 2018 (EU-27=100)
		1989=100	2005=100	2010=100	2015=100	
Poland	3.2	256	174	138	119	71.0
Bulgaria	1.0	132	147	125	115	51.1
Czech Republic	1.7	169	138	123	113	91.1
Estonia	2.1	185	138	140	119	82.2
Hungary	1.7	164	129	130	118	71.2
Latvia	0.9	132	131	134	113	69.1
Lithuania	1.1	137	147	139	115	80.8
Romania	1.7	163	162	141	122	66.1
Slovakia	2.5	207	162	127	112	73.5
Slovenia	1.8	173	129	118	115	87.5
UE-15[a]	1.3	149	118	113	108	108.2

Source: Próchniak et al. (2019); Eurostat data https://ec.europa.eu/eurostat/data/browse-statistics-by-theme (April 21, 2020); *Statistical Yearbook of the Republic of Poland*, Statistics Poland (relevant Yearbooks); own calculations.

a Weighted average.

The agriculture sector was covered by the transformation process in a particular way. As mentioned earlier, being "excessively agrarian" was a characteristic feature of CEE economies. The excessive role of agriculture was mirrored by its contribution to GDP and the proportion of labor resources engaged in it.[3] In most CEE countries, the contribution of agriculture to GDP (gross value added) in 1990 was 10–20% (Table 2.3). A significantly smaller proportion was recorded only in former Czechoslovakia. Also, in many of these countries, ca. 20% or even more (Poland, Romania) labor resources were employed in agriculture. A share below 10% was recorded in Slovenia and in the territory of today's Czech Republic (cf. Brada, 1989; Jackman, 1994). As regards agricultural production, high labor inputs were usually accompanied by large material inputs (Fabisiak & Poczta, 2011). The countries differed in how fast the farming labor force left the agricultural sector during the economic transformation. In the 1990s, the greatest reduction in employment (from 50% to 60%) was experienced in the Czech Republic and Slovakia. The changes were also relatively fast in Estonia and Hungary. In turn, in countries such as Poland and Slovenia, the decline in agricultural employment was much smaller (cf. Schiff and Montenegro, 1997;

Table 2.3 Agricultural sector's share in employment and contribution to GDP in CEE countries

Country	Agricultural employees (% share)[a]			Gross value added (% share)		
	1990	2005	2018	1990	2005	2018
Poland	25.2	17.4	9.6	11.8[b]	3.3	2.8
Bulgaria	13.5	8.9	6.6	18.0	8.6	4.2
Czech Republic	8.9	4.0	2.8	8.0	2.4	2.2
Estonia	18.0	5.2	3.3	17.0[c]	3.5	2.6
Hungary	18.2	4.9	4.8	8.0	4.3	4.3
Latvia	17.4	12.1	7.0	22.0[c]	4.3	3.8
Lithuania	18.9	14.3	7.2	27.0[c]	4.8	3.0
Romania	24.0	32.3	22.3	20.0	9.6	4.8
Slovakia	12.0	4.7	2.3	5.0[d]	3.6	3.3
Slovenia	8.1	9.1	5.4	6.0[c]	2.8	2.4

Source: Own calculations based on Eurostat data: https://ec.europa.eu/eurostat/data/browse-statis tics by-theme, *Agriculture in the European Union. Statistical and economic information.* European Commission, Brussels (relevant years); *Employment in agriculture (2005, 2018), FAOSTAT,* accessed on June 15, 2020, *Statistical Yearbook of the Republic of Poland,* Statistics Poland (relevant Yearbooks), *International Statistics Yearbook 2019,* Statistics Poland, Warsaw 2019.

a FAOSTAT data was used for the sake of comparability of agricultural employment figures. This data differs from Polish figures provided by the Central Statistical Office in Section 2.2 of this chapter due to different methodologies being used with respect to agricultural employment statistics.
b 1989 (before the transformation started, in 1990, that share dropped to 7.1%).
c Contribution to GDP.
d 1993.

Lerman et al., 2002). These processes were strictly related to the ownership form and structure of farms.

In most CEE countries, the large share of farming employees contributed to the formation of a poorly efficient agricultural sector which required large amounts of inputs and operated as state-owned farms or cooperatives (Table 2.4). Note that agricultural cooperatives were established under state coercion. Things looked different in Poland and former Yugoslavia which discontinued the forced establishment of cooperatives. Thus, individual farms remained the prevailing form of farming (cf. Sarris et al., 1999; Swinnen, 2001; Lerman et al., 2002; Majerova et al., 2009). However, although most agricultural holdings were privately held in these two countries (especially in Poland), they operated in an institutional environment (including the market) typical of a command-and-quota economy. Information on the role of what is referred to as the collectivized (socialist) agricultural sector is presented in Table 2.4.

Before the transformation, state-owned and cooperative farms used ca. 90% (or more) of agricultural land (except for Poland and Slovenia). Far-reaching changes in the structure of land ownership and use took place from the beginning of the transformation until these countries joined the EU. The transformation process included the privatization of the state sector; land

Table 2.4 Share of different forms of farm ownership in agricultural land use (%) before and after the transformation in CEE countries

Country	Before the transformation[a]				After the transformation[b]			
	Coop-era-tives	State-owned hold-ings	Com-panies	Indivi-dual farms	Coop-era-tives	State-owned hold-ings	Com-panies	Indivi-dual farms
Poland	4.0	19.0	0.0	77.0	1.9	5.5	4.7	87.9
Bulgaria	58.0	29.0	0.0	13.0	61.0	0.0	23.0	26.0
Czech Republic	61.0	38.0	0.0	1.0	32.0	1.0	43.0	24.0
Estonia	57.0	37.0	0.0	6.0	0.0	1.0	37.0	62.0
Hungary	80.0	14.0	0.0	6.0	20.5	0.0	24.4	55.1
Latvia	54.0	41.0	0.0	5.0	0.0	0.2	9.1	90.7
Lithuania	91.0		0.0	9.0	3.3	0.5	32.2	64.0
Romania	59.0	29.0	0.0	12.0	7.0	0.0	13.0	80.0
Slovakia	69.0	26.0	0.0	5.0	50.2	0.3	26.9	22.6
Slovenia	8.0		0.0	92.0	0.0	0.0	5.8	94.2

Source: Fabisiak and Poczta (2011).

a As per the last agricultural census before the economic transformation.
b As per the last agricultural census before the accession to the EU.

used by cooperatives was restituted to its owners or their descendants. State-owned farms and cooperatives supervised by the government were replaced by a developing sector of individual farms, newly established companies and cooperatives created on a voluntary basis. The transformation of ownership and the reduction of the role of agriculture in employment and its contribution to GDP were the main manifestations of how these economies (and their rural areas) shifted to non-agricultural activities. That shift provided grounds for a multipurpose development of rural areas and for non-agricultural uses of resources previously linked to agriculture. This made it possible to allocate some of the agricultural resources to other activities, including rural tourism.

2.2 Transformation of social structures

According to Wilkin (2008) and Halamska (2011), in addition to the shift to non-agricultural activities, the processes which play a key role in the rural transformation (as witnessed in recent years) include: the decline in the share of rural areas; the redefinition of social groups; and the formation of a new agricultural model reflected by a slow adjustment of agricultural production structures to the market economy. This section analyzes the transformation which is part of the redefinition of social groups, a process that affects the socioeconomic situation of Polish households. Comprehensive information regarding this topic can be derived from representative surveys on household budgets conducted every year by the Central Statistical Office. The surveys play a crucial role in analyzing the population's quality of living, and are the main source of information on how money is spent, on quantities of food consumed, revenues, and other aspects of living conditions of specific population groups. As a basic classification, the households are divided into five main socioeconomic groups of population which include farmer households in addition to employee households, self-employed households, pensioner households, and unemployed households.

The creation of privately held farms was decisive not only for what Polish agriculture and rural areas would become in the decades to come but also for how fast the socio-occupational structures could be modernized during the transformation. The 1990s saw a major slowdown in rural–urban migration; at the same time, gainful off-farm activities started to be governed by the market mechanism. The extension of social benefits was an accompanying process which involved an increase in social transfers allocated to farms. When confronted with restrictions to agricultural development and with value added being transferred from the agriculture to other sectors through the price mechanism, these dynamic changes had to result in a tremendous growth of importance of non-farming income in total incomes of farmer households (Zegar & Gruda, 2000). In this context, note that incomes of farming families are also used as a way of accumulating resources for production purposes (unlike in employee families, for instance). Faced with a radical decline in incomes, farmer households protected their consumption at the expense of accumulation, which is a common defense instrument. As Zegar and Gruda (2000) note, the share of consumption in the income distribution structure grew from 67.5% in 1990 to nearly 96% in 1998 as a consequence of that process. Therefore, the share of accumulation declined significantly, which means this sector moved in an opposite direction to the economy as a whole.

Households can be viewed as a specific micro-system which transforms under the impact of macro-structural changes. Undeniably, a relationship exists between the situation in the labor market and changes in the socioeconomic structure of households. This is because that structure is largely determined by the population's economic activity, unemployment, and type of revenue stream. Figure 2.1 shows the trends related to the structure of households grouped by the main source of income. During 1993–2002, due to structural transformation and economic fluctuations, Poland experienced severe perturbations in the operation of the labor market. This undoubtedly had an impact on the direction and pace of socioeconomic change. In the whole period, the basic group of households was composed of those which had employment as the main source of income. Their share dropped considerably in 2000 (to ca. 40%) and reached 47.6% in 2018. Pensioner households had the second largest share in the household structure; compared to 1993, it grew by nearly six percentage points. As Kowalska (2002) emphasized, that process—in addition to the declining average household size—was part of adverse demographic trends affecting Poland in the early years of socioeconomic transformation. Conversely, the consistent (though sluggish) increase in the importance of self-employed households was a positive aspect. In 2018, the share of farmer households in the total household population was half that recorded in 1993. The share of households with no steady sources of income was quasi-constant throughout the study period.

According to the 1988 National Census, 22.9% of the rural employed population had a non-farming job; 24.0% worked in agriculture (including

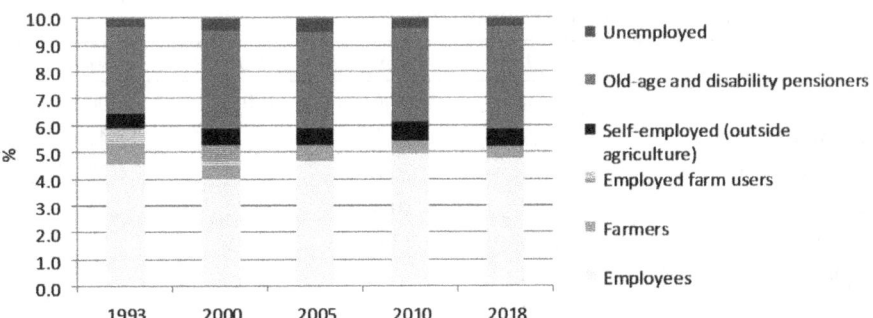

Figure 2.1 Changes in the socioeconomic[*] structure of Polish households (grouped by main source of income), 1993–2018.
Source: Own calculations and compilation based on Central Statistical Office data from representative surveys of Polish household budgets in 2003–2018: *Household budget survey*, Statistics Poland (relevant Yearbooks).

19.9% in the private sector and 4.3% in the public sector); 17.1% were unemployed; and 35.8% were dependent people (Statistics Poland, 2003). The years that followed were marked by further changes in the importance of different sectors of the rural economy and by people moving between socio-economic groups (Table 2.5). In more than 15 years, employed households became the largest group of rural households as their share increased from less than 22% to over 47%. At the same time, note that due to changes in the classification of farms by socioeconomic criteria, in 2015, that group included most farms which in 2000 formed a separate statistical category of "employed farm users." In 2000, that category accounted for one-fifth of all households. From 2000 to 2015, the percentage of white-collar households more than tripled. This indicates the emergence and growing importance of what is referred to as the rural middle class. Another finding is the increasing share of off-farm self-employed households (from 4.3% in 2000 to 6.2% in 2015).

The acceleration of processes referred to above was particularly noticeable after Poland joined the EU. What can also be noticed in the study period is a minor drop in the share of farmer households (from 12.5% in 2000 to 11.3% in 2015). That period also saw a decline in the importance of pensioner households in the socioeconomic structure of rural households.

Table 2.6 presents data on the evolution of average monthly household per capita incomes during 2010–2018, with rural households being identified as a separate category. Disposable per capita income[4] more than doubled over that period both in the general household population and in rural areas. However, the average monthly disposable per capita incomes of rural households continued to be lower than in other Polish households. The study period saw an increase in all sources of income in absolute terms. At the same time, there was a change in the structure of monthly disposable per capita incomes. Incomes from employment

Table 2.5 Changes in the socioeconomic structure of rural households (grouped by main source of income), 2000–2015 (%)

Specification	2000	2005	2010	2015
Total	100.0	100.0	100.0	100.0
Employees:	21.7	37.2	45.2	47.4
blue-collar jobs	16.3	26.6	31.4	32.3
white-collar jobs	5.4	10.6	13.9	15.1
Farmers	12.5	13.6	11.8	11.3
Employed farm users[a]	19.4	–	–	–
Off-farm self-employed	4.3	4.2	5.9	6.2
Pensioners	38.1	39.5	34.0	31.8
Unemployed	4.0	5.5	3.1	3.3

Source: Own compilation based on Statistics Poland (2017).

a See comments below Figure 2.1.

Table 2.6 Levels and structure of average monthly per-capita incomes in the total household population and in rural households in Poland, 2000–2018

Specification	Total										Rural areas									
	2000		2010		2015		2018		difference (2018–2010)		2000		2010		2015		2018		difference (2018–2010)	
	PLN	%	PLN	%	PLN	%	PLN	%	PLN	%	PLN	%	PLN	%	PLN	%	PLN	%	PLN	%
Disposable income	610.5	100.0	1,200.8	100.0	1,386.2	100.0	1,693.5	100.0	1,083.0	–	483.0	100.0	948.1	100.0	1,105.7	100.0	1,432.7	100.0	949.7	–
including actual disposable income	588.6	96.4	1,154.0	96.1	1,337.9	96.5	1,643.1	97.0	1,054.5	0.6	465.1	96.3	914.6	96.5	1,073.9	97.1	1,398.7	97.6	933.6	1.3
including: income from employment	291.0	47.7	633.7	52.8	757.9	52.8	890.0	54.7	599.0	4.9	181.8	37.6	441.7	46.6	556.3	50.3	681.9	47.6	500.1	10.0
income from off-farm self-employment	52.8	8.7	108.0	8.7	120.1	9.0	147.3	8.7	94.5	0.0	25.2	5.2	69.3	7.3	82.0	7.4	105.5	7.4	80.3	2.2
agricultural income of an individual farm	32.8	5.4	51.5	5.4	44.2	4.3	63.6	3.2	30.9	–1.6	75.7	15.7	124.3	13.1	105.2	9.5	151.3	10.6	75.6	–5.1

(*Continued*)

Table 2.6 (Cont.)

Specification	Total										Rural areas									
	2000		2010		2015		2018		difference (2018-2010)		2000		2010		2015		2018		difference (2018-2010)	
	PLN	%	PLN	%	PLN	%	PLN	%	PLN	%	PLN	%	PLN	%	PLN	%	PLN	%	PLN	%
social welfare payments including:	169.2	27.7	307.5	25.6	353.4	25.5	414.2	24.5	245.0	−3.3	150.2	31.1	236.8	31.1	281.0	25.0	321.0	22.4	170.7	−8.7
old-age and disability pensions	150.3	24.6	276.8	23.0	315.2	23.0	371.7	21.9	221.4	−2.7	134.1	27.8	213.1	22.5	248.5	22.5	284.8	19.9	150.8	−7.9
other social welfare payments	29.4	4.8	40.3	3.4	44.0	3.2	114.2	6.7	84.8	1.9	26.4	5.5	42.4	4.5	42.2	3.8	129.8	9.1	103.3	3.6
other incomes	33.7	5.5	55.4	4.6	59.8	4.3	56.8	3.4	23.1	−2.2	23.2	4.8	31.6	3.3	36.5	3.3	40.8	2.8	17.6	−2.0

Source: Own calculations and compilation based on Statistics Poland (2017) and Central Statistical Office data from representative surveys of Polish household budgets in 2000–2018: *Household budget survey*, Statistics Poland (relevant Yearbooks).

considerably strengthened their role (reaching 47.6%, which means growth by ca. ten percentage points from 2000 to 2018). The share of incomes derived from off-farm self-employment increased, too, though much less significantly. At the same time, compared to 2010, a smaller proportion of disposable incomes was derived from working on an individual farm in 2018. This is indicative of the evolution of the rural labor market towards an increased importance of off-farm activities. Such revenue streams as old-age and disability pensions also play a less significant role. The increase in importance of incomes from other social allowances is mostly related to the introduction of the 500+ program[5] in 2016.

Central Statistical Office data indicates the severity of income disparities in Polish urban and rural areas, as measured with the Gini coefficient.[6] The Gini coefficient for 2003–2018 (determined based on disposable household income per capita) suggests that disparities in household incomes are greater in rural areas than in cities and than the average level for Poland (Figure 2.2). A considerable upward deviation (an increase in rural income disparities) was recorded in 2013, and was mostly due to a deterioration in incomes of farmer households. It is important to find out whether income disparities between rural households decrease over time and whether their Gini coefficient follows a trend similar to what is recorded in urban areas and in the total household population. According to data presented in this chapter, rural and urban households followed divergent trends until 2013. Indeed, income disparities between rural households kept growing until 2013, whereas income disparities between urban households started to decline after 2005. In recent years, the Gini coefficient has clearly followed a downward trend in both household types.

According to a study by Jędrzejak and Pekasiewicz (2017) and Wołoszyn and Wysocki (2014), farmers are the most diversified socio-occupational group in terms of incomes. The authors indicate that during 2006–2014, ca. 26–28% of farmer households were at risk of poverty, whereas ca. 7–10% earned incomes above the wealth line. In the socioeconomic group of farmers,

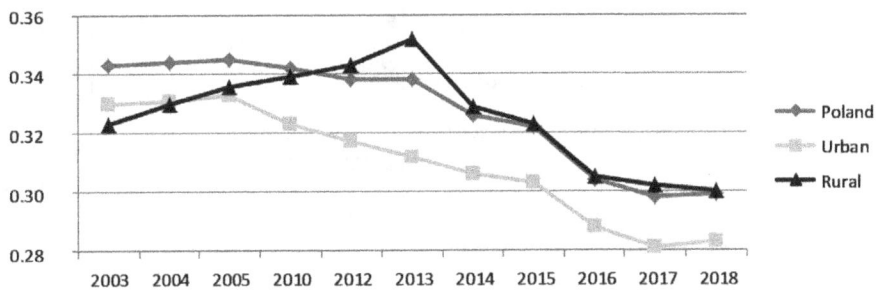

Figure 2.2 Income disparities in Poland between groups residing in different areas, as measured with the Gini coefficient, 2003–2018.
Source: Own calculations and compilation based on Central Statistical Office data from representative surveys of Polish household budgets in 2003–2018: *Household budget survey*, Statistics Poland (relevant Yearbooks).

the indicators of both poverty and wealth were much higher than the corresponding levels calculated for the entire population of Polish households.

The household survey was based on the assumption that farmer households are households whose exclusive or main means of subsistence are incomes derived from an agricultural use of an individual farm. It therefore comes as no surprise that incomes from individual farms have the largest share in the structure of disposable incomes of farmer households (Figure 2.3). However, during 2003–2018, that share followed a downward trend (in 2003, it was slightly above 72%; in 2009, it was barely 66%). The decline was mostly compensated by the increase in incomes from employment (by 11 percentage points) and in other social allowances (by ca. five percentage points). Old-age and disability pensions can also be observed to lose their importance as a source of income for farmer households.

Farmer households found themselves in a particularly difficult financial situation at the beginning of the economic transformation path. Real incomes of individual farms were on a decline: in 1990, they reached 48.6% of what they were in 1989; in 1991, they reached 73.9% of what they were in 1990. This was mainly caused by the decline in prices charged by farmers and by the deteriorating proportion between these prices and products (inputs) purchased by farmers. Two phenomena occurred concurrently: an absolute decline in incomes due to the growing crisis; and the deteriorating situation of farmers relative to other socio-occupational groups. This gave rise to a sense of injustice and exacerbated the frustration of the farming population. Such a situation is referred to as the Easterlin paradox, which means that the subjective perception of changes is no less important than their absolute (objective) dimension (Lewandowski, 1994).

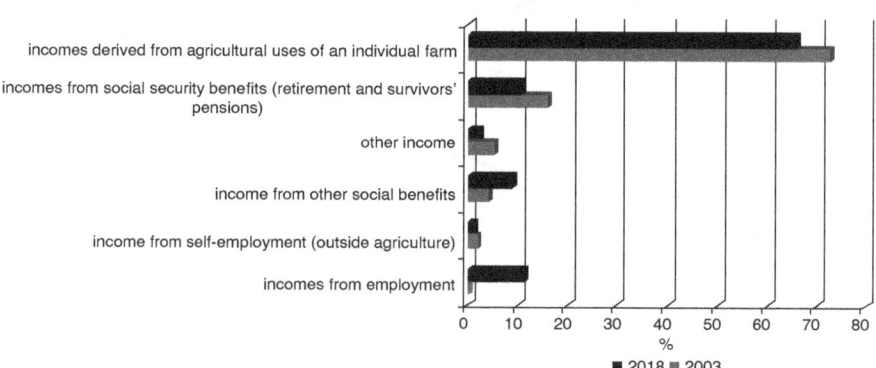

Figure 2.3 Structure of disposable incomes of farmer households in 2003 and 2018 (%). Source: Own calculations and compilation based on Central Statistical Office data from representative surveys of Polish household budgets in 2003–2018: *Household Budget Survey*, Statistics Poland (relevant Yearbooks).

Central Statistical Office data on household budgets clearly shows that following the accession (i.e. in a period when agriculture was increasingly supported with domestic and EU funds), farmers saw their real incomes grow faster than other socio-occupational groups. This was reflected by an improvement in the proportion between incomes of farmers and other groups of households.[7] As shown in Figure 2.4, during 2004–2018, the average income gap between farmers and employees was by 17.5 percentage points lower than that recorded in 2003. This can be explained by the increase in farming incomes which largely resulted from direct payments received under the Common Agricultural Policy (for more information, see Section 2.4 of this chapter).

It can be definitely concluded that agriculture is not—and will not be—able to ensure revenue streams to all (or even most) rural residents at a level that could be viewed as satisfactory. Based on broad research carried out at district level in Poland, Rosner and Stanny (2007) found that the diversification of the rural labor market had reached an advanced stage. The diversification of incomes of farming families leads to pluriactivity. The contribution of off-farm income to the budgets of farming families slowly grows. However, both the high share of farming employment at country level and the forecasted concentration processes (which will erode the demand for farm labor) suggest that attaining the desired proportion between agricultural and non-agricultural employment is among the aspects of critical importance to the rural socioeconomic development process. Therefore, there is need for further diversification of the rural economy through measures which include supporting the development of economic activity of the rural population, developing good counseling services for newly created economic operators, or supporting the development of rural infrastructure.

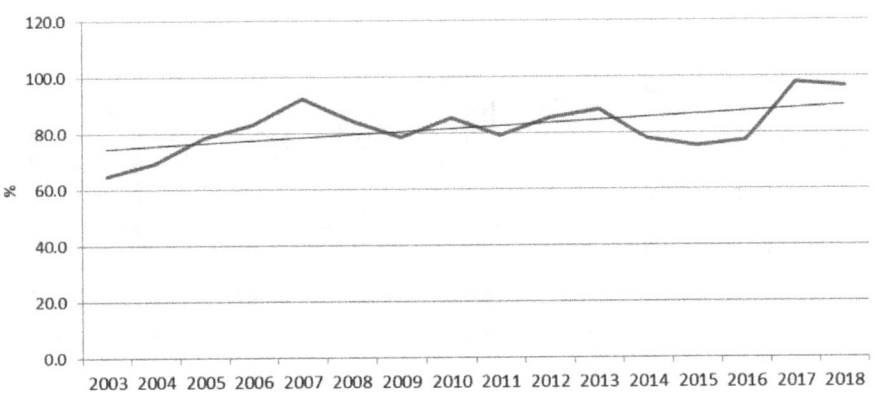

Figure 2.4 Proportion between incomes of farmer households and employed households, 2003–2018 (%).
Source: Own calculations and compilation based on Central Statistical Office data from representative surveys of Polish household budgets in 2003–2018: *Household Budget Survey*, Statistics Poland (relevant Yearbooks).

2.3 Rural shift to non-agricultural activities

2.3.1 Changes in labor, land, and capital resources

Inputs are closely related to production structures having a direct impact on the potential of farms to provide agritourism services. These problems are significant from the point of view of the potential flow of labor resources (primarily), but also of capital and land resources, to non-agricultural economic activity.

Agricultural activities strongly depend on resources of environmental productive inputs while also having a strong impact on their potential and condition. In many European regions, agriculture is the sector that formed the rural landscape; as a consequence, rural areas can be a source of public goods such as biodiversity and landscape diversity, rural vitality, high quality of air, etc. Properly functioning markets for productive inputs are an essential condition for sustainable development, both of agriculture and of rural areas (Swinnen & Knops, 2013). Also, they are the key conditions that foster growth and put the agricultural sector in a position to maintain its international competitiveness (van Bavel et al., 2009).

The fact that agriculture-related resources (especially including land) are natural is an extremely important characteristic. Indeed, land plays a particular role in agricultural production, mostly because of the spatial nature of production and of the impact on the forces of nature on farming processes. From 1980 to 2019, the area of agricultural land used by the farms was on a steady decline (Figure 2.5).

In 2019, the area of agricultural land used in Poland was by 4.3 million ha smaller than in 1980. A particularly sharp decline in agricultural land resources (by 2.5 million ha) was recorded in the 1990s. The reduction in land resources was mostly due to land being reallocated to non-farming uses.

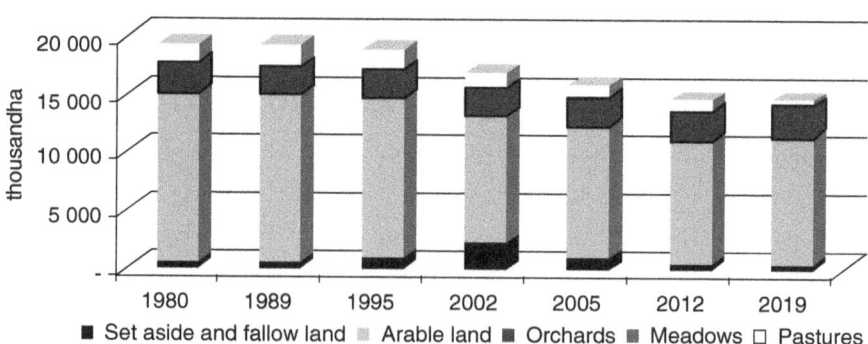

Figure 2.5 Changes in agricultural land resources in Poland (thousand ha).
Source: *Statistical Yearbook of the Republic of Poland*, Statistics Poland (relevant Yearbooks); *Statistical Yearbook of Agriculture and Rural Areas*, Statistics Poland (relevant Yearbooks); *Land Use and Sown Area*, Statistics Poland (relevant Yearbooks).

Usually, land excluded from farming uses during 1996–2002 was unsuitable for an economically viable agricultural production; this is indirectly shown by the large decline in the area under crops grown in poor-quality soils (rye, oat, potato). Also, a part of agricultural land was reallocated to non-farming uses such as residential housing, road infrastructure or leisure facilities (*Statistical Yearbook of the Republic of Poland*, 2003). The deteriorating economic condition of farms experienced after the economy switched to the market system also resulted in discontinuing agricultural production on a growing area of agricultural land. That process clearly slowed down in 2002, which can be explained by the expected accession of Poland to the European Union and by the agricultural sector being covered by CAP instruments, including area payments. The area of set-aside and fallow land located in arable land went down from 2,302 thousand ha in 2002 to 350 thousand ha in 2019, and is similar in size to what was recorded in the 1980s.[8] Although its share decreased from 77% in 1980 to 74% in 2019, arable land continues to be the dominant component of agricultural land. The area of pastures also declined (by ca. 1.2 million ha in absolute terms); that form of land use reduced its share to 2.5% (vs. 8.1% in 1980). Meanwhile, the area of orchards grew by nearly 40 thousand ha.

Reliable information on agricultural land resources maintained in good agricultural condition in farms with over 1 ha of agricultural land can be derived from data of the Agency for Restructuring and Modernization of Agriculture regarding land covered by direct payments. From 2005, these resources have been accounting for slightly more than 14 million ha, although the area declared in the first post-accession year was smaller (*Reports of the Agency for Restructuring and Modernization of Agriculture,* relevant Yearbooks). The 14 million ha can be viewed as the production potential of the Polish agriculture included in agricultural land resources.

The scope of changes to how agricultural land is used in Poland clearly shows that, as is the case in most highly developed countries, land no longer inhibits growth of agricultural production. Today's economics of land farming is based on substituting land with capital while alleviating the pressure on agricultural land. This allows some agricultural land to be set aside, especially that of low quality or of unfavorable relief or location. It can be concluded that the discontinuation of farming activities in such areas is an ineluctable process, whatever are the difficulties it poses to farmers who continue to be active there. At the same time, that process is feasible thanks to intensification measures deployed in regions with more favorable farming conditions.

As emphasized by Tomczak (2005), during the transformation, Polish rural areas and agriculture failed to find a way to tap into the production potential derived from labor resources being excessive. Although many years have passed, large employment in Polish agriculture continues to be a problem which needs to be addressed. Excessive employment in Polish agriculture not only involves extremely low labor efficiency levels in that sector (in the European context) but also, most importantly, creates a gap in the Polish GDP which could be bridged by the excessive agricultural labor force working in other sectors. The changes in how labor force has been estimated since the 2002 National Census allowed the excess resources to be put in real terms. Nevertheless, data presented in Figure 2.6 suggests that over 2.3 million people are employed in agriculture, which accounts for more than 15% of total employment. This is an extremely high level, especially when compared to EU-15 countries. As land resources decline faster than the labor force, the labor-to-land ratio deteriorates. Hence, in 2018, there were 16 persons per 100 ha of agricultural land in the Polish agricultural sector. When expressed in AWUs (Annual Work Units),[9] the situation of labor inputs seems a little bit better in absolute terms. The AWU is an indicator of labor inputs used in agricultural production; from 2002 to 2018, these inputs declined by nearly 600,000.

However, data shown in Table 2.7 suggests that the labor-to-land ratio has been unfavorable for many years, and has undergone only slight changes. The improvement in the relationship between capital inputs (total of intermediate consumption and depreciation) and labor inputs can be primarily explained by the increase in capital expenditure: in 2002, there was EUR 4,300 per AWU, compared to nearly EUR 11,000 per AWU in 2019. Similarly, the increase in capital expenditure and the decline in land resources resulted in improving the capital-to-land ratio. However, it needs to be emphasized that the relationships between agricultural productive inputs are largely inconsistent with what is witnessed in countries which—due to climate conditions—have a similar structure of agricultural production, i.e. in EU countries (especially including northern and western Europe). As a consequence, labor productivity in Polish agriculture (measured as output) is over three and nearly six times lower than the average

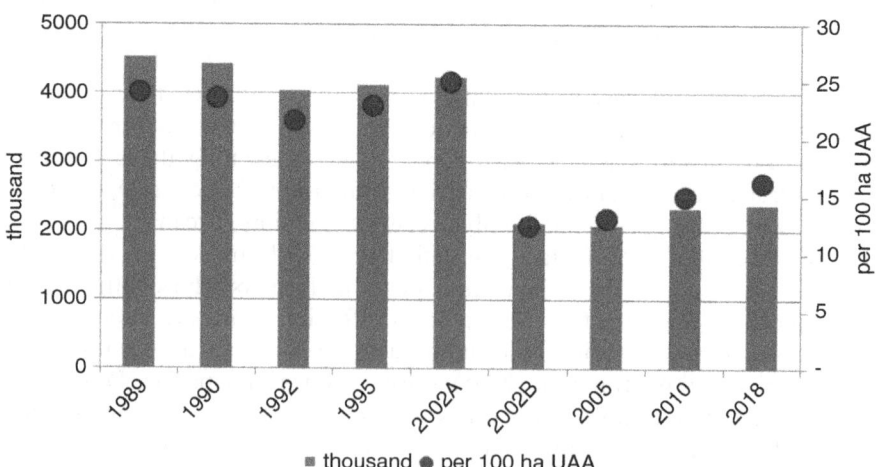

Figure 2.6 Agricultural employment, 1989–2018 (as of December 31).
Source: *Statistical Yearbook of the Republic of Poland*, Statistics Poland (relevant Yearbooks).

Table 2.7 Relationships between productive inputs in Polish agriculture, 2002–2019

Years	Area of agricultural land per AWU (ha)	Capital expenditure per AWU (EUR thousand)	Capital expenditure per ha of agricultural land (EUR)
2002	7.5	4.3	574.3
2003	7.1	3.8	541.8
2004	7.2	4.2	589.4
2005	6.9	4.5	648.4
2006	7.0	4.8	688.8
2007	7.0	5.8	824.2
2008	6.9	6.6	956.7
2009	7.1	5.4	770.7
2010	8.1	6.8	836.6
2011	7.9	7.9	1,008.3
2012	7.8	8.1	1,038.0
2013	7.6	8.2	1,082.6
2014	7.6	8.6	1,136.2
2015	7.5	8.4	1,112.1
2016	7.5	8.2	1,085.7
2017	8.7	10.0	1,146.2
2018	8.8	10.4	1,188.1
2019	8.8	10.7	1,223.5
2019/2002	117.5	250.5	213.1

Source: Own calculations based on *Statistical Yearbook of Agriculture*, Statistics Poland (relevant Yearbooks); *Systematics and Characteristics of Farms* (Statistics Poland, 2004); *Characteristics of Agricultural Holdings*, Statistics Poland (relevant Yearbooks), *Economic Accounts for Agriculture*, https://ec.europa.eu/eurostat/web/agriculture/data/database (April 4, 2020).

Note: 1 AWU (Annual Work Unit) = 2,120 working hours per year.

levels recorded in UE-28 and UE-15, respectively. Only the Romanian and Croatian agricultural sectors report lower levels of labor productivity than Poland. Another conclusion is that land resources are poorly productive, too. This can be explained by low levels of production intensity, on the one hand, and by lower levels of organization of agricultural production (a large share of cereals in the total cropping mix and a relatively low livestock density), on the other (Baer-Nawrocka & Poczta, 2016).[10]

In summary, it may be concluded that the persistently high levels of resources employed in agriculture are not conducive to making them more efficient. A similar problem is also true for several CEE members of the EU, including especially Bulgaria and Romania.[11] Low labor efficiency in agriculture gives rise to income problems which, in turn, are accompanied by development problems affecting a significant part of farms and related farming families. These issues can be solved primarily by creating off-farm rural jobs, including in the tourism business. Poczta (2010) emphasizes that if the resources employed in agriculture were moved to a non-agricultural activity, Poland could—in some aspects—benefit from the effect described in A. Lewis's development model. One significant difference could consist in triggering rural development processes without the migration of rural residents to cities.

2.3.2 The role of agriculture in the national economy

Different theories of economic development demonstrate that agriculture loses its quantitative importance as the economy develops. This is especially true for the declining share of agriculture in the social product made by an increasingly smaller number of employees. In parallel, human and animal labor inputs are replaced with capital which enables maintaining, if not increasing, the levels of agricultural output. As the farming labor force stays within the sector, it faces excessive employment (Tracy, 1993), which currently is the case in Poland. At the same time, Wilkin (2010) emphasizes that in highly developed countries, the importance of agriculture to the economy, society, and environmental condition is much greater than what could be inferred from statistics. This is because the multifunctional nature of today's agriculture is highly valued. It seems that production and market functions of agriculture are obvious and generally known. However, the modern role of agriculture is focused on what is outside the farm, and means addressing a wide variety of environmental, cultural, economic, and social needs. Society somehow expects agriculture to fulfill these functions, and is willing to pay for it (Wilkin, 2010).

The scientific literature on market and non-market functions of agriculture proposes a classification of its non-commercial functions, dividing them into four groups (Van Huylenbroeck et al., 2007):

1   Green functions, which include managing land resources to maintain their valuable properties; creating conditions for wild animals and plants; protecting animal welfare; maintaining biodiversity; improving the flow of chemicals in agricultural production systems.
2   Blue functions, related to the management of water resources (improving their quality; using water and wind as energy sources; preventing floods).
3   Yellow functions, which mean maintaining rural coherence and viability; maintaining and enhancing cultural traditions and the identity of rural areas and regions; developing agritourism and hunting activities.
4   White functions, which mean ensuring food security and food safety.

Considering the share of agriculture in the global output and its contribution to GDP, the clear trend towards reducing the quantitative importance of agriculture in the Polish economic system was particularly noticeable during 1989–2002 (Figure 2.7). This was the direct consequence

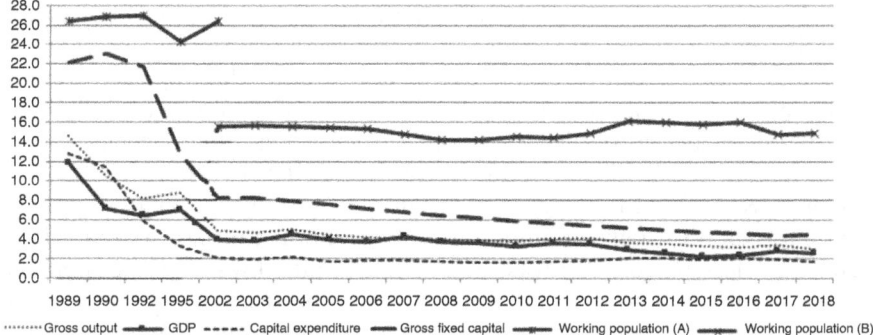

Figure 2.7 Share of the agricultural sector in the national economy (%).
Source: *Statistical Yearbook of the Republic of Poland*, Statistics Poland (relevant Yearbooks), *Employment in National Economy*, Statistics Poland (relevant Yearbooks), *Fixed Assets in National Economy*, Statistics Poland (relevant Yearbooks), own calculations.

of the economic transformation related to the transition to a market economy and to the change in the price system. In that period, the decrease in the contribution of agriculture to global output was mainly driven by the decline in agricultural prices; in turn, the decrease in the share of agriculture in GDP was additionally pushed by the deterioration in agricultural trading terms.[12] After Poland accessed the European Union, and especially in the last couple of years when the outcomes of the CAP have become evident, the situation has become relatively stable. In 2018, the share of agriculture in the global output and in GDP was 3.0 and 2.6%, respectively. Since 2002, the share of agriculture in investment expenditure has been fluctuating around 2%. However, after Poland joined the EU, the absolute value of agricultural investment expenditure nearly tripled. According to Central Statistical Office data, as a consequence of the above, agricultural gross fixed capital formation (GFCF) has increased, especially in the last couple of years. Nevertheless, in 2018, the consumption rate of these assets reached 77%; the corresponding rates for active technical fixed assets were as high as 93% (vehicles) and nearly 82% (machinery and equipment) (*Fixed Assets in National Economy*, Statistics Poland 2018).

The Polish agricultural sector continues to provide employment for nearly 15% of all employed in national economy who, however, account for only 4.5% of total GFCF. In the European context, this is an extremely high percentage of people employed in agriculture, which is largely the consequence of the relatively slow structural transformation of that sector. Factors that hampered the transformation include the demographic conditions, reflected in a relatively slow transfer of farms between generations and in a lack of demand for agricultural labor from other sectors of national economy. As mentioned in Section 2.1 of this chapter, Poland and many other Central and Eastern European countries had an excess of agricultural labor force, especially in the period preceding the transformation. In Poland, the economic transformation failed to increase the number of non-agricultural jobs. Often, it was quite the opposite; as rural industrial workers became unemployed, they had farming (usually combined with social transfers) as the sole revenue stream (cf. Trzeciak-Duval, 1999; Csaki, 2000; Gorton and Davidowa, 2004; Lerman et al., 2002; Macours & Swinnen, 2005; Pouliquen, 2001). In his analysis of the beginnings of the Polish economic transformation journey, Wilkin (1992) states that national income was on a decline, and so was national demand. The inflation rate and loan interest rates were persistently high. The sharp increase in unemployment made it virtually impossible to migrate out of agriculture, hence it could be said that Polish agriculture (especially individual farming) was a buffer for the remaining part of the economy affected by a number of difficulties. The above clearly shows that whether, and how fast, the population moves out of agriculture depends equally on that sector's structural transformation and on the development of the national economy as a whole.

When considering the role of agriculture and the increased importance of its multifunctional nature, many researchers note the meaning of its triple (territorial/spatial, natural, and socio-cultural) embeddedness (Wilkin, 2010). This is an important aspect which makes agricultural production stand apart from most other production types, which usually lost their embededness under the influence of competition, forcing them to embrace mobility. For obvious reasons, farmers and farms have a very limited capacity to move. Moreover, agriculture adapts to a slightly different extent to the changing global economic conditions than most other sectors. Therefore, as emphasized by many economists, state intervention in the agricultural sector is so important. What also makes agriculture a special sector of the economy is that people see and value its multifunctional nature. The essence of being multifunctional consists in understanding that non-commodity outputs of agriculture are usually related to its production function. This is referred to as jointness of these two function types. Wilkin (2010) states the following about non-commodity outputs and production functions: *In order for the former to be delivered, the agriculture must exist and operate in the latter area of activity, too.*

2.3.3 Evolution of production structures

The socioeconomic transformation was a major breakthrough in the formation of the agrarian structure of Central and Eastern European countries. As mentioned in Section 2.1 of this chapter, in 1989, right before the economic shift, Poland had the largest private sector of all communist countries. Its share in the economy was around a quarter of GDP and nearly half of employment, although most of them worked in agriculture and did not add much value. As Piątkowski (2019) notes:

the extended private sector was largely the legacy of the 'Polish path to socialism': after 1956, the party abandoned the agricultural collectivization process, allowed individual farmers to keep ca. 80% of land, and tolerated a relatively large private market for trade activities and services.

(p. 223)

The introduction of the market economy encouraged some CEE countries to decollectivize agriculture and give the former owners (or their successors) back the right to land previously taken away from them. The scope and path of decollectivization was aligned with the political and economic capacities and objectives of each country. An important increase in the importance of individual farming was witnessed in the Baltics (Lithuania, Latvia, Estonia), in Romania and in Hungary. As mentioned earlier, in Poland (just as in Slovenia), individual farms always had a great importance, and the decollectivization processes reinforced their role (see Table 2.2 in Section 2.1).

Note that nearly all European countries were somehow affected by the deficiencies of the agrarian structure. However, the nature, intensity and—above all—the starting time of transformation processes in the agrarian structure in post-war Europe differed across Community countries, resulting in discrepancies which persist to this day. The largest differences are particularly noticeable between Central and Eastern European countries and Western and Northern European countries, which were the first to start the transformation of the farm structure. This evolution took place parallel to changes in labor and capital resources, driven by rapid economic development. The accession of CEE countries to the European Union generated strong momentum for major changes in their agrarian structure. However, in Poland, the expected concentration of farmland is a sluggish process which can be explained by reasons such as the extended system of agricultural support. Using—or sometimes even owning—agricultural land makes individuals eligible for an economic rent in the form of area payments. This encourages many farmers to continue their activity or to postpone the sale of their land.

According to Central Statistical Office data, in 1990, Poland was home to 3.8 million economic operators active in the agricultural sector, including 2,137.5 thousand individual farms. That number decreased by 96.1 thousand from 1990 to 1996. The market mechanism resulted in the polarization of the structure of farming land. The number of the smallest farms (with an area of 1–2 ha) kept growing until 2002, and was accompanied by an increase in their area of agricultural land. This has been encouraged by the gradually extended system of diverse forms of farm aid (especially since 1994). Over subsequent years, these farms lost in importance both in absolute and relative terms, which suggests that land was gradually concentrated in larger farms (Baer-Nawrocka & Poczta, 2014). This was accompanied by a decrease in the number of agricultural producers. Between 2010 and 2018, the number of farms larger than 1 ha of agricultural land declined by nearly 83,000 on a countrywide basis (Table 2.8). A drop in the number of farms was recorded in

Table 2.8 Number of farms (thousands) and area of agricultural land owned by farms grouped by size in Poland in 2010 and 2018

Specification	Farms						Area of agricultural land					
	2010		2018		Difference (2018–2010)		2010		2018		Difference (2018–2010)	
	number	%	number	%	number	%	area (thousand ha)	%	area (thousand ha)	%	area (thousand ha)	%
Total	1,509.1	–	1,425.4	–	–83.8	–5.6	15,503.0	–	14,669.0	–	–833.9	–5.4
Up to 1 ha	24.9	–	24.1	–	–0.8	–3.1	256.4	–	16.9	–	.	.
Above 1 ha, including	1,484.3	100.0	1,401.3	100.0	–83.0	–5.6	15,246.6	100.0	14,652.2	100.0	–594.5	–3.9
1.01–1.99	300.6	20.3	285.0	20.3	–15.5	–5.2	500.4	3.3	426.9	2.9	–73.5	–14.7
2.00–4.99	489.8	33.0	449.0	32.0	–40.8	–8.3	1,688.5	11.1	1,448.7	9.9	–239.8	–14.2
5.00–9.99	346.3	23.3	315.0	22.5	–31.4	–9.1	2,503.1	16.4	2,224.2	15.2	–278.9	–11.1
10.00–14.99	151.5	10.2	142.0	10.1	–9.5	–6.3	1,849.5	12.1	1,704.4	11.6	–145.1	–7.8
15.00–19.99	72.0	4.9	70.6	5.0	–1.4	–2.0	1,244.6	8.2	1,198.7	8.2	–45.8	–3.7
20.00–49.99	97.0	6.5	105.5	7.5	8.5	8.8	2,836.3	18.6	3,122.0	21.3	285.7	10.1
50 ha or more	27.0	1.8	34.1	2.4	7.1	26.2	4,624.3	30.3	4,527.2	30.9	–97.1	–2.1

Source: Own compilation based on the *Statistical Yearbook of Agriculture* and *Characteristics of agricultural holdings*, Statistics Poland (relevant Yearbooks).

all size groups below 20 ha; the number of smaller farms (2.00–9.99 ha) declined at a faster rate. In these two size groups, the number of farms went down by a total of over 72,000, which accounts for almost 87% of the total drop in the number of farms in the group with a size above 1 ha. Meanwhile, the number of farms with a size above 20 ha went up by ca. 15,600. The growth rate was particularly fast in the group of above 50 ha. The decline in the total number of farms at the expense of the decrease in the group of small operators is a pattern which prevails in nearly all European Union member countries, though to differing degrees (Sadowski et al., 2013). However, each of these countries has groups of different sizes which act as milestones in the farm restructuring process. They are referred to as "threshold groups," i.e. groups of the physically smallest farms which were the first to report growth compared to the base year. What the "size threshold" tells us is that the farmers from the country concerned believe that having a smaller farm is economically unviable and it is therefore reasonable to discontinue farming operations (Sadowski et al., 2019). During 2010–2018, Polish agriculture saw an increase in the number of farms in the group with 20.00–49.99 ha of agricultural land. Just as in other CEE EU members, the threshold groups in Poland were much smaller than in the EU-15. For instance, in France and Germany, the threshold group is the one above 100 ha.

The changes in the structure of farms grouped by size are strictly related to changes in how these farms use their land. Hence, a reduction in size was recorded in farms with an area of up to 20 ha of agricultural land, whereas all groups of larger farms reported growth, except for farms with an area of over 50 ha which clearly reduced the area of utilized agricultural land compared to 2010. However, their share in land use remained at a level of 31% and their share in the total number of farms larger than 1 ha of agricultural land was 2.4%. Despite these positive developments, the smallest farms (with an area of up to 5 ha of agricultural land) account for more than half (52.3%) of all Polish farms, but use barely 13% of total agricultural land resources. Farms with an area of 5.00–9.99 ha (which hold over 15% of total agricultural land resources) also are a relatively important group (22.5% of the total farm population). According to research by Baer-Nawrocka and Poczta (2016), in countries with a production mix similar to that of Polish agriculture (mainly including western and northern EU members), ca. 80–90% of total agricultural land resources are usually concentrated in the largest farms (above 50 ha of agricultural land). As a consequence, the average farm size in these countries is several times that found in Poland (10.5 ha of agricultural land).

In addition to natural and soil conditions, centuries-long socioeconomic, political, and systemic processes also determine the differences in production structures between regions (Maps 2.1 and 2.2). The largest numbers of the smallest farms (defined as those with an area of up to 5 ha) are found in southeast Poland (Podkarpackie, Małopolskie, Śląskie, and Świętokrzyskie

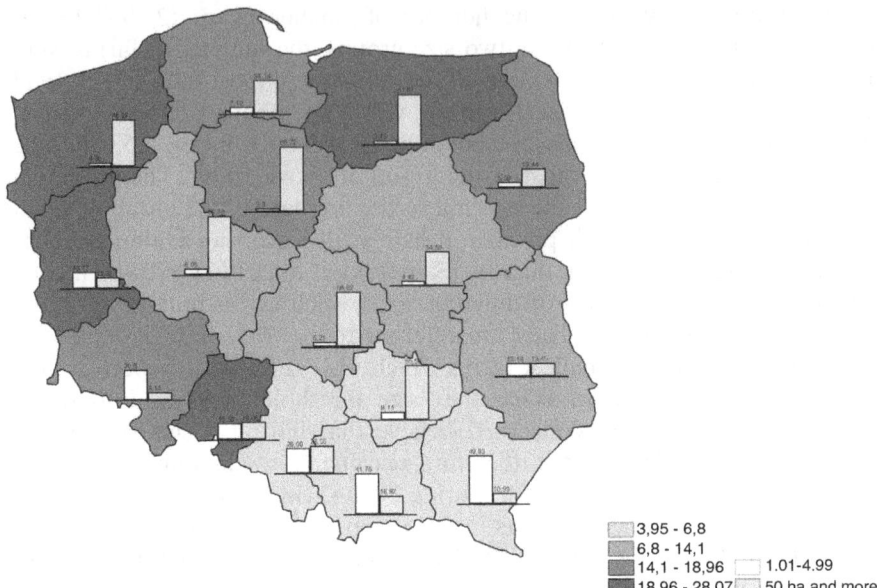

3,95 - 6,8
6,8 - 14,1
14,1 - 18,96   1.01-4.99
18,96 - 28,07   50 ha and more

Map 2.1 Average size and share of farms with an area of up to 5 ha and above 50 ha
in the total farm population in 2018 (by voivodeship).
Source: Own compilation based on *Land use and Sown Area*, Statistics Poland,
Warsaw 2019.

voivodeships). In the above voivodeships, the share of these farms in the total
population of holdings larger than 1 ha is 65–85%, i.e. considerably above the
average figure for Poland (52.4%). However, as mentioned earlier, rather than
in the distribution of farm numbers, the agricultural structure problem consists in the concentration of agricultural land in different size groups. In the
Podkarpackie and Małopolskie voivodeships, these farms hold as much as
nearly 42% and 50% of agricultural land, respectively. In the Świętokrzyskie
and Śląskie voivodeships, the situation is only slightly better as that share
falls within the interval of 26–30%. As a consequence, these voivodeships
have the smallest average area of farms, varying in the range of 4 to 7 ha.
Conversely, a much more favorable structure of farms can be found in voivodeships where former state-owned agricultural holdings gave rise to the establishment of relatively large farms (compared to those existing in other Polish
regions). These are the western and northern parts of Poland, including the
Zachodniopomorskie, Warmińsko–Mazurskie, Lubuskie, Pomorskie, Opolskie,
and Dolnośląskie voivodeships. In these regions, most agricultural land is concentrated in farms larger than 50 ha, and the average farm size is much above the
average Polish level of 10.5 ha of agricultural land.

In this context, it needs to be clearly emphasized that small farms are not
economically problematic if they hold only a small share in total agricultural

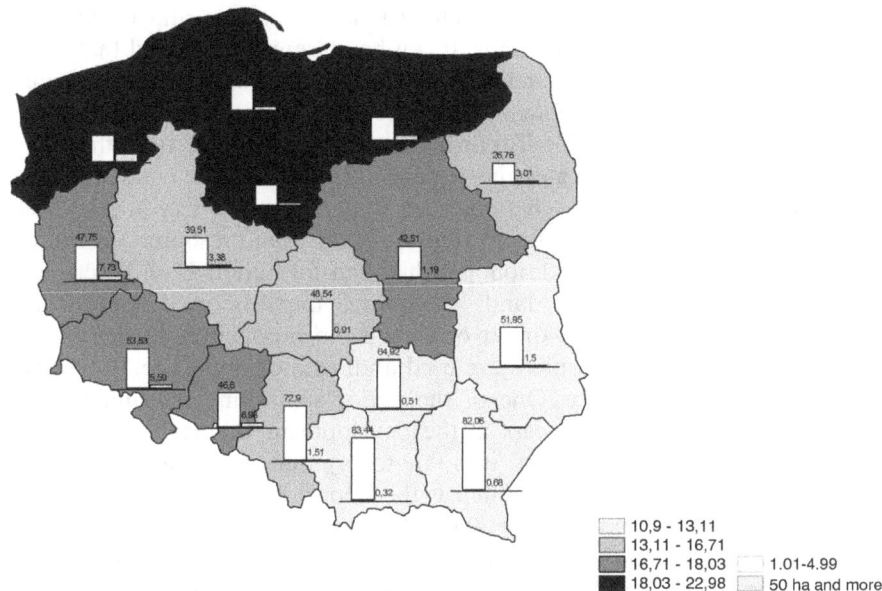

Map 2.2 Percentage of farmer households which derived incomes from economic activity; share of agricultural land held by farms with an area of up to 5 ha and above 50 ha in the total farm population in 2018 (by voivodeship).
Source: Own compilation based on *Land use and Sown Area*, Statistics Poland, Warsaw 2019; *Characteristics of Agricultural Holdings,* Statistics Poland, Warsaw 2019.

land resources, and are not socially problematic if their owners have an additional (usually main) off-farm revenue stream. This is the case in many EU countries. Conversely, problems emerge if small farms operate on a considerable part of total agricultural land resources. This was shown to be the case especially in southeast Poland. The existence of a large number of small, cost-ineffective farms is an economic and social problem. Firstly, this directly translates into low incomes and poverty among people related to these farms, especially if farming incomes are their main revenue stream. Secondly, these farms often make improper use of—and increasingly abandon—their land due to lack of funds and skills. While this is especially true for foothill and mountainous regions in the southeast part of the country, it also considerably affects other parts of Poland. In these regions, people massively quit farming or opt for extreme production extensification practices. The majority of the smallest farms (up to 5 ha) have not been investing in fixed assets for many years. Moreover, they discontinue animal production and no longer use poorest quality agricultural land; as a consequence, they are at various stages of decline. In this context, note that a large proportion of small and very small farms fuels the adverse reallocation of agricultural production space to non-farming uses. The above is exacerbated by problems related to people's mindsets, such as being reluctant to sell their family legacy, and by farmers keeping their land to access EU funds. It can be reasonably expected that these regions will be increasingly severely affected by the production and economic gap separating them from regions with a more favorable agrarian structure.

The spatial polarization of Polish farms, a persistent process, is all the more important since it also indicates the regional differences in a context of the future transformation of the national agrarian structure. Considering the potential economic strength of farms (represented by land resources) it can be concluded that the basic condition for this transformation is the increased concentration of agricultural land. Therefore, there is need for further restructuring as an important driver of changes in production efficiency and in how productive input resources are used and, as a consequence, in income and production performance. One of the major significant determinants of the agrarian restructuring process is the development of local markets for non-farming jobs. In this context, it is fair to say that the evolution of agricultural structures, strictly interrelated with support for the creation of new rural jobs, is the key challenge for agricultural and cohesion policies at EU level and for national agricultural and economic policies. Note also that these are long-term processes. As Rosner (2001) emphasizes, agricultural restructuring processes took more than 40 years in Germany, although the economic context was much more favorable. They included the concentration of land, production and capital. In Poland, agricultural restructuring (concentration) processes can be expected to be similar in many ways.

Another important aspect in the context of the agrarian and socioeconomic transformation of rural areas is that engaging in an economic activity (which is an option for the population to quit farming and, thus, to contribute to a more in-depth agrarian restructuring) is usually associated with personal characteristics (referred to as entrepreneurship) but also with the strength of the potential entrepreneurs' determination to do so. The economic capacity to undertake a production or service activity is a matter of importance, too. These conditions are linked to the socioeconomic and infrastructural development of a region, tradition of economic activity, wealth of the community, and absorptive capacity of local markets. A study by Wojewodzic (2017) found that farmers from south Polish regions (i.e. territories with a fragmented agrarian structure) did not engage in non-agricultural economic activities. Instead, they often chose employment. A similar trend can be observed in other Polish regions. In an effort to find new revenue streams, small farm owners turn to employment. Conversely, larger farms much more often diversify their on-farm economic activities to improve their income situation (see Map 2.2). Indeed, they have a large pool of productive inputs which cannot be fully used solely for farming purposes. Wilkin (2010) identifies large differences in the diversification of farm activity between EU countries and regions. In 2005, the largest shares of diversified farms were found in Finland (29%), France (25%), and the UK (24%). The most frequent forms of on-farm diversification are:

- processing of agricultural products, usually practiced in southern countries (mainly in Portugal and Italy) but also in Romania and Hungary;
- tourism services, which gained the greatest popularity in the UK, Austria, and Belgium;
- production and contractual work, which gained the greatest popularity in Bulgaria, Finland, and Greece.

2.4 Role of EU funds in the transformation in Poland with particular regard to rural areas

In the mid-1980s, European Communities started to pay attention to and implement the cohesion policy (named at that time the regional or structural policy).[13] The increased importance of cohesion among the policies implemented by the Communities was also reflected by placing a greater emphasis on structural measures under the CAP. The whole history of the CAP can be divided into two essential sub-periods. The first one ended in the mid-1990s and was the time of a policy designed to boost supply in an effort to modernize agriculture and increase agricultural output.[14] The second essential sub-period was clearly focused on demand, and had some elements of a structural policy for agricultural and rural areas which the Community started to implement in 1992, in parallel with what is referred to as the MacSharry reform.[15] This was the start of compensatory payments for farmers (now known as direct payments);[16] also, more and more attention was paid to rural development and to non-production functions of agriculture.[17] The reform was a breakthrough which strengthened the importance of the rural development policy. Since the implementation of the reform, considerable amounts of funds have been allocated to measures related to afforestation, environmental protection, preservation of the traditional rural landscape, and development and activation of local communities, including support for tourism. This was the starting point for a sustainable multi-purpose agriculture policy and for a multifunctional development of whole rural areas. Today, European agriculture, as defined in the European Model of Agriculture (EMA),[18] is a provider of many public goods (food security, climate change mitigation, biodiversity, bio-energy, soil preservation and renewal, good water management, water quality management, contribution to cultural values, formation of valuable landscape; generally speaking: participation in sustainable development). This is one of the basic reasons for supporting agriculture and rural areas with public funds (from the EU budget).

Central and Eastern European countries, including Poland, became Community members when the second phase of the CAP was already at an advanced stage; focus was no longer on agricultural modernization (the agriculture of EU-15 countries completed that stage during the first period of CAP implementation).[19] Instead, emphasis started to be placed on other, non-production functions of agriculture. As a consequence, in 2004, the Polish agricultural sector found itself in a new economic and institutional environment; it was not obvious whether it would fit there. In practice, the Polish agri-food sector benefited from the participation in the Single European Market.[20] However, support with funds from the EU budget is of fundamental importance to the entire Polish economy, including agriculture and rural areas. Table 2.9 presents information on financial transfers between the EU budget and Poland from the moment Poland joined the EU to May 31, 2020. It suggests that during that period, a total of EUR 185.4 billion was transferred from the EU budget to Poland, whereas Poland paid EUR 59.6 billion to the EU budget. Hence, Poland's net financial position was EUR 125.6 billion (taking account of what was paid back to the EU budget). Of the transfers from the EU budget to Poland, the largest amounts (EUR 118.6 billion during the period concerned) were allocated to the cohesion policy.

Poland is the largest beneficiary of Union aid. During 2014–2020, the European Union allocated a total of EUR 82.5 billion to Poland under the cohesion policy. These funds are distributed under national and regional programs. National programs include the following targets and amounts to be used:

1   Infrastructure and Environment Program EUR 27.4 billion
2   Smart Development Program EUR 8.6 billion
3   Knowledge–Education–Development Program EUR 4.7 billion
4   Digital Poland Program EUR 2.2 billion
5   Eastern Poland Program EUR 2.0 billion
6   Technical Support Program EUR 0.7 billion
                    (https://www.funduszeeuropejskie.gov.pl/strony/o-funduszach/)

The largest amounts were allocated to the Infrastructure and Environment Program which has the following priorities: low-emission economy; environmental protection; development of the country's technical infrastructure; and energy security. Other funds are allocated under 16 regional programs managed by Marshall's Offices. Support for agritourism, including rural tourism infrastructure, is covered by the Infrastructure and Environment Program and the Eastern Poland Program, but primarily by 16 Regional Operating Programs which correspond to the country's administrative division into 16 voivodeships.

Rural areas and agriculture play a particular role in the EU policy and budget. Upon accessing the EU, Poland became one of the largest beneficiaries of the CAP and the largest beneficiary of funds disbursed under what is referred to as the 2nd pillar of the CAP. Funds transferred to agriculture and rural areas under the CAP account for EUR 60.7 billion. For the sake of relevance, it would be reasonable to increase that amount with transfers from the European Agricultural Guidance and Guarantee Fund and the Financial Instrument for Fisheries Guidance which were formally positioned as structural funds under the cohesion policy in the 2004–2006 Financial Perspective. Therefore, as of May 31, 2020, the agri-food sector has been supported with a total of more than EUR 62 billion from the EU budget. This means that the share of the CAP in all transfers from the EU budget to Poland was 33.5%

Table 2.9 Summary of financial transfers between the EU budget and Poland (EUR)

Specification	Amounts transferred since Poland became an EU member state (total from May 1, 2004 to May 31, 2020)
I. Transfers from the EU to Poland,[a] including:	185,408,798,027
1. Cohesion Policy	118,558,099,986
ISPA Fund	1,968,089,608
Fund for European Aid to the Most Deprived (FEAMD)	318,985,997
Cohesion Fund	37,119,996,084
Within the 2004–2006 financial perspective	2,961,675,332
Within the 2007–2013 financial perspective	22,387,151,159
Within the 2014–2020 financial perspective	11,771,169,594
Structural Funds	79,151,028,297
Within the 2004–2006 financial perspective	8,561,029,831
European Social Fund	2,038,181,591
Financial Instrument for Fisheries Guidance	*178,399,777*
European Regional Development Fund	5,155,405,682
European Agricultural Guarantee Fund	*1,189,042,781*
Within the 2007–2013 financial perspective,	44,798,808,847
European Social Fund	10,088,161,808
European Regional Development Fund	34,790,647,038
Within the 2014–2020 financial perspective	25,791,189,619
European Social Fund	6,233,049,686
European Regional Development Fund	19,558,139,933
2. Common Agricultural Policy	*60,723,417,506*
Direct payments (ARMA)	*37,874,848,484*
Market interventions (ADA)	*1,807,035,878*
Rural Development Program until 2013	*16,049,975,136*
2014–2020 Rural Development Program	*3,937,488,125*
Other CAP transfers	*165,249,153*
European Fisheries Fund until 2013	*710,441,822*
2014–2020 European Maritime and Fisheries Fund	*178,378,907*
3. Connecting Europe Facility (CEF)	1,611,331,542
4. Funds transferred under Heading 3a of the EU budget	187,421,682
Migration Funds	209,314,620
5. Other transfers	865,454,563
II. Contributions to the EU budget	59,631,777,671
Gross National Income (GNI)	39,941,803,581

(*Continued*)

Table 2.9 (Cont.)

Specification	Amounts transferred since Poland became an EU member state (total from May 1, 2004 to May 31, 2020)
Value Added Tax (VAT)	8,433,805,724
Traditional own resources (TOR)	7,255,956,016
Discounts	4,000,212,350
III. Repayments to the EU budget	170,652,925
IV. Balance: PL – EU (I – II – III)	125,607,550,518

Source: *Ministry of Finance*, http://www.mf.gov.pl/ministerstwo-finansow/dzialalnosc/unia-europejska/transfery-finansowe-polska-ue (June 30, 2020).

a A cash-based approach was used: the transfers are registered as they credit the relevant account in the National Bank of Poland.

(italicized items in Table 2.9). It needs to be emphasized that the transfers alone from the EU budget to Poland exceed the Polish contribution to the EU budget (more than EUR 62 billion vs. a contribution of less than EUR 60 billion). This proves the important role of Polish rural areas and of Polish agriculture as beneficiaries of funds allocated in the EU budget.

Rural areas and agriculture are supported under the 1st and 2nd pillar of the CAP. Under the 1st pillar of the CAP, agricultural support is mostly implemented through area payments (direct payments). Support under the 2nd pillar is allocated to the agricultural sector and rural areas, including environmental protection. The implementation of the 2nd pillar of the CAP is reflected by adequate rural development programs which are a structural policy. The European Commission believes that the structural policy for agriculture should be aligned with current problems of each region. Therefore, rural development programs can be partially designed at national or regional level. As a consequence, when creating their rural development programs, member countries can adjust them to specific rural and agricultural problems found in each country or region by selecting appropriate measures and distributing the funds as needed while complying with the restrictions imposed by Union authorities.[21] This contributes to a flexible implementation of structural CAP measures and helps making the decision on their form at national or regional level.

As Poland accessed the EU, Polish farmers became eligible for direct aid which was supplemented with funds from the national budget. Polish farmers attained the target payment level in 2010. In 2013, just as in other countries who joined the EU in 2004, that level was attained only based on payments from the EU budget. In 2004, Polish farmers filed 1,400.4 thousand applications for payment. Afterwards, that number increased, but in subsequent years, it followed a downward trend; as the number of farms decreased, it went down to ca. 1,350 thousand. The total amount of payments received by

Table 2.10 Amount of direct payments made within the 2004–2016 campaign (as at March 31, 2017) and estimations until 2020 (PLN billion)

Campaign																	Total
2004	2005	2006	2007	2008	2009	2010	2011	2012	2013	2014	2015	2016	2017	2018	2019	2020	Total
6.3	6.7	8.2	8.3	8.6	12.2	12.6	14.1	13.7	14.1	14.2	14.3	14.6	14.6	14.6	14.9	14.9[a]	207.0

Source: *Agency for Restructuring and Modernization of Agriculture (ARMA)* (May 26, 2020), data from the Ministry of Agriculture and Rural Development, own calculations.

a Estimation.

Table 2.11 Funds available under rural and agricultural development programs in Poland, 2004–2013 and 2014–2020

Program	Payments effected (PLN million)
Total for 2004–2006	17,646.2
Agriculture Sectoral Operating Program	6,876.7
2004–2006 Rural Development Program	10,769.5
2007–2013 Rural Development Program	67,499.0
Total for 2004–2013	85,145.2
2014–2020 Rural Development Program[a]	59,163.9[b]
Total for 2004–2020	144,309.1

Source: Own compilation based on data of the Agency for Restructuring and Modernization of Agriculture, the Ministry of Agriculture and Rural Development, and the 2014–2020 Rural Development Program (June 20, 2020) http://www.arimr.gov.pl/pomoc-unijna/wdrazane-program y-i-dzialania-dane-liczbowe/program-rozwoju-obszarow-wiejskich-2007-2013.html

a Amount of payments planned.
b EUR 13,513.3 million, calculated as per the ECB exchange rate as of September 30, 2019 (EUR 1 = PLN 4.3782).

farmers each year is shown in Table 2.10. Until the end of April 2019, the farmers obtained a total of nearly PLN 192.1 billion under this scheme. It can be reasonably estimated that from 2004 to 2020, Polish farmers will receive a total of PLN 207 billion (which is close to EUR 50 billion) in the form of direct (income) support.[22] That amount had a crucial impact on farmer incomes; in larger farms, it also indirectly contributed to modernization and development processes related to both agricultural and non-agricultural production.

In Poland, during 2004–2020, projects designed to improve agricultural structures, in broad terms, and to modernize agriculture were or are implemented under the following structural programs focused on agricultural and rural development:

1 2004–2006 Sectoral Operating Program "Restructuring and modernization of the food sector and rural development";
2 2004–2006 Rural Development Plan;
3 2007–2013 Rural Development Program;
4 2014–2020 Rural Development Program.

After joining the EU, Polish agriculture became the largest beneficiary of structural support of all member countries. High levels of support under the 2nd pillar of the CAP are of fundamental importance to the Polish agriculture and rural areas. This is because both Polish rural areas and Polish agriculture experience a series of deficiencies which are and can be alleviated or offset with the use of rural development funds. From 2004 until the end of the 2007–2013 perspective, the Polish agri-food sector and Polish rural areas received a total of more than PLN 85 billion under rural and agricultural development programs. In the current perspective, payments under the 2nd pillar of the CAP are supposed to exceed PLN 59 billion. Hence, during 2004–2020, the estimated amount of payments should total to PLN 144 billion, which is approximately EUR 33 billion (Table 2.11).

CAP structural funds are allocated to the development of agriculture and of the whole agri-food sector, and to support for environmental measures and rural development. While these goals compete with one another for funds allocated under the 2nd pillar of the CAP, they can also create synergies. Indeed, the development of agriculture and of the food industry fosters rural development. In turn, based on the creation of non-agricultural jobs, infrastructural development etc., rural development simulates modernization processes and creates demand for agricultural products. Environmental well-being encourages rural development in both the agricultural and non-agricultural dimension, including the development of the tourism function. However, one can try to estimate the amount of support provided under these rural programs. An attempt to identify more "rural" measures among those supported with EU funds shows that a total of nearly PLN 13 billion was allocated to rural development until the 2007–2013 programming period. In the 2014–2020 financial perspective, a total of less than PLN 8 billion is supposed to be allocated to rural development from the RDP (i.e. from CAP funds), which is much less than what was transferred in the 2007–2013 perspective (over PLN 12 billion). This is caused by the decrease in support under the 2nd pillar of the CAP at EU budget level, and by the decision made at national level to reallocate 25% of RDP funds to strengthen the 1st pillar of the CAP (to increase the direct payments budget). In turn, rural development support was planned to be supplemented with an amount of ca. PLN 22.8 billion available under the cohesion policy (Table 2.12).

Previous support with (structural) funds disbursed under the 2nd pillar of the CAP allows to conclude that from the Polish perspective, the need for increasing the competitiveness of agriculture and of the entire agri-food sector, and the need for improving labor productivity and the agrarian structure have been matters of fundamental importance to the implementation of that policy since the very moment Poland accessed the EU. However, as the financial perspectives succeed each other, new conditions emerge for the implementation of the Union's agricultural policy. They are related to aspects such as environmental protection needs, conditions of rural living, and promotion of diversification of rural activities, including the development of rural tourism functions. The policy implemented under the 2nd pillar of the CAP is increasingly adapted to these new conditions as it shifts from agricultural producer support to a policy for a broader support of rural development. This becomes even more important from the perspective of development of the whole country.

A major aspect consists in making a synergic use of funds available under the common agricultural policy and the cohesion policy. More detailed studies on the impact of the cohesion policy on the development of rural tourism are available for the 2007–2013 Perspective. Also, there is incomplete data on the

Table 2.12 Rural development support with EU funds until 2020

Specification	Amount (PLN million)
Sectoral Operating Program "Restructuring and modernization of the food sector and rural development"	816.4
Measures:	
• Rural renewal; preservation and protection of cultural heritage	
• Diversification of agricultural and similar activities in order to ensure diversity of activities or alternative revenue streams	
• Leader+ pilot program	
2007–2013 Rural Development Program	12,172.6
Measures:	
• Diversification into non-agricultural activities	
• Establishment and development of micro-enterprises	
• Basic services for the economy and rural population	
• Rural renewal and development	
• Implementation of local development strategies	
• Implementation of cooperation projects	
• Functioning of a local action group; acquisition of skills and mobilization.	
Total for programs implemented until 2013	12,989.0
2014–2020 Rural Development Program	7,924.7[a]
Measures:	
Basic services for and renewal of rural areas	
Support for local development under the LEADER initiative	
Total funds allocated to CAP implementation until 2020	20,913.7
Planned support for rural development with funds of the cohesion policy	22,766.6[a]
including:	
• water and sewerage infrastructure	5,253.8
• water management	2,189.1
• development of entrepreneurship	6,567.3
• social revitalization; revitalization of the rural infrastructure	8,756.4
Total support for rural development until 2020 (under the CAP and cohesion policy)	43,680.3

Source: Beba et al. (2016); own compilation based on data from the Agency for Restructuring and Modernization of Agriculture and the Ministry of Agriculture and Rural Development.

a Calculated as per the ECB exchange rate as of September 30, 2019 (EUR 1 = PLN 4.3782).

initial period of the 2014–2020 perspective (Institute of Rural and Agricultural Development, 2019) (Tables 2.13 and 2.14). This data shows that—mostly because of their geographic features—rural areas were the greatest beneficiary of most measures and operations addressing the tourism industry. Tourism development in rural areas is among the essential stimulators of multipurpose rural development and a condition for a shift to non-agricultural activities, which means making rural development no longer solely dependent on agriculture. The authors of *Wpływ polityki spójności na rozwój obszarów wiejskich* (Impact of the cohesion policy on rural development) conclude that:

Table 2.13 Summary of indicators reflecting the outcomes of intervention in the area of cultural heritage and tourism in the 2007–2013 perspective

Indicator	Total	Including in rural areas
Length of reconstructed tourist trails	2,886	2,237
Length of tourist trails built	5,649	4,468
Number of new tourism products	1,116	675
Number of reconstructed tourism and recreation facilities	935	505
Number of tourist information lefts created	304	2,192
Number of tourist information portals created	92	53
Number of tourism and recreation facilities built	2,133	1,244
Number of modified tourism products	215	124
Length of tourist trails covered by intervention measures	8,535	6,704
Number of tourism products covered by intervention measures	1,331	799
Number of cultural facilities covered by intervention measures	339	207

Source: Institute of Rural and Agricultural Development (2019b).

Table 2.14 Summary of indicators reflecting the outcomes of intervention in the area of cultural heritage and tourism in the 2014–2020 perspective

Indicator	Total	Including in rural areas
Length of newly opened tourist routes	650.0	650.0
Length of newly opened educational trails	42.8	42.8
Length of renewed tourist trails (km)	114.5	38.5
Length of educational trails	42.8	42.8
Length of tourist trails created (km)	528.3	435.4
Number of tourism and recreation facilities covered by support	4.0	0.0

Source: Institute of Rural and Agricultural Development (2019b).

it is necessary to better synchronize the Common Agricultural Policy with the Cohesion Policy. The development needs and capacities differ strongly across rural areas. This should encourage the implementation of a diversified policy mix which suits well the needs and must include funds of the 1st and 2nd pillar of the CAP as well as other funds and instruments of the cohesion policy

and that there is need to

increase the role of local government in planning, financing and implementing projects designed to improve cohesion. This is obviously in consonance with what the EU requires and plans for the upcoming programming period (2021–2027). Generally, local government units are highly aware of their region's most urgent needs. They ensure a greater stability of institutions and human resources than central-level authorities and are more efficient in using the available funds.
(Institute of Rural and Agricultural Development, 2019a, pp.122–123)

In summary, it needs to be emphasized that Polish rural areas became a highly important beneficiary of Poland's membership of the EU. One of the ways of supporting rural areas was through the huge amount of funds transferred from the EU budget. The basic role was played by funds provided under the EU's cohesion policy and Common Agricultural Policy. They resulted in agricultural development and an increase in farmer incomes, on the one hand, and stimulated the rural shift to non-agricultural activities and multipurpose rural development, on the other. Polish rural areas are no longer agricultural areas. Instead, they become an increasingly attractive place for working, living, and relaxing. Also, the delivery of tourism services becomes a growing and promising axis of multipurpose rural development.

Notes

1 It was largely applicable to the agricultural sector and agri-food processing, as well as to subsidizing the prices of finished food products. This had a great impact on the situation of the agricultural sector in the transformation period.
2 The plan included ten acts adopted by the body referred to as the "Contracted Parliament" on December 27 and 28, 1989, and signed by President W. Jaruzelski on December 31, 1989.
3 For more information on the evolution of the role of agriculture in the Polish economy, see Section 2.3.2.
4 As defined by the Central Statistical Office, disposable income "is the total current income of households derived from different sources, less personal income tax, less other income and ownership taxes paid by self-employed, and less social and health insurance contributions."
5 The social allowances program has been in place in Poland since April 1, 2016, and consists in paying a monthly parental benefit of PLN 500 for each child in a family.
6 The Gini coefficient falls within the interval of [0; 1]. The higher the value of the coefficient, the greater is the concentration and heterogeneity of incomes.

7 The relationship between annual incomes of farmers and other socioeconomic groups is among the key problems of agriculture in highly developed countries. Note that research on income proportions faces certain methodological barriers. They mostly result from the way farming incomes are determined based on the Economic Accounts for Agriculture (EAA), data from the farm accounting system (FADN) or household budget data. According to a number of studies, the conclusions can vary depending on the data source. The study by the authors of this chapter suggests that income disparity exists in nearly all member countries. Based on sector-level data, the average income disparity in the EU agriculture has been fluctuating around 40–44% over recent years.

8 Note, however, that due to available statistics (especially the changes in how land resources are counted and how land uses are defined), there is limited comparability of long-term data. Moreover, it is not possible to fully identify the methods and forms of using land which is no longer agricultural land held by farms.

9 According to the Central Statistical Office, 1 AWU in agriculture corresponds to 2,120 hours of work a year.

10 The issue of productivity of inputs in Central and Eastern European countries was addressed by authors such as Trzeciak-Duval (1999), Chaplin et al. (2004), Gorton and Davidova (2004), Latruffe et al. (2008).

11 Poor productivity of farming labor inputs in CEE countries is a problem reported by Alexandri et al. (2008), Bedrač and Cunder (2012), Koteva et al. (2008), Krisciukaitiene et al. (2012), Melnikeine and Krisciukaitiene (2008).

12 In Poland, both the prices of agricultural supplies and sales prices were subject to direct regulations until 1989. Hence, farming incomes were largely co-determined by the government's price policy for that sector. Switching to a market economy involved the deregulation of prices, opening the agricultural market to international competitors and withdrawing most state subsidies for agricultural and food products and productive inputs. This is why agricultural producers saw a significant deterioration in their position after the introduction of the market economy.

13 However, the European Regional Development Fund was established already in 1975 (as a consequence of pressures related to the British accession) and was supposed to support less developed regions to promote the economic and social cohesion across the CEE.

14 As a consequence, the Community attained a high level of food self-sufficiency.

15 An Irish politician, Ray MacSharry was the Commissioner for Agriculture at the European Commission during 1989–1993.

16 The introduction of direct payments allowed farming incomes to be maintained at a stable level while gradually reducing institutional prices and driving increased internal and external demand for the Community's agricultural products.

17 As a consequence, two funds have been in place since the EU 2007–2013 Financial Perspective. The European Agricultural Guarantee Fund was designed to finance market activities (including direct payments, refunds for the exportation of agricultural products to third countries, intervention measures to regulate agricultural markets). Funds disbursed by the European Agricultural Fund for Rural Development were allocated to measures that promote rural development and restructuring.

18 The European Model of Agriculture is a term which emerged in 1998 in the Explanatory Memorandum, and was subsequently used in 1999, in the Agenda 2000 (Wilkin, 2009). However, the EMA concept is related both to the arrangements made in the Treaties of Rome and to amendments to the CAP initiated by the MacSharry reform (Tomczak, 2009).

19 In earlier years, CEE countries also experienced some earlier agricultural modernization processes, and used industrial productive inputs in the agricultural sector. However, firstly, this was done outside an environment characteristic of a free-market economy. Secondly, in most of these countries, the sectoral

concentration processes were forced by political decisions. Moreover, Poland, Slovenia, and Croatia (as former Yugoslavian republics) did not implement modernization processes witnessed in Western European countries.

20 Note that the Polish agri-food sector has had a resounding success as a member of the Single European Market. Since the accession of Poland to the EU, agri-food exports have more than sextupled, from EUR 5.2 billion in 2005 to EUR 31.4 billion in 2019, with 80% delivered to the EU market. At the same time, positive net agricultural exports have increased from EUR 0.8 billion to over EUR 10 billion.

21 In the current perspective, examples include the institutional restrictions to the financing of certain measures, as provided for in Regulation No. 1305/2013, which have a moderate impact on allocation decisions made by member countries. These restrictions consist in allocating no less than 5% of the total EAFRD contribution to the LEADER approach, and no less than 30% of the total EAFRD contribution to environmental measures and combating climate change. The propositions of the European Commission for the next Perspective go even further.

22 After Poland accessed the EU, transfers of CAP-related public funds have accounted for ca. 50% of farmer incomes.

References

Agriculture in the European Union. Statistical and Economic Information. European Commission, Brussels (relevant Yearbooks).

Alexandri, C., Lucia, L., Tudor, M., & Voicilas, D. M. (2008). The dynamics of farms in Romania-factors of influence, economic implication and perspectives. In W. Józwiak (ed.), *Farms in Central and Eastern Europe – today and tomorrow* (pp. 105–120). Institute of Agricultural and Food Economics – National Research Institute, Warsaw.

Åslund, A. (2002). *Building Capitalism. The Transformation of the Former Soviet Bloc.* Cambridge University Press, Cambridge.

Åslund, A., & Orłowski, W. M. (2014). *The Polish Transition in a Comparative Perspective.* mBank – CASE Seminar, Proceedings No. 133/2014, CASE, Warsaw.

Åslund, A., Boone, P., & Johnson, S. (1996). how to stabilize: lessons from Post-communist countries. *Brookings Papers on Economic Activity*, no. 1, 1–97.

Baer-Nawrocka, A., & Poczta, W. (2014). Transformation of agriculture. In W. Poczta & I. Nurzyńska (eds.), *Rural Poland 2014. Rural Development Report* (pp. 85–122). Scholar Publishing House.

Baer-Nawrocka, A., & Poczta, W. (2016). Polish agriculture vs. agriculture in the European Union. In J. Wilkin & I. Nurzyńska (eds.), *Rural Poland 2016: The Report on the State of Rural Areas* (pp. 67–87). Scholar Publishing House.

Balcerowicz, L. (1997). *Socjalizm, kapitalizm, transformacja. Szkice z przełomu epok.* Wydawnictwo Naukowe PWN, Warszawa.

Beba, P., Poczta, W., & Kiryluk-Dryjska, E. (2016). Ewolucja kierunków wsparcia wsi i rolnictwa w Polsce środkami Wspólnej Polityki Rolnej o charakterze strukturalnym [Evolution of directions of sport for agriculture and rural areas in Poland by CAP's structurak funds]. *Rocz. Nauk. SERiA [Annals of The Polish Association of Agricultural and Agribusiness Economists]*, 18 (5), 9–16.

Bedrač, M., & Cunder, T. (2012). Implementation of Rural Development Programme 2007–2013 in Slovenia and challenges for the future. In Z. Floriańczyk & D. Civjjsnovic (eds.), *Rural Development Policies from the EU Enlargement Perspective* (pp. 107–120). Rural Areas and Development Vol. 9, European Rural Development Network.

Blanchard, O. (1997). *The Economics of Post-Communist Transition*. Oxford University Press, New York.

Brada, J. C. (1989). Technical progress and factor utilization in Eastern European economic growth. *Economica* 56, 433–448.

Chaplin, H., Davidova, S., & Matthew, G. (2004). Agricultural adjustment and the diversification of farm households and corporate farms in Central Europe. *Journal of Rural Studies* 20, 61–77. doi:10.1016/S0743-0167(03)00043-3.

Characteristics of Agricultural Holdings, Statistics Poland (relevant Yearbooks).

Csaba, L. (1995). *The Capitalist Revolution in Eastern Europe. A Contribution to the Economic Theory of Systemic Change*, Edward Elgar, Aldershot.

Csaki, C. (2000). Agricultural reforms in Central and Eastern Europe and the former Soviet Union. Status and perspectives. *Agricultural Economics*, 22 (1), 37–54. http s://doi.org/10.1016/S0169-5150(99)00039-0.

Economic Accounts for Agriculture (2020). April 4. https://ec.europa.eu/eurostat/web/a griculture/data/database

Employment in National Economy, Statistics Poland (relevant Yearbooks).

Employment in Agriculture (2005, 2018). Faostat.

Eurostat. Data. https://ec.europa.eu/eurostat/data/browse-statistics-by-theme

Fabisiak, A., & Poczta, W. (2011). *Adaptacja sektora rolnego krajów Europy Środkowej i Wschodniej w procesie integracji z Unią Europejską*. Wyd. Uniwersytetu Przyrodniczego w Poznaniu. Poznań.

Fixed Assets in National Economy, Statistics Poland (relevant Yearbooks).

Gorton, M., & Davidova, S. (2004). Farm productivity and efficiency in the CEE applicant countries: a synthesis of results. *Agricultural Economics* 30 (1), 1–16. doi:10.1016/j.agecon.2002.09.002.

Halamska, M. (2011). Transformacja wsi 1989–2009. Zmienny rytm modernizacji [The Polish countryside in the process of transformation 1989–2009: A differing pace of modernization]. *Studia Regionale i Lokalne* [*Regional and Local Studies*], 2 (44), 5–25.

Household Budget Survey, Statistics Poland (relevant Yearbooks).

Institute of Rural and Agricultural Development (2019a). *Wpływ polityki spójności na rozwój obszarów wiejskich*. Polish Academy of Sciences, Wolański sp. z o.o., Raport końcowy, Warszawa.

Institute of Rural and Agricultural Development (2019b). *Wpływ polityki spójności na rozwój obszarów wiejskich*. Polish Academy of Sciences, Wolański sp. z o.o., Załącznik 1 do Raportu końcowego, Warszawa.

*International Statistics Yearbook*2019. Statistics Poland, Warsaw.

Jackman, R. (1994). Economic policy and employment in the transition economies of Central and Eastern Europe: What have we learned? *International Labour Review* 133 (3), 327–345.

Jarmołowicz, W., & Piątek, D. (2013). Polska transformacja gospodarcza. Przesłanki – przebieg – rezultaty. In S. Owsiak & A. Pollok (eds.), *W poszukiwaniu nowego ładu ekonomicznego*, Warszawa: Wyd. PTE.

Jędrzejak, A., & Pekasiewicz, D. (2017). Nierówności dochodowe gospodarstw domowych rolników na tle innych grup społeczno-ekonomicznych w Polsce w latach 2006–2014 [Income Inequality of Households of Farmers Compared with Other Socio-economic Groups in Poland in the years 2006–2014]. *Problemy Rolnictwa Światowego*, 17 (3), 166–176. doi:10.22630/PRS.2017.17.3.63.

Kornai, J. (2006). The great transformation of Central Eastern Europe: Success and disappointment. *Economics of Transition*, 14 (2), 207–244. https://doi.org/10.1111/j. 1468-0351.2006.00252.x.

Koteva, N., Mladenova, M., Rissina, M., & Bashew, H. (2008). Structural changes in agricultural farms in the EU CAP conditions. In W. Józwiak (ed.), *Farms in Central and Eastern Europe – today and tomorrow* (pp. 121–140). Institute of Agricultural and Food Economics – National Research Institute, Warsaw.

Kowalska, A. (2002). Zmiany struktury społeczno-ekonomicznej gospodarstw domowych i ich uwarunkowania. *Polityka Społeczna*, no. 4, 8–14.

Krisciukaitiene, I., Melnikeine, R., & Galnaityte, A. (2012). Competentives of Lithuanian farms and their agriculture production from present to medium – term perspectives. *European Union Food Sector after Last Enlargements – Conclusions for the Future CAP*. Institute of Agricultural and Food Economics – National Research Institute, Warsaw, 145–164.

Land use and Sown Area. Statistics Poland (relevant Yearbooks).

Latruffe, L., Davidova, S., & Balcombe, K. (2008). Productivity change in Polish agriculture: an illustration of a bootstrapping procedure applied to Malmquist indices. *Post-Communist Economies*, 20 (4). https://doi.org/10.1080/14631370802444708.

Lerman, Z., Csaki, C., & Feder, G. (2002). *Land Policies and Evolving Farm Structures in Transition Countries*. Policy Research Working Paper No. 2794. World Bank, Washington, DC.

Lewandowski, J. (1994). Rolnictwo we współczesnej gospodarce Polski [Agriculture in the present-day Polish economy]. *Ekonomista*, no. 4, 481–491.

Macours, K., & Swinnen, J. F. M. (2005). Agricultural labour adjustments in transition countries: the role of migration and impact on poverty. *Review of Agricultural Economics*, 27 (3), 405–411.

Majerova, V., Marikova, P., & Herova, I. (2009). The structural changes in the rural areas and agriculture: the case of the Czech Republic. In *The Structural Changes in the Rural Areas and Agriculture in the Selected European Countries* (pp. 31–46). Institute of Agricultural and Food Economics – National Research Institute, Warsaw.

Matyska, W., Potocka, M. & Samecki, P. (2019). *30 lat transformacji ekonomicznej w Europie Środkowej*. Wyd. Centrum Europejskie Natolin, Warszawa.

Melnikeine, R., & Krisciukaitiene, I. (2008). The competitiveness of Lithuanian farms in the EU. In W. Józwiak (ed.), *Farms in Central and Eastern Europe – Today and Tomorrow* (pp. 25–37). Institute of Agricultural and Food Economics – National Research Institute, Warsaw.

Ministry of Finance. http://www.mf.gov.pl/ministerstwo-finansow/dzialalnosc/unia-europejska/transfery-finansowe-polska-ue (June 30, 2020).

Orłowski, W. M. (2019a). Rozwój gospodarczy Polski w latach 1989–2018 na tle regionu Europy Środkowo-Wschodniej. In M. Belka & W. M. Orłowski (eds.), *Gospodarka polska – szanse i zagrożenia*. Raport przygotowany dla Europejskiego Funduszu Rozwoju Wsi Polskiej, Warszawa.

Orłowski, W. M. (2019b). Trajectories of the Economic Transition in Central and Eastern Europe. In G. Gorzelak (ed.), *The Regional Dimension of Transformation in Central Europe*. Routledge, Abingdon.

Piątkowski, M. (2019). *Europejski lider wzrostu [Europe's growth champion]*. Poltext Sp. z o.o.

Poczta, W. (2010). Wspólna Polityka Rolna po 2013 roku – uzasadnienie, funkcje, kierunki rozwoju w kontekście interesu polskiego rolnictwa [EU Common Agricultural Policy after 2013 – substantiation, functions and directions of development in the context of interests of Polish agriculture]. *Wieś i Rolnictwo* [*Village and Agriculture*], No. 3, 38–55.

Pouliquen, A. (2001). *Competitiveness and Farm Incomes in the CEEC Agri-food Sectors. Implications before and after Accession for EU Markets and Policies*. Independent Study for the European Commission. Office for Official Publications of the European Communities, Luxembourg.

Próchniak, M., Lissowska, M., Maszczyk, P., Rapacki, R., & Sulejewicz, A. (2019). Wyrównywanie luki w poziomie zamożności między Europą Środkowo–Wschodnią a Europą Zachodnią. In *Europa Środkowo-Wschodnia wobec globalnych trendów, gospodarka, społeczeństwo i biznes*. Raport Szkoły Głównej Handlowej w Warszawie. Gazeta SGH, Krynica.

Rosner, A. (2001). Społeczno-ekonomiczne uwarunkowania przemian strukturalnychw rolnictwie. In I. Bukraba-Rylska & A. Rosner (eds.), *Wieś i rolnictwo na przełomie wieków* (pp. 47–61). Polish Academy of Sciences, Institute of Rural and Agricultural Development, Warsaw.

Rosner, A., & Stanny, M. (2007). Zróżnicowanie poziomu rozwoju gospodarczego obszarów wiejskich w Polsce. In A. Rosner (ed.), *Zróżnicowanie poziomu rozwoju społeczno-gospodarczego obszarów wiejskich a zróżnicowanie dynamiki przemian* (pp. 47–114). Polish Academy of Sciences, Institute of Rural and Agricultural Development, Warsaw.

Sadowski, A., Czubak, W., Poczta, W., & Rowiński, J. (2019). Struktury obszarowe i ekonomiczne rolnictwa oraz innych państw unijnych. In W. Poczta & J. Rowiński (eds.), *Struktura polskiego rolnictwa na tle Unii Europejskiej* (pp. 47–84). Wyd. CeDeWu, Warszawa.

Sadowski, A., Baer-Nawrocka, A., & Poczta, W. (2013). *Gospodarstwa rolne w Polsce na tle gospodarstw Unii Europejskiej*. Wyd. Główny Urząd Statystyczny, Warszawa.

Samecki, P. (2012). Adaptability through change: from misdevelopment to a successful transition in Central Europe. In G. Gorzelak, G. Chor-Ching, & K. Fazekas (eds.), *Adaptability and Change: The Regional Dimensions in Central and Eastern Europe* (pp. 80–93). Scholar Publishing House.

Sarris, A. H., Doucha, T., & Mathijs, E. (1999). Agricultural restructuring in central and eastern Europe: implications for competitiveness and rural development. *European Review of Agricultural Economics* 26 (3), 305–329. https://doi.org/10.1093/erae/26.3.305.

Schiff, M., & Montenegro, C. E. (1997). Aggregate agricultural supply response in developing countries. A survey of selected issues. *Economic Development and Cultural Change*, 45 (2), 393–410.

Reports of Agency for Restructuring and Modernisation of Agriculture (ARMA) (relevant years).

Statistical Yearbook of Agriculture and Rural Areas, Statistics Poland (relevant Yearbooks).

Statistical Yearbook of Agriculture, Statistics Poland (relevant Yearbooks).

Statistical Yearbook of the Republic of Poland, Statistics Poland (relevant Yearbooks).

Statistics Poland (2003). *Population and Households. The State and the Socio-economic Structure*, Warsaw.

Statistics Poland (2004). *Systematics and Characteristics of Farms*, Warsaw.

Statistics Poland (2017). *Socio-economic Situation of Households in 2000–2015. Urban-Rural Diversity*, Warsaw.

Swinnen, J. F. M. & Knops, L. (2013). Factor Markets: Diversity under a Common Policy. In J.F.M. Swinnen & L. Knops (eds.), *Land, Labour and Capital Markets in European Agriculture. Diversity under a Common Policy*. Brussels: Centre for European Policy Studies (CEPS). 1–12.

Swinnen, J. F. M. (2001). *Implications of EU Enlargement for Agri-food Markets and Policy*. Materials for the conference Outlook for Agriculture. Agribusiness and the Food Industry in Central and Eastern Europe, Budapest, May 18, 2001.

Tomczak, F. (2005). *Gospodarka rodzinna w rolnictwie. Uwarunkowania i mechanizmy rozwoju*. [*Family agricultural economy: mechanism and development determinants*]. Institute of Rural and Agricultural Development Polish Academy of Sciences, Warsaw.

Tomczak, F. (2009). *Ewolucja wspólnej polityki rolnej UE i strategia rozwoju polskiego rolnictwa*. Institute of Agricultural and Food Economics – National Research Institute, Warsaw.

Tracy, M. (1993). *Food and Agriculture in a Market Economy. An Introduction to Theory, Practice and Policy*. Prague: V. Krigl, Polygrafická a nakladatelská činnost.

Trzeciak-Duval, A. (1999). A decade of transition in Central and Eastern European agriculture. *European Review of Agricultural Economics*, 26 (3), 283–304https://doi.org/10.1093/erae/26.3.283.

Van Huylenbroeck, G., Vandermeulen, V., Mettepenningen, E., & Verspecht, A. (2007). Multifunctionality of agriculture: a review of definitions, evidence and instruments. *Living Reviews in Landscape Research* 1 (3), 1–43. http://www.livingreviews.org/lrlr-2007-2003.

van Bavel, B., De Moor, T., & van Zanden, J. L. (2009). Introduction: Factor markets in global economic history. *Continuity and Change* 24 (1), 9–21. doi:10.1017/S0268416009007000.

Wilkin, J. (1992). *Rolnictwo – hamulec czy motor transformacji polskiej gospodarki?* In A. Kwieciński (ed.), *Znaczenie rolnictwa w procesie transformacji gospodarczej w Polsce*. Polish Policy Research Group, Warsaw University.

Wilkin, J. (2008). Ewolucja paradygmatów rozwoju obszarów wiejskich [Evolution of the Paradigms of rural development]. *Wieś i Rolnictwo* [*Village and Agriculture*], No. 3, 18–28.

Wilkin, J. (2009). Uwarunkowania rozwoju polskiego rolnictwa w kontekście europejskimi i globalnym. Implikacje teoretyczne i praktyczne. In U. Płowiec (ed.), *Polityka gospodarcza a rozwój kraju*. Wyd. PTE, Warszawa.

Wilkin, J. (2010). Wielofunkcyjność rolnictwa – nowe ujęcie roli rolnictwa w gospodarce i społeczeństwie. In J. Wilkin (ed.), *Wielofunkcyjność rolnictwa. Kierunki badań, podstawy metodologiczne i implikacje praktyczne* [*Multifunctional agriculture. Research trends, methodolical basis and practical implications*] (pp. 17–40). Institute of Rural and Agricultural Development. Polish Academy of Sciences, Warsaw.

Winiecki, J. (1999). Formalne i nieformalne reguły postępowania w warunkach postkomunistycznej transformacji. In E. Adamowicz (ed.), *Gospodarka w okresie przemian*. Oficyna Wydawnicza SGH, Warszawa.

Winiecki, J. (2012). *Transformacja postkomunistyczna. Studium przypadku zmian instytucjonalnych*. Wydawnictwo C.H. Beck, Warszawa.

Wojewodzic, T. (2017). *Procesy dywestycji i dezagraryzacji w rolnictwie o rozdrobnionej strukturze agrarnej [Divestment and disagrarisation processes in farms of fragmented agrarian structure]*. University of Agriculture in Krakow.

Wołoszyn, A., & Wysocki, F. (2014). Nierówności w rozkładzie dochodów i wydatków gospodarstw domowych rolników w Polsce. [Inequalities in income and expenditure distributions of farmers' households in Poland]. *Roczniki Naukowe SERiA [Annals of The Polish Association of Agricultural and Agribusiness Economists]* XVI (6), 535–540.

Zegar, J. S., & Gruda, M. (2000). *Relacje dochodów ludności chłopskiej i pozarolniczej. Wskaźnik parytetu dochodów.* Institute of Agricultural and Food Economics – National Research Institute, Warsaw.

3 Key changes in rural tourism development in Poland between 1989 and 2019

Lucyna Przezbórska-Skobiej

The development of rural tourism may be investigated in terms of several development stages (phases), in which it has developed as a result of growing demand and supply. Moreover, tourism-related operations have changed, new trends and forms have appeared, while concepts and definitions for rural tourism have evolved. This chapter also proposes various concepts for tourism regionalization in rural areas of Poland and proposes information concerning legal regulations having a significant effect on the development of rural tourism in Poland.

3.1 Origins and development phases of rural tourism

According to the literature on the subject, rural tourism and agritourism are nothing new to rural areas both in Poland and around the world (OECD, 1994, p. 7; Lane, 1994; Sikora, 1995; Drzewiecki, 2001; Kutkowska, 2003, p. 59). Poland has seen these forms of tourism at least since the 19th century (Sikora, 1995; Maciąg, 1996; Przezbórska, 1998; Dębniewska, 2000, p. 17; Roberts & Hall, 2001, 151; Bański, 2006, p. 129; Balińska, 2009; Zawadka, 2010a) when "local regionalist movements supported the initiatives for a socioeconomic revival of rural areas through tourism" (Sikora, 2012, p. 75; cf. Maciąg, 1996; Wojciechowska, 2009). However, as emphasized by Dębniewska and Tkaczuk (1997) and Marzejon-Frycz (2012), these were the beginnings of rural tourism rather than of agritourism which, as defined today, requires a number of measures to be taken to make farms suitable for hosting tourists (Gurgul, 2005). People and events that contributed to the development of rural tourism in the 1800s include doctors who wanted to change the bad life habits of big city residents as well as industrial revolution and urbanization processes (Wojciechowska, 2006; Marcinkiewicz, 2013; Surdacka, 2017). Moreover, the impoverishment made city dwellers look for cheaper forms of leisure in Polish rural areas (Surdacka, 2017) in an attempt to address their ambitions and put themselves on a more or less equal footing with the wealthy bourgeoisie and landholders who visited international spa and seaside resorts in the 1800s (Marcinkiewicz, 2013). Marcinkiewicz (2013, p. 24) wrote that in 1936, the voivodeships included in what then was the Polish territory were home to 792 holiday resorts which a total of over

209,000 visitors. In 1938, holiday centers located near Warsaw hosted nearly one-fifth of the total Polish tourist population.

The rapid development of rural tourism in the early 1900s, especially in the two decades between the wars when Poland regained its independence, was interrupted by the outbreak of World War II (Marzejon-Frycz, 2012). In the postwar era, war damages and the shift to another economic system were factors that initially inhibited this kind of activity in Poland. The pre-war rural tourism initiative resurged only in the mid-1950s (Sikora, 1995; Marzejon-Frycz, 2012). In the 1950s, the organization of rural holiday facilities was strictly controlled by the government; unregistered hosts were only allowed to rent out up to three rooms (Sikora, 2012; Surdacka, 2017). Hence, at that time, these were totally different tourism processes. As late as in 1995, Sikora (1995) wrote that in Poland, the long-standing tradition of rural holidays involved visiting friends and family rather than paying for a farm accommodation; however, "that tradition also means the organization of what is referred to as summer holiday stays and village holiday resorts." Traveling to rural destinations is a trend that became widespread in the 1960s; contributing factors were the reactivation of "Gromada,"[1] the Tourism and Holiday Cooperative (in 1957) (Drzewiecki, 1992; Sikora, 1999; Mikuta & Żelazna, 2004), and the activity of other institutions, such as the National Paid-Leave Fund (Passaris, Sokólska, & Vinaver, 2002; Marzejon-Frycz, 2012; Marcinkiewicz, 2013). That period saw the creation of tourism villages and the emergence of a rural tourism movement guided by the "the Jabłońskis welcome the Matysiaks"[2] slogan; this was the name of the best holiday village competition launched by the editorial team of the *Zielony Sztandar* (Green Banner) magazine. According to Drzewiecki (2009, p. 8), staying in a holiday village was supposed to be "different both from the urban working environment and from crowded resorts [...] whereas the organization of holiday villages was generally intended to provide financial support" both for the employees (cheap holidays) and for rural service providers (additional revenue streams). In the 1950s and 1960s, these forms of leisure were supported and co-financed by different state institutions (Pieńkos & Tymińska, 2013). In the 1960s, Gromada already had an extensive network of collaborators, i.e. farmers who hosted urban dwellers in their homes during the summer season (Drzewiecki, 1992; Sikora, 1999); also, Polish rural areas were becoming increasingly multifunctional (Drzewiecki, 2009). Some even called for "establishing a formal and legal framework for what is referred to as 'summer holiday stays' by statutorily providing holiday villages with a special legal status similar to what is enjoyed by spa resorts" (Drzewiecki, 2009, p. 9; cf. Drzewiecki, 1983). However, in the context of a centrally planned economy, these achievements were lost in the 1970s when the responsibility for "summer holiday stays" was transferred to the newly created Voivodeship Tourism Enterprises. These institutions were given the monopoly for intermediation in the rental of private rural accommodation. Moreover, high taxes were imposed on accommodation providers, resulting in a pronounced slowdown in the development of that form of leisure (Drzewiecki, 2002; Sikora, 2012; Marcinkiewicz, 2013). The recession was strongly exacerbated in the martial law period (Drzewiecki, 2002; Marzejon-Frycz, 2012). However, in the late 1980s/early 1990s, the political and economic shift in Poland produced essential changes in the operation of holiday villages. State monopoly was phased out, whereas farmers and other rural residents were allowed to engage in any economic activity (Drzewiecki, 2009). Nevertheless, in Poland, the 1990s were a tough period of socioeconomic transformation, and the situation of Polish farmers at that time can be compared to problems faced by the French, Italians or Austrians in the 1950s, 1960s or 1970s. However, Western European rural dwellers could rely on considerable interest and financial support from regional or central authorities because rural tourism and agritourism—as components of multipurpose rural development—were supported at all government levels. Conversely, in Poland, agritourism remained outside the interests of the authorities, at least in the initial stage of the economic transformation (Marzejon-Frycz, 2012).

The development of rural tourism and agritourism in their modern meaning and form started in the 1990s after Poland embarked on the market economy path and as farmers and other rural residents were allowed to offer private accommodation without restrictions (Drzewiecki, 2005; Wojciechowska, 2006; Sikora, 2012). At that time, rural tourism became a part of the rural policy as a factor in the restructuring of the agricultural sector, as a driver of the restructuring and modernization of regions, especially those with a poorly developed agricultural sector (Kielesińska, 2014), and as a way to earn additional incomes and fight against rural unemployment (Strzembicki & Kmita, 1994; Bieńkowski, 2001; Roberts & Hall, 2001; Gaworecki, 2003b, p. 216; Chądrzyński, 2009; Wojciechowska, 2009; Sikora, 2012). Several development stages (phases) of rural tourism can be identified (Mikuta & Żelazna, 2004; Wojciechowska, 2005, 2006, 2009; Drzewiecki, 2009; Sikora, 2012). In addition to an increase in supply and demand, rural areas also witnessed a change in how tourism operates. New trends and forms have emerged, and the terms and definitions related to rural tourism have evolved (Drzewiecki, 1998; Wojciechowska, 2006, p. 115; 2009, p. 17–22). The evolution of the rural tourism terminology in the Polish literature on the subject is shown in Figure 3.1.Wojciechowska (2009, pp. 18–19) explains it by the differences in the approach to rural tourism: the terms used in the initial period were "semantically related to tourists" (people); afterwards, they referred to a tourism destination (a place: village, rural area; rural tourism) and, ultimately, to the function of the destination (function and specialization; agritourism). Today, "rural tourism," a term which once again refers to the location of tourism activities, is used increasingly often.

The evolution of the rural tourism terminology in the Polish literature, as shown in Figure 3.1, resulted in the emergence of terms used today, including tourism in rural areas, rural tourism and agritourism. However, many authors of the relevant literature point out the chaos in the terminology relating to different kinds and forms of tourism which develop in a rural setting (Szwichtenberg, 1998; Przezbórska, 1998; Gaworecki, 2003a; Jalinik, 2005a; Sznajder & Przezbórska, 2006, p. 17; Wojciechowska, 2006, p. 114; Wojciechowska, 2007a;

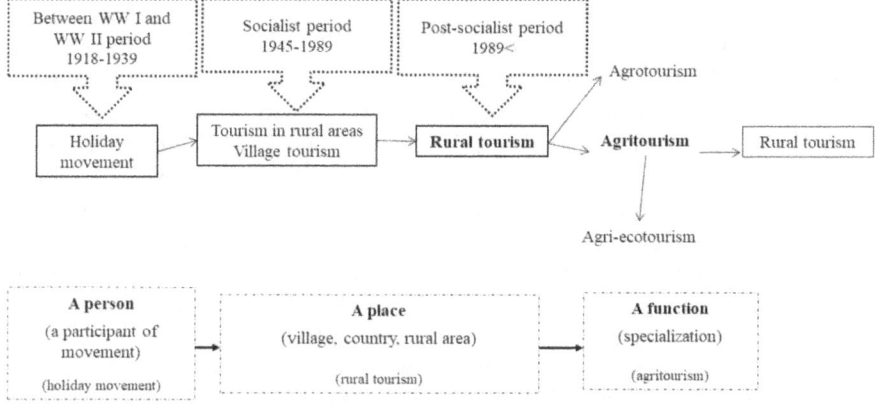

Figure 3.1 Evolution of terms describing tourism in rural areas based on the Polish literature.
Source: Own study based on Wojciechowska (2006, pp. 115–116).

Mazurski, 2009; Wojciechowska, 2009, pp. 34–44). According to Mazurski (2009, p. 7) the "terminological chaos," which affects tourism, too, is "among the major problems of today's science [...] which complicates communications not only between representatives of different disciplines but also between researchers active in the same scientific domain." This is what makes it so important to orchestrate the terms related to rural tourism development.

Tourism in rural areas is the most widespread of the terms listed above (Iwicki, 2001). It covers any form of tourism which takes place in rural areas (Bański, 2006, p. 128), including farm stay (agritourism), participation in folkloric events, eco-tourism,[3] and visiting national parks, landscape parks or nature reserves (Wyrwicz, 1996; Szwichtenberg, 1998; Iwicki, 2001; Wojciechowska, 2009). Depending on how rural areas are defined, this can have a narrower meaning (a rural area that meets the criteria of dispersed settlement patterns and a non-urbanized landscape) or a broader meaning (a rural area delimited based on an administrative decision) (Roberts & Hall, 2001; Majewski & Lane, 2001).

Rural tourism[4]

> includes the forms of tourism which take place in a non-urbanized rural environment characterized by: traditional social structures; a local atmosphere; small-scale accommodation facilities; personal contacts; and a low intensity of recreational activities which do not require large investments and do not have any major environmental impacts.
>
> (Majewski, 1997, p. 52)

That definition includes a small boarding house, a forest lodge, a small hotel, and private accommodation, i.e. diverse forms of tourism found in what is referred to as true rural areas (Drzewiecki, 1993, 1995; Kożuchowska, 2000; Gaworecki, 2003a, p. 98; Wojciechowska, 2009, p. 22; Majewski, 2010; Kiper & Özdemir, 2012) as well as activities "related to nature, hiking, health tourism and landscape, cultural and ethical exploration [...] with a direct use of rural resources and values" (Drzewiecki, 1995; Gaworecki, 2003b; Wojciechowska, 2009, p. 22). Lane (1994), author of one of the definitions most commonly quoted in the international literature, sees rural tourism simply as "tourism activities which take place in rural areas." Similar definitions can be found in international publications, e.g. in a report by the Organization for Economic Co-operation and Development (OECD, 1994) and elsewhere (e.g. Sharpley & Sharpley, 1997; Andrzejewska, 1999; Iakovidou, Partalidou, & Manos, 2000; Yeoman, 2000). In turn, Majewski and Lane (2001, p. 40) identified five characteristics which should be inherent of rural tourism "in its purest form": it should be located in rural areas; be functionally rural (based on small entrepreneurship; open spaces; contact with nature and cultural heritage; and traditional communities and habits); have a rural scale (both the facilities and the towns should be small); have a traditional nature (follow a slow development path; be related to local communities); and be multifaceted (reflect the diversity and complexity of the rural environment, economy, history and location).

Note that no commonly accepted Europe-wide definition of "rural tourism" has yet been developed (Bott-Alama, 2005, p. 14). Even the European Commission stated in 1990 in the "Community Action to Promote Rural Tourism" (CEC, 1990, p. 11) that "There is no precise definition in Europe of rural tourism, the forms and interpretations of which vary considerably from country to country [...] but is very often labeled as cheap tourism" (cf. Drzewiecki, 1998, p. 25; Bott-Alama, 2005, p. 14).

The chaos in rural tourism terminology largely results from difficulties in defining rural areas and from different criteria being used in their delimitation (CEC, 1990, p. 11; Durydiwka, 2012, p. 48).[5] This issue is faced both at country level an in international institutions (e.g. OECD, 1994).

In turn, *agritourism*, [6] less frequently referred to as agro-tourism or agrarian tourism,[7] includes all forms of tourism related to farming activities and/or agricultural premises (Drzewiecki, 1992, 1995; Wiatrak, 1995, 1996; Dębniewska & Tkaczuk, 1997; Philip, Hunter, & Blackstock, 2010). Hence, agritourism can be concluded to be a kind of rural tourism which is related to "agriculture and an active farm or equivalent holding (i.e. an animal, fish or horticultural farm etc.)" (Majewski, 1995, 1997; Bański, 2006; Balińska, 2009; Płocka, 2009; Strzembicki, 2009). The key difference between rural tourism and agritourism boils down to the fact that the latter is closely related to agriculture and that "vegetable production and animal husbandry are among the attractions offered to tourists during their farm stay" (Majewski, 2000a, p. 15; cf. Jalinik, 2005b, p. 85). "Agritourism, a combination of two words (agro and tourism)"[8] (Sznajder & Przezbórska, 2006, p. 15) provides a highly precise reflection of what this activity actually is (Dębniewska & Tkaczuk, 1997; Wicks and Merrett, 2003; Kiper & Özdemir, 2012, p. 128). As emphasized by

Strzembicki (2009, p. 1), "a direct relationship exists between agritourism and an active farm" which consists in that "different farm resources are used in addressing tourist needs." Hence, agritourism cannot be equated with rural tourism (Strzembicki, 2009).

In analyzing the term "agritourism," one also has to note that it was mainstreamed "by supply-side operators who represent the interests of farms providing tourism services"[9] (Sznajder & Przezbórska, 2006, p. 15). As a consequence, that term was considerably extended to cover all kinds of activity related to services offered not only to tourists but also to holidaymakers. This is most likely the reason why "agritourism" means something different to agritourism service providers (a broader approach)[10] than to tourists themselves (a narrower approach)[11] (Sznajder & Przezbórska, 2006, p. 15). According to Drzewiecki (2009, p. 45) this is because Polish agritourism "entered a new development phase [...] characterized by a transition from experiments and trials to a professional activity." Hence, the development and progressing commercialization of agritourism make it increasingly reasonable to repeat the question asked by Sznajder, Przezbórska, and Scrimgeour (2009): "is today's agritourism still what it used to be?"[12] In his practical findings from research on modern Polish agritourism, Drzewiecki (2009, pp. 46–47) presented a highly radical proposition to no longer use the term "agritourism" because the share of farm owners in the group of agritourism service providers is on a consistent decline and "the society sees a false image of agritourism due to that name being abused by accommodation providers, publishers and Internet operators." Already in 2005, Kłodziński (2005) found that term to be highly popular yet abused; he noted that agritourism "is only a part of what the rural environment can offer to tourists" (Kłodziński, 2005, p. 10, after: Pająk, 2002). A similar conclusion regarding the abuse of "agritourism" can be drawn from a study by Bański et al. (2012, p. 75). They observed that in a group of 800 tourism service providers (accommodation providers) based in rural areas, only 17% allowed the tourists to participate in farm works, although as many as 80% of the facilities surveyed were referred to as "agritourism" establishments by their owners (which means they should be intrinsically related to a farm). Kmita-Dziasek (2004, p. 135) noted that "the differences between rural tourism and agritourism become blurred,"[13] while Majewski and Lane (2001, p. 43) openly write about the myth that "rural tourism essentially means agritourism." This could be contrary to the expectations of customers who "choose an agritourism offer due to its specific cognitive values" (Kmita-Dziasek, 2004, p. 135) and because they can be in an active farm, have a direct contact with farm activities and live an authentic agricultural experience (Philip, Hunter, & Blackstock, 2010, p. 756). A similar process was spotted by Tyran (2010, p. 204, after: Paluszek, 2008) who wrote that competition in the market for agritourism services and the resulting diversification of agritourism services brought a major revamping of products and services offered by agritourism farms. "Today, agritourism is rarely seen as 'laying under a pear tree' and assisting in field works. Modern agritourism farms offer such activities as horse riding lessons, equine-assisted therapy, handicraft workshops and special events for families and businesses" (Tyran, 2010, p. 204), i.e. activities which increasingly often have less and less to do with agriculture. This is typical not only of Poland; similar problems in defining and perceiving agritourism were indicated by Roberts and Hall (2001); Philip, Hunter, and Blackstock (2010); Dionysopoulou, Katsoni, and Argyropoulou (2014).

The scope and hierarchy of rural tourism-related terms (referred to as the tourism pyramid) is shown in Fig. 3.2.

It is therefore worthwhile to go back to these terms and recall the differences between them, all the more so since their meanings also evolve as a consequence of the development of tourism processes in rural areas. This is so important because an accurate definition of both rural tourism and agritourism—and the subsequent consistent use of these terms—is necessary due to cognitive reasons and in order to structure the terminology; but furthermore, it is also essential for public employees and planners in their day-to-day work (Drzewiecki, 1992; Getz & Page, 1997; Majewski & Lane, 2001, p. 34; Jalinik, 2005a; Drzewiecki, 2009, p. 41) and for farmers themselves and other rural residents engaged in the delivery of tourism services, for instance due to tax reliefs for agritourism farm owners and the availability of European funds (Kutkowska, 2003, p. 59; Kachniewska, 2011, pp. 54–58; Wiatrak, 2014, p. 30).

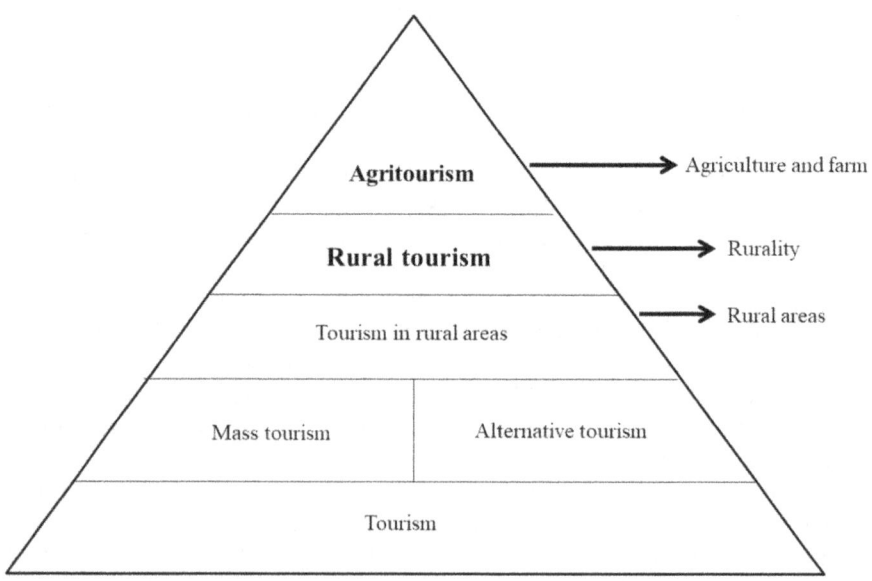

Figure 3.2 Scope and hierarchy of concepts connected with tourism in rural areas (tourism pyramid).
Source: Przezbórska-Skobiej (2015b, p. 105).

The review of publications presented in this chapter also shows how much tourism processes have changed in rural areas. According to Wojciechowska (2006, p. 99), the development phases of rural tourism in Poland are closely related to the origins and evolution of the relevant terminology but also to the organizational formation of tourism, including at different state administration and local government levels and at different stages of Polish history (Wojciechowska, 2012). She used three criteria (key factors of tourism infrastructure development; key development drivers of tourism traffic; and organizational characteristics of tourism traffic) to identify five development phases of tourism in Polish rural areas, including one pioneering (initial) phase which ended in 1873 and four essential phases which correspond to "specific milestones in Polish history" (Wojciechowska, 2009, p. 11). Wojciechowska (2006) used 1873 as the threshold year, marked by the establishment of the Galician Tatra Association (which evolved into the Tatra Association in 1874 and into the Polish Tatra Association in 1919). It is believed to be the first Polish tourism organization (Izydorczyk, 1975) whose numerous goals included promoting the development of tourism in rural areas (Wojciechowska, 2009). The other four phases were described as follows by Wojciechowska (2006):

- Phase 1 (1873–1918): rural leisure was available only to a narrow intellectual and financial elite of urban residents, and involved relaxation, sightseeing, and therapies offered in mansions, manors or spa centers, mostly located close to big cities (cf. Mikuta & Żelazna, 2004; Majewski & Zmyślony, 2014; Matczak, 2015).
- Phase 2 spanned the two decades between the wars (1918–1939). That period saw accelerated development and territorial concentration of tourism traffic around big cities and in remote areas attractive to tourists (e.g. Subcarpathia, coastal area, great lake regions). This was also the beginning of scientific research on tourism traffic; statistics started to be recorded, and terms such as tourism traffic or holiday resort went mainstream in the scientific community (Wojciechowska, 2006, 2009; Surdacka, 2017).
- Phase 3 (1945–1989) was a period of the socialist state; the government intervened in the development and organization of mass supply of and individual demand for rural tourism (which had a restrictive effect on its development). Recreational traffic and private accommodation are the terms which started to be used in official statistics. The relevant literature referred to concepts such as summer holiday stays, rural tourism and tourism in rural areas.
- Phase 4 started in 1989, lasts to this day, and is related to subsequent economic and political changes experienced in Poland. It is characterized by the coexistence of two styles of rural tourism: the old and the new style. The new style means rapid development of agritourism which became a trendy activity (Sikora, 1995). That fact, combined with substantive organizational and legislative support from relevant

institutions, resulted in a quite rapid formation of tourism demand and supply. Furthermore, in this period, the new forms of tourism have given rise to doubts as to their names and the content and scope of their respective definitions. These years also saw a rebound in the development of rural tourism in suburban (metropolitan) areas (Majewski & Zmyślony, 2014).

In turn, according to Drzewiecki (2005, p. 49) the following three stages of agritourism development can be identified in Poland:

- 1st stage (which ended in the early 1990s): a period when farmers rented out simple, low-standard accommodation facilities;
- 2nd stage (2003–2005): the farmers improved the sanitary and accommodation conditions of their agritourism farms (usually up to a "standard" level) while developing supplementary services;
- 3rd stage (2003–2005): considerable quality improvements, diversification of recreation options offered by agritourism farms, and rapid growth in the number of agritourism accommodation facilities in Poland.

At the same time, Zawadka (2010b) identified four stages of agritourism development: 1st stage: early 1990s; 2nd stage: from mid-1990s to 2000, a year marked by organizational changes in the Polish tourism sector; 3rd stage: from 2000 to mid-2000s, i.e. until the implementation of measure 2.7 of the 2004–2006 Sectoral Operating Program "Restructuring and modernization of the food sector and rural development" (the "Pilot LEADER+ program'). According to Zawadka (2010b), in its early stages, agritourism was the core of rural tourism and was often equated with it. The next stages saw the enhancement of the rural holiday offer and the emergence of other providers of tourism in rural areas. This resulted in a decreasing importance of agritourism in rural tourism as a whole.

Wojciechowska (2006, 2009, 2012) and other authors (Drzewiecki, 2009; Sikora, 2012; Marcinkiewicz, 2013; Majewski & Zmyślony, 2014; Matczak, 2015) also emphasize that both rural tourism and agritourism evolved and changed during each phase; their development path changes in a continuous way, as evidenced by demand and supply and by the terminology used in describing this process. After World War II, rural tourism in Poland "evolved from a small economic activity to a sector which often dominates the rural economy" (Matczak, 2015, p. 79). The boundaries of that evolution were delimited by changes in Polish rural areas and in tourism itself.

In summary, attention needs to be paid to the conclusions from the "Rural tourism and agritourism: new paradigms for the 21st century," an international conference on the future of rural tourism and agritourism held in Kielce in 2015. In the report on conference discussions, Majewski & Zawistowska (2015) wrote that in the 2035 perspective, rural tourism and agritourism can continue to develop in line with either an optimistic or a pessimistic scenario. "The optimistic scenario assumes that these forms of tourism will strengthen or at least maintain their current important position, measured with their market share. The pessimistic scenario means losing their current position" (Majewski & Zawistowska, 2015, p. 304). The adverse phenomena which can affect the tourism market include: a decline in both supply and demand; strong competition; absence of successors in the next generations; and/or a decrease in cost-efficiency of rural tourism operators, especially small-scale businesses. Hence, if these processes

> are accompanied by a deterioration in the quality of services offered, it will translate into lower demand and [...] major changes in the market for rural tourism and agritourism. Some operators active in that market will evolve into large-scale companies which will make rural tourism less attractive.
>
> (Majewski & Zawistowska, 2015, p. 304)

These problems can be partly solved by changes in strategic planning and in marketing and organizational activities, and through creativity and collaboration "in all dimensions and at all levels: from service providers to government authorities, in the tourism sector and beyond" (Majewski & Zawistowska, 2015, p. 304).

3.2 Regionalism in rural tourism development

The attractiveness to tourists varies strongly across the Polish territory due to different environmental conditions and uneven levels of tourism development (Lijewski, Mikułowski, & Wyrzykowski, 2002). This, in turn, is reflected by the numerous attempts made to define Polish tourist regions based on different methods and criteria for regionalization and delimitation (e.g. Mileska, 1963; Bajcar, 1969; Kruczek & Sacha, 1994; Gołembski, 1999; Lijewski, Mikułowski, & Wyrzykowski, 2002; Liszewski, 2009; Durydiwka, 2012). Regionalization, defined as dividing a territory in regions based on specific characteristics, or as a process of separation of regions, i.e. "relatively uniform territories which differ in specific natural or acquired characteristics from adjacent territories" (*Encyklopedia* [Encyclopedia], PWN Publishing House), means reducing the infinite number of individual locations on the globe to a finite number of territorial units (Szymla, 2000). Thus, regionalization consists either in separating certain larger territories (regions) composed of smaller areas which share similar characteristics subject to evaluation (zonal regions or regions centered around a hub, i.e. nodal regions) (Szymla, 2000) or in dividing a territory into smaller units based on certain criteria (Kruczek & Zmyślony, 2010). According to Chojnicki (Kruczek & Zmyślony, 2014, p. 7, after: Chojnicki, 1966, pp. 7–43), region is a concept which boils down to two dimensions:

- the analytic dimension: the region is considered a "unified area exhibiting a specific feature or a combination of features of relevance to a defined research problem";
- the subjective dimension: the region is referred to as a "real social object [...] which forms a spatially separate entity that needs to be discovered rather than created" (Kruczek & Zmyślony, 2014, p. 7).

According to Mazurski (2009, p. 7) the term "tourist region" itself continues to be debatable, as reflected in the literature on the subject (Wysocka, 1975; Drzewiecki, 1980; Biderman, 1981; Jackowski, 1981; Matczak & Suliborski, 1984; Warszyńska, 1985; Faracik, 2006; Mazurski, 2009; Saja, 2009; Kruczek & Zmyślony, 2010; Faracik, 2011; Derek, 2008; Durydiwka, 2012; Falkowski, 2016).

Delimitation is defined as "the action of fixing the boundary of something" (*Słownik Języka Polskiego* [Polish language dictionary], PWN Publishing House). Hence, delimitation of regions boils down to defining their borders. Both processes, i.e. regionalization and delimitation, are of great theoretical and practical importance. Furthermore, regionalization and (especially) delimitation are difficult and complicated terms (Gołembski, 2002, p. 11; Kruczek & Zmyślony, 2014, p. 12).

While regionalization—as a procedure used to identify regions—can rely on different methods, it is essentially carried out in two ways: "by dividing an area into regional units based on defined criteria" (a top-down approach) or "by merging smaller adjacent units which share similar environmental characteristics" (a bottom-up approach) (Kruczek, Zmyślony, 2010, p. 12). Mazurski (2009) identified four types of determinants of tourist regions and their borders, namely:

- *explorative criteria* (the whole geographic area covered by the analysis is divided into regions),
- *"purely" tourist criteria* (some areas can remain non-attributed to any region; this criterion involves the introduction of an additional factor which specifies the type of tourism);
- *mixed (tourist and explorative) criteria*: the whole geographic area is divided into regions;
- *attractiveness of an area to tourists* (depending on how much is the area saturated with tourist attractions).

Kruczek & Zmyślony (2014, p. 12) noted that the literature on the subject lacks the description of unequivocal, widely accepted delimitation criteria. Physical and geographic, economic and spatial or administrative criteria are the ones most commonly referred to. The widespread administrative criterion allows the use of the

applicable administrative division and the term "region" defined as a strictly delimited part of the national territory which forms an economic complex with a specific profile or specialization and with a separated administrative and economic center which plays a decisive role in the development and functioning of the whole region.

(Kruczek & Zmyślony, 2014, p. 12)

Importantly, the regional division of a country based on the economic and administrative criterion is a "reference point for the regional policy and a basis for the collection of socioeconomic data" (Kruczek & Zmyślony, 2014, p. 12). The administrative criterion is often used in practice because it provides the ability to access relevant data. However, it is not a flawless approach, and its major downside "results from the fact that the boundaries of tourist regions usually do not coincide with administrative boundaries" (Gołembski, 2002, p. 11).

The criterion of complementarity between tourism values, tourism development, and tourism traffic (according to Mileska, 1963) provides a basis for identifying *real regions* and *potential regions*. Moreover, *rural tourism regions* and *agritourism regions* can be distinguished based on the criterion of the dominant tourism type (according to Gaworecki, 2010). Territorial tourism units have a hierarchical structure, with regions as higher-order territorial units and sub-regions as lower-order units. Tourist regions and sub-regions are home to a wide variety of settlement units (tourism towns) with tourism functions (e.g. tourism destinations, spa or sightseeing centers) (Czerwiński, 2011).

In the late 1980s/early 1990s, based on research on Polish rural areas, Drzewiecki (1985, 1992) identified the rural recreational space as an area which witnesses the development of different forms of tourism, especially including rural tourism and agritourism. Drzewiecki (1992, p. 74) defined it as "a part of a country's recreational space which includes areas located "outside cities, urbanized areas [...], specialized areas [...] and restricted areas," with a predominantly agricultural economy and a prevalence of natural components, with no signs of degradation, and which "demonstrates characteristics that are conducive to relaxation [...] and processes that enable humans to recover their mental and physical strength." Poland is widely believed to be a country offering a very favorable context for agritourism and rural tourism development (Roman, 2018). According to a study carried out by Drzewiecki in the early 1990s (Drzewiecki, 1992), the conditions for becoming a tourism location were met at that time by as much as 66% of all Polish communes (i.e. 1369 of them, with an area of 207,700 km^2); in each commune, tourist space accounted for less than 10% to over 90% of the local territory (Drzewiecki, 2009). In the late 1980s/early 1990s, employees of the Institute of Tourism assessed the opportunities for rural tourism development in Polish communes. They found that only one commune out of three had appropriate conditions for a tourism-driven activation (Kmita, 1994). In turn, according to research by Przezbórska and Lira (2011) on the rural natural environment's attractiveness and suitability for the development of rural tourism and agritourism at district level, of the 314 Polish land districts, only 50 (15.9% of the total number) offered excellent or very good natural conditions for the development of rural tourism and agritourism. Other analyses carried out by Przezbórska-Skobiej and Lira (2012) based on a defined synthetic valuation index for the Polish agritourism space suggest that land districts located in west, northwest, northeast, and southeast Poland exhibited higher quality of agritourism space and were home to the largest population of agritourism farms. All studies referred to above showed that Polish rural areas considerably differ in how suitable they are for the development of the recreational and tourism function; this means that the rural recreational space is highly heterogeneous. Note, however, that agritourism farms exist nearly everywhere in Poland, and their development factors can result from both exogenous (e.g. attractiveness of the area) and endogenous conditions (referring to the properties and characteristics of farms or farm owners themselves; this means, for instance, the attractiveness of the product and service offering) (Sikora & Jęczmyk, 2005). This was pointed out by Drzewiecki (1992, p. 45; cf. 2009, p. 40) who wrote that recreation, including tourism, "can exist in any part of geographical space" while agritourism can develop in any location not affected by natural and anthropogenic "prohibitive factors," e.g. health risks (a highly polluted environment), or "restrictive factors," i.e. nuisances (including an excessive concentration of livestock production, minor environmental pollution or noise) (Drzewiecki, 1998; Sznajder & Przezbórska, 2006; Drzewiecki, 2009). Indeed, according to economic theory, the location of this kind of activity is chosen freely (Drzewiecki, 2009, p. 40). Based on his research, Drzewiecki (2005, p. 47) identified seven regions of agritourism development: Western Pomerania, Pomerelia (including the coastline), Warmia–Masuria, Greater Poland, Sudetes, Carpathians, Lublin Upland, and Podlachia. The literature on the subject also proposes other ways of extracting rural tourism and agritourism regions in Poland. One of the first divisions of the national territory in rural tourism regions was presented by Sikora (1999) who identified seven rural tourism development regions: Pomerania, Masuria, Greater Poland, Central Poland, Eastern Poland, Sudetes, and Carpathians. In turn, based on her own definition of agritourism space, Wojciechowska (2009, pp. 140–143) proposed the Polish voivodeships to be divided into groups and sub-groups in accordance with pre-established development levels of agritourism accommodation resources. In her paper, she identified twelve regions with an outstanding agritourism offer, including: the coastal region; Pomerania; Kashubia–Masuria–Suwałki; Mazovia; Greater Poland; Łódź; the western region; the Świętokrzyskie region; the eastern region; Sudetes; and Subcarpathia (Wojciechowska, 2009).

According to a study by Przezbórska-Skobiej (2015b), the northeast part of the Polish territory (Pisz, Mrągowo, Hajnówka, Węgorzewo, Białystok, Giżycko, and Sejny districts) offers more favorable conditions for the development of agritourism. The same is true for some areas located in west (Nowa Sól and Wschowa districts) and southwest Poland (Jelenia Góra and Kłodzko districts).

Conversely, the least favorable conditions for rural tourism development were identified in central Poland (in the Kujawsko–Pomorskie voivodeship, e.g. Inowrocław and Nakło districts) and in urbanized areas of Silesia and Lesser Poland (e.g. Myślenice, Wodzisław or Bieruń-Lędziny districts). As shown by the analyses, northern Poland territories have a better potential for rural tourism development because they demonstrate the highest concentration of stimulating factors (and are home to the largest group of agritourism and rural tourism facilities). As part of the same study, Polish rural territories were subject to a valuation procedure with a distinction between rural tourism and agritourism development regions. The k-means method was used to identify clusters of districts and communes with similar conditions for the development of rural tourism. Combined with an analysis of the distribution of rural and agritourism accommodation facilities, findings from this study suggest that Poland witnesses the emergence of rural tourism and agritourism development regions (Przezbórska-Skobiej, 2015b). The identified district clusters include:

- the eastern (Podlachia–Lublin) region;
- two northern regions (Pomerania and lakelands);
- northwest regions and northeast lakelands;
- three mountain regions (Carpathian, Świętokrzyskie, and Lower Silesia–Opole);
- central Poland region (Greater Poland and Łódź).

At commune level, it was proposed to identify as many as 16 actual regions of rural tourism development. Rural tourism facilities and agritourism farms based in the identified regions can provide the tourists with a diversified product and service offering underpinned by the characteristics of local rural areas. The study referred to above also included using a logit model in assessing the conditions for the development of agritourism farms and the likelihood of creation of new farms (Przezbórska-Skobiej, 2015b). According to that analysis, the best conditions for the development of agritourism farms exist in territories with abundant tourism resources, i.e. in the area which extends from west and northwest Poland to the northeast (coastal areas and lakelands) and in southern Poland (the belt formed by the Sudetes, Carpathians, Carpathian foothills, and Świętokrzyskie Mountains), as well as around large urban agglomerations (similar conditions were noted by Durydiwka, 2005).

Note that the distribution of agritourism farms across the Polish territory is highly uneven due to a number of different factors (Drzewiecki, 1992, 2009; Przezbórska, 2007) related, for instance, to the heterogeneity and quality of natural assets in each part of the country. The uneven distribution of agritourism farms becomes particularly evident when comparing their numbers between districts and communes. For instance, in 2011, as many as 1,200 agritourism farms (19.3%, i.e. almost one-fifth of all agritourism farms active in Poland) were located in as little as ten districts (0.5% of the total number of Polish districts). In the same year, there were 908 communes in Poland with not even a single agritourism farm, and 394 communes with one agritourism farm each. On the other hand, there were communes with tens of active farms; for instance, the Solina commune (Lesko district) was home to as many as 85 of them (Przezbórska-Skobiej & Sobotka, 2016).

Based on an analysis of the distribution of agritourism accommodation facilities in Polish rural and urban-rural communes, Przezbórska-Skobiej and Sobotka (2016) proposed a classification into twelve real agritourism regions, including five large ones: Baltic Coastal Zone; Pomerania Lakeland; Masuria Lakeland; Sudetes (mountains and foothills); and Carpathians (mountains and foothills), divided into agritourism sub-regions (Figure 3.3).

The Carpathians (mountains and foothills) have the largest number of agritourism accommodation facilities of all regions identified. The 53 communes constituting that region were home to as many as 706 agritourism farms, including as many as 408 in the Bieszczady sub-region. In turn, the Masurian Lakeland (the geographically largest of the regions identified) is home to 520 agritourism farms active in 37 communes. It was divided in two sub-regions referred to as Olsztyn Lakeland and Great Masurian Lakeland for convenience. A similar number of agritourism farms was recorded in the Pomerania

Legenda

Pobrzeże Bałtyku
1. woliński
2. koszaliński

Pojezierze Pomorskie
3. Pojezierze Kaszubskie
4. Pojezierze Bytowskie
5. Bory Tucholskie
6. Pojezierze Drawskie
7. Pojezierze Wałeckie

Pojezierze Mazurskie
8. Pojezierze Olsztyńskie
9. Kraina Wielkich Jezior Mazurskich
10. Pojezierze Suwalskie
11. rajgrodzki
12. białowieski
13. Pojezierze Poznańskie
Sudety i Pogórze Sudeckie
14. jeleniogórski
15. kłodzki

16. głuchołaski
17. kroczycki
18. Góry Świętokrzyskie
19. Roztocze
Karpaty i Pogórze Karpackie
20. żywiecki
21. małopolski
22. gorlicki
23. bieszczadzki

0   60 km

Figure 3.3 Polish agritourism regions and sub-regions in 2011.
Source: Przezbórska-Skobiej and Sobotka (2016, p. 189).

Lakeland (543 farms in 40 communes), divided in five agritourism sub-regions. However, the identification of Pomerania Lakeland as a separate region may give rise to doubts. Indeed, when defining and describing the Polish rural recreational space, Drzewiecki (1992, pp. 74–75), concluded that specialized sub-regions, such as recreational sub-regions and areas well-equipped in tourism facilities do not constitute rural recreational space. Therefore, "just like urban areas (urban communes), seaside rural and urban-rural communes at the highest level of development of the tourism function in Poland" should not be counted as rural tourism (including agritourism) development areas; similar problems were noted by Derek (2008) and Bański et al. (2012).

An analysis by Przezbórska-Skobiej and Sobotka (2016) additionally confirmed that no agritourism regions had been formed in central Poland (outside Greater Poland), a region which also exhibits the greatest dispersion of agritourism farms (as mentioned in 2006 by Sznajder and Przezbórska).

The analyses also discovered the emergence of smaller territorial units (potential regions) which are likely to become agritourism regions in the future, e.g. three communes in the south of the Małopolskie voivodeship (Poronin, Bukowina Tatrzańska, and Łapsze Niżne) which have been home to a total of 35 agritourism farms since 2011, or three communes in the north of the Warmińsko–Mazurskie voivodeship (Lidzbark, Górowo Iławeckie, and Bartoszyce) which were home to 26 agritourism farms (Przezbórska-Skobiej & Sobotka, 2016). From the perspective of the quality of the geographic environment and other factors of importance to agritourism development, Polish rural areas still hold considerable reserves which can be used in developing agritourism (Drzewiecki, 2009).

In summarizing the findings from research by different authors on the regionalization of Polish agritourism and rural tourism development in Poland, it has to be emphasized that rural tourism facilities are distributed unevenly and are geographically dispersed. Regions and sub-regions of rural tourism and agritourism development, as identified in research by different authors, exhibit high levels of territorial concentration and are strongly related to areas with outstanding natural values; in these locations, tourists are mainly offered accommodation and leisure services either during the summer season (coastal areas, lakelands) or all year round (mountains and foothills). Another important aspect is that plans and strategic documents of most Polish communes should not assume that agritourism has any major impact on rural economic development. This is caused by the absence (or small number) of active agritourism farms and rural tourism facilities; they have not been created thus far due to lack of appropriate assets (resources) or to small interest shown by tourists (Drzewiecki, 2005; Durydiwka, 2005; Drzewiecki, 2009, Sikora & Jęczmyk, 2005).

3.3 Players supporting rural tourism development

Support from the institutional environment is an important factor affecting the development of tourism services in Polish rural areas, especially including agritourism services (Table 3.1). Almost since the very beginning, different entities,

Table 3.1 Operators active in supporting the development of rural tourism and agri-tourism in Poland

Direct support providers		Indirect support providers	
Local community		Support from district-level authorities	District offices
Commune (local authority)			Organizations, e.g. local tourism organizations
Consulting	Agencies, foundations	Support from region-level authorities	Voivodeship offices
	Agricultural consultancy (Agricultural Consultancy lefts, Agricultural Advisory Centre in Brwinów (the state-controlled entity))		Marshall's offices
Industry authorities	Agritourism associations		Organizations, e.g. regional tourism organizations
	Agricultural and tourism chambers	Support from national authorities	Ministries, e.g. Ministry of Economy (1999–2005), Ministry of Sports and Tourism (2007–2019), Ministry of Development, Ministry of Agriculture and Rural Development
	Agricultural chambers		
	Tourism chambers, e.g. Polish Tourism Chamber (since 1990)		
	Agricultural and tourism societies		Offices, e.g. Office of Physical Culture and Tourism (1991–2000)
	Tourism associations		
	Associations of accommodation providers		Organizations, e.g. Polish Agency for Tourism Development (1993–2014), Polish Tourism Organization (since 1999)
	Rural tourism associations	International support	International tourist organizations (e.g. the World Tourism Organization, European Travel Commission, Committee on Tourism of the European Union)
	Agri-ecotourism associations		

Source: Own study based on Mikuta and Żelazna (2004); Wojciechowska (2007b); Wiatrak (2009); Wojciechowska (2009); Cichowska and Klimek (2010); Panansiuk (2019).

organizations, and institutions, including without limitation agricultural consultancy centres, agritourism associations, and local government units, have been involved in the development of Polish rural tourism and agritourism (Roman & Niedziółka, 2017; Knecht, 2009; Wojciechowska, 2009; Balińska, 2016; Bednarek-Szczepańska, 2017).

In the context of the four development phases of Polish rural tourism, as identified by Wojciechowska (2007a) (the pioneering phase which ended in 1873, and the four essential phases which correspond to specific milestones in Polish history: phase 1, from 1873 to 1918; phase 2, from 1918 to 1939; phase 3, from 1945 to 1989; and phase 4, from 1989 to this day), it can be noted that the role of supporting actors evolved from one phase to another. Phase 1 (which exhibited clear organizational characteristics) saw the emergence of the first organizations in charge of supporting tourism in Poland. For instance, the Galician Tatra Association, considered to be the first Polish tourism organization, was established in 1873 and was followed by the Polish Sightseeing Association (1906) and the Poznań Sightseeing Association (1913) (Izydorczyk, 1975). In phase 1, supporting organizations did not play any major role in rural tourism development. However, phase 2 saw the establishment of the General Tourism Department in the Ministry of Communications (in 1932), followed by the Tourism Support League created on the initiative of the Ministry in 1935 to supervise tourism activities in Poland. Its role was all the more important since it showed interest in the organization of holiday resorts and in the development of holiday villages. Special tourism instances, referred to as holiday and tourism committees, were appointed within district, commune and "gromada" councils on the initiative of the League to provide support and advice related to holiday and tourism actions. The tasks and composition of the committees at all administrative levels were strictly provided for by the law (Łazarek, 1972). Supporting actors established during phase 3 include the National Paid-Leave Fund and "Gromada," the Tourism and Holiday Cooperative referred to in Section 3.1, reactivated after World War II. At that time, the development of rural tourism and the organization of rural holidays was strictly controlled by the government. In the 1970s, the General Committee of Physical Culture and Tourism published the "statute of the rural village"; accordingly, a rural village was required to offer no less than 50 beds in double to quadruple rooms. Holiday villages and stays started to be promoted in that period, e.g. in *Wieś letniskowa* (The Informator, a holiday village newsletter) published in 1975 by the Center for Tourism Information and Advertising (Wojciechowska, 2006). However, the strongest support for rural tourism and agritourism development was provided only in phase 4, i.e. after 1989.

The period 1989–2019 saw an evolution of the role of the aforementioned entities such as agricultural consultancy centers, agritourism associations, and local government units. The forms of mobilization measures changed, too (Wojciechowska, 2009; Wiatrak, 2009). Wojciechowska (2009, cf. Kmita, 1999; Sikora, 1999; Wilkin, 2002) divided their activities into five basic groups: consultancy, information measures, integration measures, education, and promotion. Then, she analyzed the evolution of these activities in two periods: the early 1990s and after 2000. Financial measures are not included as part of the supporting measures because Wojciechowska (2007b) considered them to be covered by other measures, namely consultancy, integration, and promotion (for a description of financial measures, see Section 3.5). Importantly, institutional support includes the instruments provided by each institution, on the one hand, and their use in developing rural tourism and managing the available resources, on the other. Furthermore, these instruments and related measures should be coupled with each other to deliver synergies (Wiatrak, 2009). Based on the relationships between entities designed to be active at three (local, regional or national) levels, Wojciechowska (2009) also presented the types of the organizational structure of the Polish agritourism sector. She defined them as the supporting structure (implemented in two stages, 1991–1993 and 1994–1995), the consolidating structure (implemented 1996–1999), and the destabilizing structure (implemented since 2000).

In the initial phases of agritourism and rural tourism development, information measures were focused on owners of agritourism farms and rural tourism facilities and on tourists (promotion activities in the press; lectures; brochures; shows). Simple consultancy measures—such as encouraging the farmers to engage in agritourism—were of much greater importance. Information measures lost their popularity over the years. Later on (after 2000), consultancy on matters such as finance, loans and law delivered by advisors, professionals, leaders, and experts (especially in EU programs and projects) gained in importance as a consultancy measure. In the first period, a major role was also played by educational measures (courses, trainings, conferences and seminars). After 2000, education measures became more sophisticated or specialized. Hence, it needs to be highlighted that agricultural consultancy centers played a special role in the initial stage of rural tourism and agritourism development (Knecht, 2009; Cichowska & Klimek, 2010; Sikora, 2012; Balińska, 2016). In the early 1990s, farmers gained new experience, overcame some barriers and shifted their mindsets to embrace new activities and open themselves to tourists (Zawadka, 2010b). Mikuta and Żelazna (2004, p. 288, cf. Zawadka, 2010a) noted that "the activity of the rural population in undertaking a tourism activity was triggered by agricultural consultancy centers" which started to provide information, trainings and consultancy in the early 1990s on behalf of the Ministry of Agriculture and Food Economics. Similar functions were, and continue to be, fulfilled by the Agricultural Advisory Centre in Brwinów, a public organizational unit reporting directly to the Minister of Agriculture and Rural Development (called the National Consultancy Center for Agriculture and Rural Development until 2005) (Sikora, 2012).

The development of the local tourism economy is also impacted by the activity of local government units which, in addition to development planning, should also take an active part in improving and expanding the tourism offer (Knecht, 2009; Balińska, 2016). At a local level, it is hard to overestimate the support from local government units (Wojciechowska, 2009). Indeed, as indicated by Kosmaczewska (2007), strong relationships exist between tourism development (including agritourism and rural tourism), a

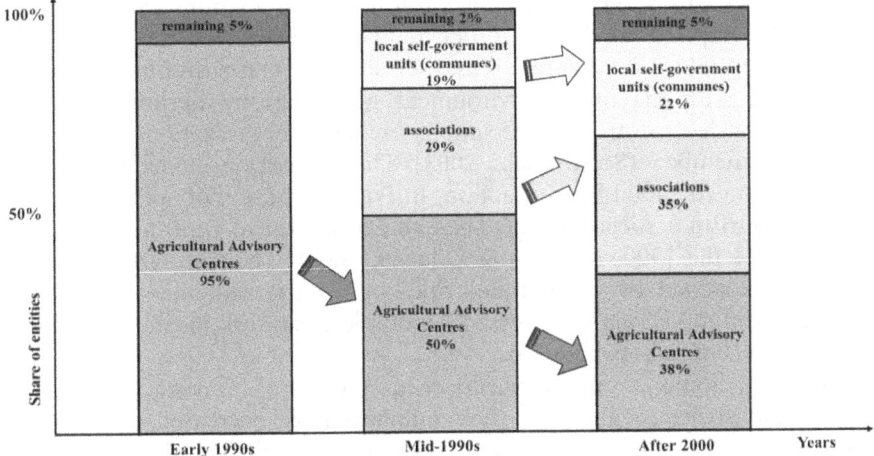

Figure 3.4 Model of the evolving participation of different entities in measures taken to support the development of rural tourism and agritourism in Poland.
Source: Wojciechowska (2007b, 2009, p. 107)

country's land use patterns, and measures taken by local government units acting as public operators. She believes local tourism development to be the combined effect of exogenous conditions (which are beyond the control of the commune, e.g. tourism-related measures taken by central or voivodeship-level authorities) and endogenous conditions which are directly or indirectly controlled by the commune. Meanwhile, research suggests that the small activity of local government units is among the barriers to rural tourism and agritourism development in some parts of Poland (Karolczak et al., 2002; Sikora & Jęczmyk, 2005). The literature emphasizes that, for instance, commune-level development strategies clearly "lack a comprehensive approach to agriculture; one which would indicate it has the potential to contribute to the tourism infrastructure and to create a branded local tourism product underpinned by agritourism farm resources" (Sikora, 2012).

The 1990s also saw the emergence of agritourism associations structured in a form which allows them to promote agritourism farms and operators. They are established to pursue their statutory goals, as defined by their founders, and are financed by the founders. They also can engage in an economic activity, provided that the profits are used in pursuing their statutory goals. Objectives of an association can include providing information on a region's tourism values, and designing, publishing, and distributing maps, brochures, and guides. Ever since they were established, agritourism associations have played an important role in promoting tourism activities in Polish rural areas. Agritourism farms join an association primarily in order to make their promotion efforts more efficient. Also, it is a way to enter into cooperation with the authorities and local government units. Foryś (2016, 2017, p. 112) carried out a detailed study on the role of agritourism associations Poland. In his findings, he emphasized that "in order for the associations to make a full use of their potential vis-à-vis rural areas, they should engage, and be engaged, in cooperation with the broad environment going beyond agritourism stakeholders." Polish agritourism associations are small organizations and usually have 20–30 members (Strzembicki, 2003). Their activity is guided by a statute which sets out the main lines of action. In Poland, the activity of regional and local agritourism associations has been an accelerator of rural tourism development since the 1990s (Mikuta & Żelazna, 2004). Their importance can be primarily explained by the particularities of the tourism offer which is co-created by multiple small, dispersed economic operators located away from the markets (Pałka-Łebek, 2016).

Poland is home to many agritourism and rural tourism associations, unions, and societies, although their numbers have decreased over the last years. The number of agritourism organizations active in promoting agritourism in selected years from the 1991–2020 interval (grouped by type) is shown in Table 3.2. The location of agritourism associations is presented in Figure 3.5. According to data of the "Hospitable Farms" Polish Federation of Rural Tourism, there are 28 of them, although Sikora (2012) claims that over 150 agritourism organizations were active in Poland in 2010, including more than 70 members of "Hospitable Farms." Perhaps some of them have suffered the same fate as the Suwałki Agricultural and Tourism Chamber, one

Table 3.2 Agritourism organizations active in promoting agritourism in selected years from the 1991–2020 interval

Organization types identified based on their names	Number of organizations in selected years							
	1991	*1993*	*1994*	*1995*	*1999*	*2001*	*2004*	*2020*
Agritourism associations	0	2	4	14	27	23	31	22
Agricultural and tourism chambers	1	1	1	1	1	1	1	0
Agricultural and tourism societies	0	0	0	1	2	2	1	0
Tourism associations	1	2	3	3	2	2	6	2
Associations of accommodation providers	0	0	0	1	1	3	1	1
Rural tourism associations	0	0	1	1	1	2	4	2
Agri-ecotourism associations	0	0	0	0	1	1	2	1
Total	2	5	9	21	35	34	46	28

Source: compilation based on Wojciechowska (2009) and own calculations.

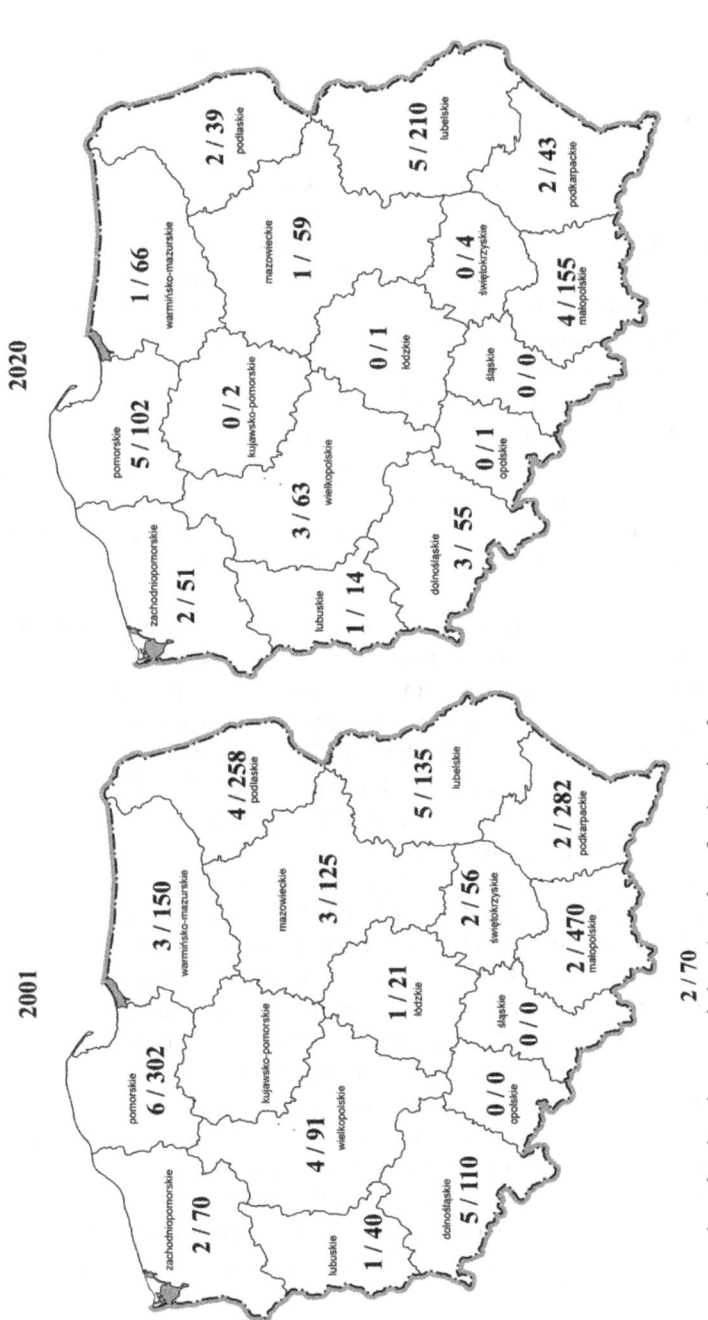

Figure 3.5 Number of agritourism associations and agritourism farms grouped in associations registered as members of the "Hospitable Farms" Polish Federation of Rural Tourism in 2001 and 2020.

Source: Own calculations based on data from www.agritourism.pl (accessed on July 20, 2020).

of the oldest organizations established to promote the development of rural tourism and agritourism.

Established in 1991, the Chamber was one of the most active supporters of rural tourism and agritourism in northeast Poland (Sznajder & Przezbórska, 2006). It was active in the Podlaskie and Warmińsko–Mazurskie voivodeships as a non-profit non-government organization. Composed of ca. 160 owners of agritourism farms and boarding houses, the Chamber was very active in a broad scope of domains, primarily including information and consultancy services, tourism training, classification of rural tourism accommodation facilities, and market research and analysis. It also hosted a Tourism Information and Regional Promotion Center together with an accommodation booking system, handled public relations and performed other promotion and advertising activities. As part of cross-border cooperation with Lithuanian partners, the Chamber made efforts to develop and commercialize a cross-border tourism product. It also collaborated with organizations dedicated to supporting the disabled in order to provide them with an opportunity to relax and recover in the region's natural environment. Unfortunately, the Chamber accumulated debts and did not stand the test of time. It was declared bankrupt in 2016 (https://www.radio.bialystok.pl/wiadomosci/suwalki/id/166836, accessed on June 20, 2019).

According to Sikora (2012), one characteristic feature of Polish agritourism and rural tourism associations is their dispersion. Since their beginnings, this has been the cause of many difficulties affecting their activities and collaboration, making them weak and passive. Therefore, a countrywide federation of rural tourism ("Hospitable Farms") was established in 1996 to reinforce the position of agritourism associations. This was the consequence of TOURIN II, the Tourism Industry Development Program implemented during 1995–1997, financed under the European Union's PHARE fund (for a broader description, see Section 3.5). The "Hospitable Farms" Polish Federation of Rural Tourism was registered by the Voivodeship Court in Warsaw on May 24, 1996. It has legal authority and operates pursuant to the Associations Act of April 7, 1989 (as amended). The Federation owns a classification system for rural accommodation facilities and has a registered logo[14] which can be used by all classified member farms. The Federation takes measures to promote rural holidays and support the wide adoption of tourism activities in rural areas (taking account of the rural cultural heritage and of the conditions for sustainable development of rural tourism). It also carried out other promotional activities, including the publication of the countrywide directory of agritourism farms *Polska. Atlas Turystyki Wiejskiej* (Poland. The rural tourism atlas) (Sikora, 2012). In 1997, the "Hospitable Farms" Polish Federation of Rural Tourism became a member of EuroGites, the European Federation of Rural Tourism (previously known as the European Federation for Farm and Village Tourism) composed of 35 rural tourism associations located in 27 European countries (based on information from https://www.eurogites. org/ accessed on July 20, 2020).

In the recent years, in addition to agritourism and rural tourism associations, an important role has also been played by Local Action Groups (LAGs) which can take different legal forms (including associations, foundations or unions) and operate in one or several communes (Sikora, 2012). The LAGs significantly help preserving local cultural heritage, and make a major contribution to education and the development of skills and interests, e.g. in old handicraft techniques, regional products, and natural or anthropogenic assets (Sikora, 2012).

Other operators engaged in supporting rural tourism and agritourism in Poland include national institutions such as the Ministry of Agriculture and Rural Development and the Agency for Restructuring and Modernization of Agriculture. The ministry, as a government institution, helps rural tourism operators through relevant strategic documents which are essential in accessing financial support from public funds and from European Union funds. In turn, the Agency for Restructuring and Modernization of Agriculture supervises the programs financed with EU rural development funds and acts as the paying agency. Earlier, it supported agritourism development with soft loans subsidized by state budget (Sikora, 2012). The Ministry of Agriculture and Rural Development also takes measures to promote rural tourism, such as "*Odpoczywaj na wsi ze smakiem*" (A tasteful rural holiday, a project implemented in 2019) or "*Odpoczywaj na wsi*" (Have a rural holiday, 2016–2020). The purpose of these campaigns is to make rural areas viewed as a tourism market that offers diverse attractions all year round. Expert supervision over the campaigns was the responsibility of the "Hospitable Farms" Polish Federation of Rural Tourism. The campaign also included the development of Agroturystyka, a mobile application coupled with an agritourism farm search engine for smartphones. It offers a directory of 1,000+ rural tourism and agritourism facilities located around the country (based on information posted at https://odpoczywa jnawsi.pl/#kampania, accessed on June 14, 2020). At national level, rural tourism development is supported by the Ministry of Sport (2007–2019, the Ministry of Sport and Tourism) and the Ministry of Development (1999–2005, the Ministry of the Economy; 2005–2013, the Ministry of Regional Development; and 2013–2015, the Ministry of Infrastructure and Development). The decisions made by these institutions mostly relate to setting the development targets for tourism, including rural tourism and agritourism.

A mention should also be made of the Polish Tourism Organization established in 1999, one of the 100+ national tourism organizations from all over the world. It is a state organization appointed to cooperate with industry players and local government units which deal with tourism issues, including rural tourism and agritourism. Together with its regional and local structures (Regional Tourism Organizations and Local Tourism Organizations, respectively), the Polish Tourism Organization is mostly active in promoting the diverse dimensions of tourism by creating a positive image of agritourism and rural tourism (e.g. they deliver "the Best Tourism Product" certificate of the Polish Tourism Organization) (Wojciechowska, 2009; Zawadka, 2010b; Sikora, 2012).

Summarizing modern methods and organizations supporting rural tourism in Poland, "it is essential to realize that establishing institutional environment was initiated already in the 19th century", when rural tourism started to develop. The development of contemporary structure of rural tourism support is the result of a number of mutually complementary and co-related processes, including processes in such aspects as: political and institutional, legislative, financial, self-organization, transfer of good practices, scientific and educational, the media (Bednarek-Szczepańska, 2017, p. 41). Bednarek-Szczepańska (2017) described this system as a multidimensional system of external support, which was formed around rural tourism (Figure 3.6).

All active institutions and organizations can have a considerable impact on the development of rural tourism and agritourism by contributing to improvements in service quality, promoting holidays in Polish rural areas among the urban population or addressing the financial, organizational, educational or promotional needs of farmers and other rural residents.

3.4 Legal and organizational conditions for rural tourism development

Rural tourism in Poland has undergone major changes in the past 30 years, and the external conditions for tourism development in the country have significantly changed as a result of the European integration and related adjustment to the new rules imposed by the European Union (Table 3.3, References to European

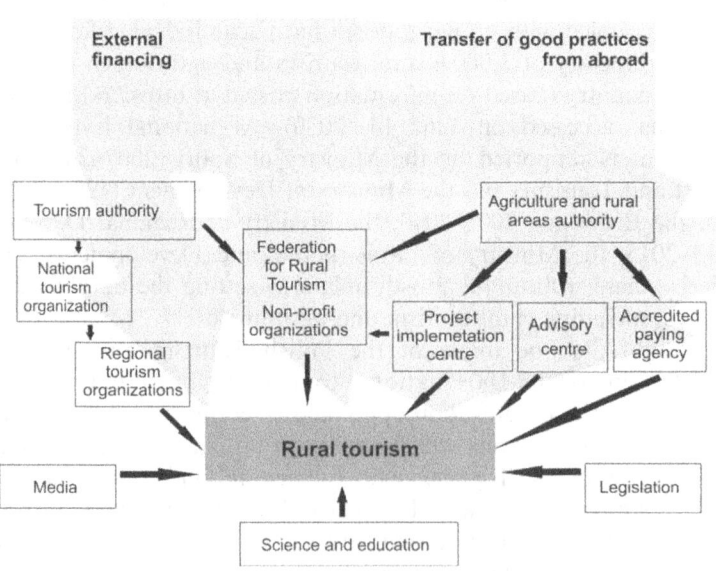

Figure 3.6 Polish model for supporting rural tourism after 1990.
Source: Bednarek-Szczepańska (2017, p. 41).

directives), and due to the ongoing economic development of Poland (Faracik et al., 2014). Legal and organizational conditions for rural tourism development have also been a very important factor influencing agritourism and rural tourism development.

As late as the mid-1990s, Sikora (1995, p. 96) wrote that "neither agritourism activity in general nor the organization of rural holidays by farm owners are regulated by dedicated legal acts. Therefore, agritourism is mostly governed by the same provisions that apply to the organization of holidays in general." Today, agritourism and rural tourism activities are regulated under a number of diverse provisions, although some manifestations of this activity are yet to be addressed in a precise legal context. Generally, agritourism and rural tourism are governed by the same provisions that apply to the organization of holidays in general (Brelik, 2015; Sikora, 2012). This section will present the key legal provisions and related amendments adopted in Poland between 1989 and 2019.

To begin the analysis of legal regulations, it has to be noted that even the terms "agritourism" and "rural tourism" themselves have an ambiguous meaning in Polish law because no widely adopted definition exists for them. The same is true for "agritourism farm" and other terms related to agritourism and rural tourism. So far, the applicable Polish regulations have not provided for any legal definition of terms such as "agritourism," "agritourism farm," "rural tourism" (Trzpioła, 2008, p. 212). This is the reason why the word "agritourism" (or related adjectives) is used by a large group of facilities which have nothing to do with farms and agricultural production (aspects which are imperative for an agritourism or rural tourism experience). Neither the Act of August 29, 1997 on hospitality services and services of tour managers and tourist guides (initially: the Tourism Services Act of August 29, 1997 on tourism services), nor the Ordinance of the Minister of the Economy and Labor of August 19, 2004 on hotel establishments and other facilities used in the provision of hospitality services, as amended, include a definition of an agritourism farm. In these legal acts, agritourism farms and private accommodation facilities are classed as "other accommodation facilities." The only definition of an agritourism activity can be found in the Ordinance of the Minister of Agriculture and Rural Development of October 17, 2007 on detailed conditions and procedure for granting financial aid under the "Diversification into non-agricultural activities" measure covered by the 2007–2013 Rural Development Program. It defines agritourism as "renting out rooms in a residential building and selling home-made meals or providing other services related to tourists staying in a farm" (Section 3.1, Item 2b). However, the absence of accurate definitions of agritourism and rural tourism and of legal regulations applicable to both of these forms of tourism is a problem faced not only in Poland. According to Matias et al. (2011), Italy is among the very few European Union countries to have a legal definition of, and dedicated legal regulations for, agritourism (Kapała, 2008). Moreover, the Polish law does not include an unambiguous definition of a farm (different definitions are provided in: the Agricultural Tax Act of November 15, 1984 [Article 2.1], as amended; the Act of December 20, 1990 on social insurance for farmers [Article

6, Item 4], as amended; and the Civil Code Act of April 23, 1964 [Article 55], as amended). Even the Community legislation fails to define agritourism; it usually uses "rural tourism" as a term extending over any manifestations of tourist activity in rural areas (Kapała, 2008).

However, like all business activities, agritourism activity is subject to a range of legal regulations, which must be abided by. Agritourism may be affected by a wide variety of different regulations, including various taxes (personal income tax, business income tax, sales tax, property tax, etc.). Fees have to be carefully examined, and permits are required in some cases (Sznajder et al., 2009). Legal problems involved in the organization and operation of agritourism and rural tourism establishments in Poland have been considered in many scientific papers published over the last 30 years (e.g. Sikora, 1995; Strzelczak & Cieślak, 1995; Filar, 1996; Hayden, 1996; Jędrzejczyk, 1996; Pomajda, 1996; Strzembicki, 1997; Raciborski, 2000; Kobyłecki, 2003; Kutkowska, 2003; Żelazna, 2004; Raciborski et al., 2005; Tyran, 2005; Firlej, 2006; Sznajder & Przezbórska, 2006; Niedziółka, 2007; Kapała, 2008; Sznajder et al., 2009; Kubal & Mika, 2012; Sikora, 2012; Brelik, 2015; Przezbórska-Skobiej, 2015a). However, "these studies mostly involved the analysis and interpretation of legal acts and legal requirements that the owner of a farm must fulfill in order to start and to run an agritourism enterprise" (Kubal & Mika, 2012, p. 5).

According to the above authors, the key legal problems related to an agritourism or rural tourism activity are as follows:

- duties related to setting up and registering a business activity;
- obligations under regulations on tourism services;
- the requirement to settle the income tax levied on revenue from tourism activities and value-added tax (VAT, a tax levied on the added value that results from the exchange of goods and services; it is an indirect tax, in that the tax is collected from someone other than the person who actually bears the cost of the tax);
- sanitary requirements related to renting out rooms, catering, and selling food products;
- fire protection and construction-related requirements;
- registration, collection of statistics, and other duties;
- eligibility for financial assistance in starting and running an activity.

In general, all the regulations can be divided into three groups (Center for Profitable Agriculture, 2005):

- general regulations which are likely to have an impact on most agritourism or rural tourism enterprises (business taxes, sales tax collection, etc.);
- employment regulations which have an impact on enterprises employing staff (fair labor standards, occupational safety and health regulations, income tax, etc.);

- permits and licenses for specific types of enterprises which have an impact on certain types of entities or types of attractions (food service permit and inspections, retail food store permits and inspections, winery licensing, fee-fishing regulations, etc.).

Table 3.3 presents a compilation of key legal regulations (legal acts) for rural tourism and agritourism activities, as applicable in Poland between 1989 and 2019.

An agritourism activity and a rural tourism activity are forms of an economic activity which, in the study period, were governed by economic activity acts: the Economic Activity Act of December 23, 1988, as amended (before 2001); the Economic Activity Law Act of November 19, 1999 (2001–2004); the Act of July 2, 2004 on the freedom of economic activity, as amended (2004–2018); and the Entrepreneurs Law Act of March 6, 2018, as amended (from 2018).

Any economic activity must be entered into the relevant register (register of entrepreneurs) at commune level, and must comply with a number of other requirements provided for in the aforementioned acts and other legal acts. However, already in the Economic Activity Act of December 23, 1988, as amended, the legislator stated that "an economic activity run personally (to the extent set forth in Para. 2) by a natural person who derives an additional income from it (a secondary gainful activity) is not required to be entered to the relevant register." This included "farmers renting out rooms or on-farm tent space to tourists, selling meals and providing the related services" (Article 9). Hence, agritourism services, i.e. tourism services delivered by farmers as an occupation secondary to their farming activity (Sikora, 1995; Kapała, 2008) are excluded from the scope of the Act. The legislator continued that policy in the next legal acts of 1999, 2004, and 2018 concerning the pursuit of an economic activity in Poland. Therefore, in accordance with these provisions, an agritourism activity is not considered to be an economic activity, and the farmers engaged in it are not considered to be entrepreneurs (Brelik, 2015). According to the tax legislation, agritourism activities supplement the incomes derived from farm activities. Thus, each farm owner can run an agritourism activity on favorable terms, provided that the aforementioned conditions are met (i.e. the farmer rents out rooms to tourists, sells homemade meals, and provides other on-farm services related to tourist accommodation) (Dorocki et al., 2012). At the same time, incomes derived from an agritourism activity are entirely exempt from personal income tax pursuant to the amendment of December 2, 1994 to the Personal Income Act of July 26, 1991 (Journal of Laws [Dz.U.] of 1991, No. 80, Item 350) (Sikora, 1995) provided, however, that the following cumulative conditions are met:

- the rooms rented out are located in residential buildings which belong to the farm;
- the rooms are rented out to holidaymakers;

Table 3.3 Legal regulations for rural tourism and agritourism activities, as applicable in Poland between 1989 and 2019

Scope of legal regulations	Legal acts	European Directives (number and reference nos.)	Consequences for agritourism / rural tourism
Undertaking and reporting an economic activity	1. Economic Activity Act of December 23, 1988, as amended (Journal of Laws [Dz. U.] of 1988, No. 41, Item 324), repealed on January 1, 2001 2. Economic Activity Law Act of November 19, 1999 (Journal of Laws [Dz. U.] of 1999, No. 101, Item 1178), repealed on August 21, 2004 3. Act of July 2, 2004 on the freedom of economic activity, as amended (Journal of Laws [Dz. U.] of 2004, No. 173, Item 1807), repealed on April 30, 2018 4. Entrepreneurs Law Act of March 6, 2018, as amended (Journal of Laws [Dz. U.] of 2018, Item 646), currently in force	Four European Directives: 32004L0038, 32002R0178, 32008R0800, 32006L0123	• exclusion of the renting out of rooms, sale of home-made meals and delivery of other on-farm services related to tourist accommodation by the farmers; • exclusion of farmers' activity consisting in the sale of non-industrially processed vegetable and animal products
Tourism services	Tourism Services Act of August 29, 1997, as amended; currently: Act of August 29, 1997 on hospitality services and services of tour managers and tourist guides (Journal of Laws [Dz. U.] of 1997, No. 133, Item 884), currently in force Ordinance of the Minister of the Economy and Labor of August 19, 2004 on hotel establishments and other facilities used in the provision of hospitality services, as amended (Journal of Laws [Dz. U.] of 2004, No. 188, Item 1945), currently in force	One European Directive: 31990L0314.	Pursuant to the Act: • Hospitality services can be provided in hotel establishments which meet the conditions for size, equipment and scope of services, as defined for the type and category of the establishment concerned, as well as sanitary, fire protection and other conditions set out in separate regulations; • Other facilities used in the provision of hospitality services means rooms and tent sites rented out by farmers in their farms, if compliant with the requirements for minimum equipment and with sanitary, fire protection and other requirements set out in separate regulations;

- Before starting the provision of hospitality services in a facility [...], the farmer who intends to provide hospitality services in his/her farm shall enter that facility to the register [...] kept by the commune head (town mayor) having territorial competence over the location thereof;

- If a farmer ceases the provision of hospitality services in his/her farm, he/she is required to inform the competent authority thereof

Pursuant to the Ordinance:

- [...] if another facility used in the provision of hospitality services is entered into the register by the farmer, he/she shall specify his/her tax ID and his/her National Court Registration Number or his/her Business Registration Number (where applicable);

- [...] a farmer who provides hospitality services shall inform the authority who keeps the register of other facilities used in the provision of hospitality services of any event resulting in a temporary interruption of service provision;

- The authority who keeps the register of other facilities used in the provision of hospitality services shall remove a facility from the register if [...] the farmer [...] ceased to provide hospitality services for a period longer than one year;

(Continued)

Table 3.3 (Cont.)

Scope of legal regulations	Legal acts	European Directives (number and reference nos.)	Consequences for agritourism / rural tourism
			• The Ordinance sets out the minimum requirements for equipment of other facilities used in the provision of hospitality services (Appendix 7 to the Ordinance); • A farm can be exempted from the requirements for equipment and scope of services applicable to on-farm campsites provided that access is ensured to drinking water and sanitary facilities
Registration requirement	1. Act of April 10, 1974 on population registers and identity cards, as amended (Journal of Laws [Dz.U.] of 1974, No. 14, Item 85), repealed on March 1, 2015 2. Act of September 24, 2010 on population registers, as amended (Journal of Laws [Dz.U.] of 2010, No. 217, Item 1427), currently in force 3. Act of December 7, 2012 amending the Act on population registers and identity cards and certain other acts (Journal of Laws [Dz.U.] of 2012, Item 1407), currently in force	Ten European Directives: 32016R0679, 32001L0055, 32003L0009, 32003L0086, 32003L0109, 32003L0110, 32004L0038, 32004L0081, 32004L0083, 32005L0085.	The registration requirement for holidaymakers and tourists was abolished on January 1, 2013. However, the requirement to register for a permanent stay is applicable to persons staying in a hospitality establishment for more than 30 days

| Preparing and serving meals | 1. Act of November 25, 1970 on health conditions for food and nutrition, as amended (Journal of Laws [Dz.U.] of 1970, No. 29, Item 245), repealed on September 23, 2001
2. Ordinance of the Minister of Land Management and Environmental Protection and of the President of the General Committee of Physical Culture and Tourism of July 26, 1973 on hygiene and sanitary requirements for hotels and other accommodation facilities (Journal of Laws [Dz.U] of 1973, No. 33, Item 196), repealed on September 23, 2001
3. Ordinance of the Prime Minister of January 29, 1999 on hotel establishments and other facilities used in the provision of hospitality services (Journal of Laws [Dz.U] of 1999, No. 10, Item 87), repealed on June 9, 2001
4. Act of May 11, 2001 on health conditions for food and nutrition, as amended (Journal of Laws [Dz.U.] of 2001, No. 63, Item 634), repealed on October 28, 2006
5. Act of August 25, 2006 on food and nutrition safety, as amended (Journal of Laws [Dz.U] of 2006, No. 171, Item 1225), currently in force | 88 European Directives:
32003R1829, 32003R1830, 32003R1946 32003R2065, 32004R0852, 32004R0882, 32004R1935, 32005R0396, 32005R1895, 32005R2073, 32005R2074, 32006L0125, 32006L0141, 32006R0401, 32006R0509, 32006R0510, 32006R0627, 32006R1881, 32006R1882, 32006R1883, 32006R1924,
32006R1924R(01),
32006R1924R(08),
32006R1925, 32006R1981, 32006R2023, 32007L0042, 32007R0333, 32007R0834, 32007R0884, 32009R0984, 31978L0142, 31980L0766, 32011R1169, 31981L0432, 31981L0712, 31982L0711, 31984L0500, 31985L0572, 31987R3954, 31989L0108, 31989L0108, 31989L0369, 31989R0944, 31992L0002, 31992L0052, 31993L0005, 31993L0011, 31993R0315, 31996L0003, 31996L0008, 31996R2232, 31997R0258, 31998L0028, 32008L0060, | Facilities which provide catering services to tourists are considered to be mass caterers, and shall be entered to the relevant register kept by the national district-level sanitary inspector or national border sanitary inspector, both of them subject to official checks by bodies of the National Sanitary Inspection Authority |

(Continued)

Table 3.3 (Cont.)

Scope of legal regulations	Legal acts	European Directives (number and reference nos.)	Consequences for agritourism / rural tourism
	6. Ordinance of the Minister of Health of April 17, 2007 on the collection and storage of food samples by non-commercial mass caterers (Journal of Laws [Dz.U.] of 2007, No. 80, Item 545), currently in force	32008L0084, 32008L0128, 32008R0282, 32008R0353, 32008R0733, 32008R0889, 32008R1235, 32008R1243, 32008R1331, 32008R1332, 32008R1333, 32008R1334, 32009L0032, 32009L0039, 32009L0054, 32009R0041, 32009R0124, 32009R0450, 32009R0470, 32009R0669, 32009R0901, 32009R0953, 32009R0975, 32009R0983, 31999L0002, 31999L0003, 31999L0021, 31999L0021, 32000R1565, 32001L0015, 32002L0046, 32002L0063, 32002L0072, 32002R0178, 32003L0040.	
Providing the guests with access to TV and radio sets	Act of February 4, 1994 on copyrights and related rights, as amended (Journal of Laws [Dz.U.] of 1994, No. 24, Item 83), currently in force	Seven European Directives: 31991L0250, 31992L0100, 31993L0083, 31993L0083R(01), 31993L0098, 31996L0009, 32012L0028	If the hosts provide their guests with access to TV and radio sets, they should pay copyright charges to the Association of Authors and Composers (ZAiKS)

Farmers' social insurance scheme	1. Act of December 14, 1982 on the social security scheme for individual farmers and their family members, as amended (Journal of Laws [Dz.U.] of 1982, No. 40, Item 268), repealed on January 1, 1991 2. Act of December 20, 1990 on the farmers' social security scheme, as amended (Journal of Laws [Dz.U.] of 1991, No. 7, Item 24), currently in force	Two European Directives: 32004R0883, 32009R0987	
Social insurance scheme for non-farmers	1. Act of November 25, 1986 on the organization and financing of social insurance (Journal of Laws [Dz.U.] of 1986, No. 42, Item 202), repealed on January 1, 1991 2. Act of October 13, 1998 on the social security scheme, as amended (Journal of Laws [Dz.U.] of 1998, No. 137, Item 887), currently in force	Four European Directives: 32013R1304, 32013R1303, 32009R0987, 32004R0883	
Personal income tax	Personal Income Tax Act of July 26, 1991, as amended (Journal of Laws [Dz.U.] of 1991, No. 80, Item 350), currently in force	Five European Directives: 32012R0236, 32013R1407, 32001R2157, 32014R0651, 32010L0024	Incomes derived from renting out guest rooms in residential buildings located on a farm in a rural setting to holidaymakers, and incomes derived from catering services provided to them shall be exempt from the income tax if no more than five rooms are rented out
Corporate income tax	1. Corporate Income Tax Act of January 31, 1989 (Journal of Laws [Dz.U.] of 1989, No. 3, Item 12), repealed on January 1, 1992 2. Corporate Income Tax Act of February 15, 1992, as amended (Journal of Laws [Dz.U.] of 1992, No. 21, Item 86), currently in force	Ten European Directives: 31990L0434, 31990L0435, 32003L0049, 32003L0123, 32004L0066, 32004L0076, 32005L0019, 32006L0098, 32011L0096, 32013L0013	

(*Continued*)

Table 3.3 (Cont.)

Scope of legal regulations	Legal acts	European Directives (number and reference nos.)	Consequences for agritourism / rural tourism
Value-added tax (VAT)	1. Act of January 8, 1993 on the value-added tax and excise duty, as amended (Journal of Laws [Dz.U.] of 1993, No. 11, Item 50), repealed on May 1, 2004 2. Value-Added Tax Act of March 11, 2004, as amended (Journal of Laws [Dz. U.] of 2004, No. 54, Item 535), currently in force	Nine European Directives: 32006L0112, 32011R0282, 31987R2658, 32015R2447, 32015R2447R(01), 32015R2446, 32015R2446R(01), 32013R0952, 32013R0952R(01)	Accommodation services set out in Section 55 of the Polish Classification of Products and Services (hospitality services, i.e. lodging and related services provided by hotels, motels and boarding houses; services of tourism accommodation facilities and short-stay accommodation facilities; services provided by tent sites and campsites) are subject to an 8% tax rate; horse riding lessons are subject to a 23% tax rate; catering services are subject to an 8% tax rate (however, the sale of mineral water, coffee or tea is subject to the general rate of 23%); discothèque and dance room services (except for facilities where drinks are served) are subject to the base rate of 23%
Property tax, local tax, resort tax	1. Act of March 14, 1985 on local taxes and fees (Journal of Laws [Dz.U.] of 1985, No. 12, Item 50), repealed on January 30, 1991 2. Act of January 12, 1991 on local taxes and fees, as amended (Journal of Laws [Dz.U.] of 1991, No. 9, Item 31), currently in force	Two European Directives: 31999L0062, 32006R1998	Renting out guest rooms to tourists in residential buildings located in rural areas by permanent residents of the relevant commune is not considered to be an economic activity under the Act if no more than five rooms are rented out

Fire protection	1. Fire Protection Act of June 12, 1975 (Journal of Laws [Dz.U.] of 1975, No. 20, Item 106) 2. Fire Protection Act of August 24, 1991, as amended (Journal of Laws [Dz.U.] of 1991, No. 81, Item 351), currently in force 3. Ordinance of the Minister of Interior and Administration of June 7, 2010 on fire protection of buildings, other structures and land (Journal of Laws [Dz.U.] of 2010, No. 109, Item 719), currently in force	One European Directive: 32016R0679
Use of buildings (Construction Law)	1. Construction Law Act of October 24, 1974 (Journal of Laws [Dz.U.] of 1974, No. 38, Item 229), repealed on January 1, 1995 2. Construction Law Act of July 7, 1994, as amended (Journal of Laws [Dz. U.] of 1994, No. 89, Item 414), currently in force	Six European Directives: 31989L0391, 31992L0057, 32010L0031, 32009L0028, 32001L0077, 32003L0030
Third-party liability; customer agreements	Civil Code Act of April 23, 1964, as amended (Journal of Laws [Dz.U.] of 1964, No. 16, Item 93), currently in force	One European Directive: 32000L0031

Source: Own study based on the Online Legal Acts System (ISAP): https://isap.sejm.gov.pl/

- the rooms rented out are located in residential buildings;
- the rooms are rented out in residential buildings located in rural areas;
- up to five rooms are rented out (excluding the dining room, living room, porch, and other accommodation accessible to the public) (Raciborski et al., 2005; Trzpioła, 2008; Brelik, 2015).

The exemption is also applicable to incomes derived from catering services for guests staying in the rooms rented out to them but does not apply to additional tourism attractions such as horse riding, sleigh rides or guided tours which are not included in the accommodation price. This means the farmer is required neither to register an activity which consists in renting out rooms and providing catering services for his/her guests nor to report it in the tax return. Hence, he/she avoids a number of formalities which are essential in running another economic activity. According to surveys carried out with farmers engaged in an agritourism activity, they incur "a quite limited fiscal and other burden, including charges related to keeping documentation and accounting records and complying with formal requirements; this should be viewed as a positive aspect" (Roman, 2014, p. 69). However, due to the fact that agritourism activities must be carried out in an active farm, they usually will only be an additional business and a secondary activity to the basic agricultural undertaking. This makes it reasonable to relax the regime applicable to the way it is established and pursued (Kapała, 2008).

Nevertheless, in all other cases (e.g. if more than five rooms are rented out; if catering services are provided not only to tourists staying in the farm; or if catering services are provided by the non-farming rural population), it becomes an economic activity (Brelik, 2015) governed by the provisions of the Entrepreneurs Law Act, and therefore must be registered. The tax exemption is not applicable to incomes derived from renting out rooms in non-residential buildings or cabins and from renting out space for tents and camper trailers (Trzpioła, 2008). The provision of rural tourism services by other rural residents outside a farm (e.g. with the use of buildings and land non-related to a farm) has always required the entrepreneur to register his/her activity, report it to the Tax Office and choose a form of taxation[15] (Raciborski, 2000; Kutkowska, 2003; Brelik, 2015). Each time, the choice of the form of taxation requires a business profitability analysis performed on a case-by-case basis (Roman, 2014).

In practice, the law on tax exemption is often abused. Tourism service providers either exceed the maximum number of rooms they rent out or simulate running a farm (e.g. buy or lease agricultural land for the sole purpose of avoiding taxation) (Bednarek-Szczepańska, 2017). For many years, there was a

> discussion on the legitimacy of this regulation due to the growth in the informal (grey) economy and restricted competition (as non-farmers and urban residents providing similar services cannot avoid taxation). However, repealing this regulation would have an adverse impact on rural

tourism. It would contribute to many farmers discontinuing the rental of rooms, especially those deriving little income from it, including undertakings active in non-tourist areas. Rather than solving the grey economy issue in tourism, the abolishment of the tax exemption could even aggravate the problem.

(Bednarek-Szczepańska, 2017, p. 40)

The renting out of rooms in combination with serving home-made meals and providing other services related to tourist accommodation is also subject to VAT at a rate of 8%. However, a VAT exemption is applicable to a small-scale activity, i.e. to taxpayers whose sales revenue did not exceed PLN 200,000 in the previous fiscal year. In that case, agritourism farm owners are required to keep daily sales records. Also, the owner of an agritourism farm or a rural accommodation facility must remember that revenues from different activities are added together; this means that a farmer who provides agritourism services while also running an economic activity which consists in, e.g. offering transport services, must add the revenues derived from all of these activities to compare them against the VAT exemption limit. If, in addition to offering accommodation and meals, an agricultural farm provides services not included in the overnight price, such services shall be subject to VAT at the applicable rate (in most cases, 23%).

Local taxes and fees, i.e. property tax, local tax or resort tax[16] (in the 42 Polish communes holding the status of health resort communes), are also of major importance to the functioning of an agritourism farm or a rural tourism facility. Since 1996, the renting out of rooms in rural residential buildings has been excluded from the definition of economic activity under the Act on local taxes and fees if the number of rooms does not exceed five. Such rooms are subject to taxation just like residential premises. Nevertheless, people who rent out more than five rooms or rent out rooms outside residential buildings and outside rural areas, must realize the need to reallocate them to commercial uses which involves a considerable increase in tax rate. The local tax is only charged in communes with particular tourism or health-promoting values confirmed by the voivodeship governor. On behalf of the commune, the tax is charged by the service provider to people staying in the commune for more than one day for tourism, leisure or training purposes. The local tax is introduced under a resolution adopted by the Commune Council (Act of January 12, 1991 on local taxes and fees, as amended, Journal of Laws [Dz. U.] of 1991, No. 9, Item 31).

The Polish legal system still lacks precise detailed regulations for rural tourism and agritourism. As mentioned earlier, there are no legal definitions for agritourism and rural tourism; and no precise conditions have been formulated for the use of the aforesaid reliefs and exemptions. Furthermore, no regulations exist to govern the administrative, legal, fiscal, insurance, construction, fire protection, hygiene and sanitary aspects of this type of activity (Brelik, 2015). As Brelik wrote (2015, p. 151), "while the legislator does not impose any additional burden on the farmer, they do not make it easier e.g. for disabled people to enjoy leisure in smaller accommodation facilities, including in agri-tourism farms" and in rural tourism facilities.

In addition to legal aspects listed above, pursuant to the Act of October 13, 1998 on the social security scheme, as amended, and to the Act of December 20, 1990 on the farmers' social security scheme, as amended, all employees and people performing an economic activity are compulsorily insured under the social insurance scheme by the Social Insurance Institution (non-farmers) or by the Farmers' Social Insurance Institution (farmers). The obligatory social insurance premium paid to the Farmers' Social Insurance Institution (KRUS) is a constant financial burden. For agritourism farm owners, a serious problem is that once they start a non-agricultural economic activity which must be reported to the authorities or is an additional business, they are required to pay a compulsory insurance contribution to the Social Insurance Institution (ZUS) (Roman, 2014). The amendment of May 2, 2004 to the act on the farmers' social insurance scheme modified the requirements which must be met by farmers engaged in a non-agricultural economic activity to remain insured by KRUS. Pursuant to the simplified regulations, the taxation form of a non-agricultural economic activity carried out concurrently with an agricultural activity does not impact an individual's eligibility for KRUS insurance. Thus, a farmer or a farmer's household member who starts (or enters into cooperation with) a non-agricultural business can remain insured by KRUS under the following conditions:

– a farmer has been fully and continually insured by KRUS pursuant to the Act for at least three years, and continues to pursue an agricultural activity or to work on a farm larger than one hectare of arable land or a specialized farm;
– within 14 days of starting (or entering into cooperation with) a non-agricultural business, the farmer provides KRUS with a statement of intent to remain covered under that insurance scheme.

(Roman, 2014, pp. 74–75)

According to the Act of April 10, 1974 on population registers and identity cards (Article 10), "an individual who stays in the territory of the Republic of Poland shall register his/her place of permanent or temporary residence (if longer than 3 months) no later than on the 30th day following his/her arrival thereto." Earlier, until 2018, the registration requirement was applicable to tourists coming to Poland for a temporary stay, including: foreign nationals (within 48 hours of arriving to their destination and no later than 48 hours of crossing the border), other tourists (within 24 hours of arriving to an accommodation facility), and tour participants (registering all participants was the responsibility of the tour manager). This aspect of activity also involves compliance with the principles set out in Regulation (EU) No. 2016/679 of the European Parliament and of the Council of April 27, 2016 on the protection of natural persons with regard to the processing of personal data and on the free movement of such data, and repealing Directive 95/46/EC (referred to as GDPR), applicable in Poland since May 25, 2018.

Even the Tourism Services Act of August 29, 1997 (current name: Act of August 29, 1997 on hospitality services and services of tour managers and tourist guides) fails to specify many important aspects of rural tourism and agritourism activities, although it regulates the responsibilities involved in organizing tourism events, hospitality services, tourist guide services, and other tourism services. Before the 1998 amendment to the Tourism Services Act, its provisions were not applied in detail to agritourism activities. The Act laid down the conditions for the provision of tourism services by entrepreneurs as defined in the Economic Activity Law Act of November 19, 1999, and therefore "its provisions were not applicable to farmers renting out rooms or tent space, selling home-made meals and providing other on-farm services to tourists" (Raciborski, 2003, p. 101). However—in accordance with Article 3 of the Act on hospitality services and services of tour managers and tourist guides—hospitality services mean short-term, commonly available renting of houses, flats, rooms, lodging for the night and sites for tents and trailers, and providing services related therewith. Hence, the farmer's activities in this area are also considered as hospitality services (Roman, 2014). The Act identifies two groups of hospitality facilities (Raciborski, 2005, p. 15):

- buildings using legally protected brand names and being subject to a categorization system;
- other facilities which guarantee minimum security levels, including rooms rented out by farmers and on-farm tent sites (as provided for in Article 35, Item 3).

Hospitality services can also be provided in other facilities if they meet the minimum requirements, including size, equipment, and sanitary, construction, and fire protection criteria (Chapter 5, Article 35, Item 2 of the Act on hospitality services and services of tour managers and tourist guides). According to the Act, both rural tourism accommodation facilities and agritourism farms are other hospitality facilities, shall comply with minimum requirements for equipment and with sanitary, fire protection and other requirements provided for in separate regulations, and shall be entered into the register of other hospitality facilities kept by the commune head (town mayor) having territorial competence over the location thereof (Article 38, Item 3). As regards construction law and fire protection regulations, agritourism farms and rural tourism accommodation facilities are not subject to any separate provisions. It is only required that facilities (residential and service buildings) used in the delivery of hospitality services shall comply with minimum requirements laid down in the Construction Law Act[17] and shall be designed, used and maintained in a manner to protect them against fire.[18] According to the Act on hospitality services and services of tour managers and tourist guides, the categorization of "other hospitality facilities" is not compulsory. They can be rated within a voluntary assessment scheme for rural accommodation resources, as established by the "Hospitable Farms" Polish Federation of Rural Tourism (Kmita, 1997; Majewski, 2000b, pp. 41–44; Niedziółka, 2007; Roman, 2014).

Another crucial aspect for rural accommodation providers are the hygiene and sanitary provisions which have long remained imprecise with respect to agritourism farms and guest rooms (Raciborski, 2000). Only upon entry into force of the Act of May 11, 2001 on health conditions for food and nutrition and, subsequently, of the Act of August 25, 2006 on food and nutrition safety, it was laid down that all facilities which provide catering services to tourists are considered to be mass caterers, and shall be entered to the relevant register kept by the national district-level sanitary inspector or national border sanitary inspector, both of them subject to official checks by bodies of the National Sanitary Inspection Authority. "The provision of food services on the agritourism farm or rural tourism entity is treated as an additional economic activity and requires special attention due to the rules of hygiene in the preparation of meals" with the appropriate authority and to ensure compliance at all stages of production, processing and distribution of food hygiene requirements (Kubal & Mika, 2012, p. 6). Moreover, unlike the case of, e.g. France, Croatia or Spain (Brelik, 2015), Polish providers of rural accommodation are not allowed to directly sell farm-made alcoholic beverages and spirit drinks without permit, although wine tasting is allowed in eno-tourism farms. However, the wine must carry the excise duty label, and be officially and legally produced as per the procedures provided for by the law. Most importantly, it must be produced within the excise duty system, and tax must be paid on it (as confirmed by the excise duty label) (Froń, 2019). Entrepreneurs are still the only ones permitted to sell it (Kapała, 2010).

Legal aspects of a rural tourism activity also include the liability of the agritourism farm or rural accommodation facility owner for damage to or loss of guests' property and the responsibility to ensure security during their stay (Sikora, 1995; Sikora, 2012). Farmers involved in agricultural activities are also required to hold liability insurance and to insure their buildings against fire and other disasters. Moreover, in order to limit the liability of agritourism farm owners, it is recommended that they buy optional liability insurance policies which extend the farmers' compulsory liability insurance.

To summarize the considerations on legal and organizational conditions for rural tourism development, it needs to be concluded that many things changed in this respect between 1989 and 2019. However, much is yet to be changed, detailed or specified, starting with the definition of many terms related to an agricultural activity (Budzinowski, 2003) and tourism activity, to finish with precise, detailed legal regulations which enable the farmers and other rural residents to provide attractive tourism services, on the one hand, and the tourists to have a safe, attractive stay in rural areas, on the other.

3.5 Financing for rural tourism development

During the development of the market economy, the Polish Government—assisted by the Commission of the European Communities (CEC)—carried out a series of development programs to provide financial support for tourism (including rural tourism) development (Airey, 1994). During 1989–2019, the priorities of multi-purpose rural development in Poland included initiatives related to the development of tourism and recreation. In the study period, rural tourism followed a rapid development path in countries where it was financially supported under diverse schemes, including European Union programs focused on regional, economic, and social development, with particular emphasis on structural funds combined with the resources of the Cohesion Fund (Przezbórska & Hegarty, 2004). In EU countries, measures for the promotion of rural tourism and agritourism were co-financed and carried out collaboratively by institutions of different levels. The sources of financing included the central state budget, regional and local budgets, the private sector, EU resources (pre-accession funds before the integration, and structural funds thereafter), and other international and domestic funds. Therefore, the sources of financing for rural tourism development can be divided into two basic groups (Przeorek-Smyka, 2009):

- *European Union funds*, including structural funds and the European Commission budget (e.g. the European Agricultural Fund for Rural Development, Community programs);
- *domestic funds*, including the state budget, national resources, regional and local budgets, funds from the private sector, and own funds of operators active in the tourism industry.

As shown in publications of various authors, before Poland's integration into the EU, investments in rural tourism and agritourism farms (renovation, reconstruction, extension, purchase of equipment, etc.) were mainly financed from their own resources (Dębniewska & Tkaczuk, 1997; Przezbórska, 2002; Firlej & Niedziółka, 2007). A number of measures, such as loans (Table 3.4), pre-accession aid from the EU (SAPARD), and afterwards the Sectoral Operational Program (SOP) "Restructuring and Modernization of the Food Sector and Rural Development 2004–2006" (i.e. external sources of financing) were employed to support rural tourism and agritourism farms (MRiRW, ARiMR, 2005a). However, only a minor proportion of enterprises have accessed it (Firlej & Niedziółka, 2007).

In most cases, agritourism is a small-scale activity, and the owners rely only on their own financial resources in starting and pursuing it. The capital they use mostly consists of their savings. Before joining the European Union, agritourism farms based their development plans on a strict calculation of incomes, and financed most of their investments using their own resources whereas loans were used to a minor extent (Dębniewska & Tkaczuk, 1997; Kurtyka, 2000; Talarowska, 2005; Sikora, 2012).

Table 3.4 Repayable financing for the development of rural tourism in Poland during 1989–2019

Source of financing	Implementation period	Actions	Source of information
Loans from the Foundation for the Development of Polish Agriculture (FDPA)	From 1993	Loans for rural residents, especially for people starting or expanding a business, including in agritourism	Kutkowska, 2003; Przezbórska, 2006
Soft loans subsidized by the Agency for Restructuring and Modernization of Agriculture (ARMA)	1994–2007	"Agrolinia 2000": • investment loans with a long repayment period; intended for the implementation of investment projects related to agriculture, agri-food processing and services for the agricultural sector (code: IP—agritourism on a farm); • educating rural leaders, including rural tourism development leaders and local development leaders; • support for rural entrepreneurship (e.g. financing for the agritourism directory and map); • small grants for bottom-up initiatives (for communes, foundations, associations, local development agencies, agricultural chambers, agricultural advisory centers);	Sikora, 1995; Dębniewska & Tkaczuk, 1997; Długokręcka, 1998; Budzich-Szukała, 2000; Kutkowska, 2003; Przezbórska, 2006; Balińska, 2016
Rural Development Foundation (RDF)	(1) 1996–1999 (2) 1995–2000	1 Soft micro-loans for micro-enterprises active in rural areas and towns, granted in collaboration with banks and regional agritourism associations (micro-loans were available for entrepreneurs from some districts) 2 Training program for rural residents planning to engage in an agritourism activity 3 Publishing (e.g. agritourism guides)	Domański, 2000; Kutkowska, 2003; Przezbórska, 2006

(*Continued*)

Table 3.4 (Cont.)

Source of financing	Implementa-tion period	Actions	Source of information
Rural Development Program (RDP) (co-financed from the World Bank, national budget, local government units and other stakeholders); coordinated by the Foundation of Assistance Programs for Agriculture (FAPA)	2000–2005	• Micro-loans and trainings for rural entrepreneurs, delivered in cooperation with the Ministry of Labor and Social Policy and the Ministry of Economy (available for entrepreneurs located in voivodeships affected by high unemployment rates or in areas which were formerly home to state-owned farms); • financial support for infrastructural development of rural areas, delivered in cooperation with the Ministry of Agriculture and Rural Development (resources for local government units)	Andrychowicz, 2000; Przezbórska, 2006
European Fund for the Development of Polish Rural Areas (Counterpart Fund)	1990–2006	1 Loans for rural investment projects related to agritourism 2 Micro-loans for the development of non-agricultural small businesses and for the creation new jobs and alternative rural revenue streams 3 Investment loans for commune– and district-level government units, intended to support the development of the technical and social infrastructure in rural areas	Sikora, 1995; Kutkowska, 2003; Europejski Fundusz Rozwoju Wsi Polskiej, 2005; Przezbórska, 2006

Source: Own study.

In the pre-accession period, soft loans were offered to those interested in starting an agritourism activity in Poland. Farm owners mostly used these funds to extend their holdings (Talarowska, 2005).

The Foundation for the Development of Polish Agriculture (FDPA), created in 1988 under co-financing of the Rockefeller Brothers Fund, played a significant role in the development of Polish agritourism. Through measures which included the rural enterprise promotion program, the Foundation granted loans on relatively favorable terms to rural entrepreneurs (farms) with up to ten employees to finance their commercial, production, service, and agritourism projects.

Through measures which included the rural enterprise promotion program, the Foundation granted loans on relatively favorable terms to rural entrepreneurs (farms) with up to ten employees to finance their commercial, production, service, and agritourism projects. Resources available under this fund were used to finance the TOURIN Tourism Industry Development Program, implemented in two main stages as TOURIN I, 1992–1995 (National Tourism Product Development Plan) (Kmita, 1997; Augustyn, 1998; Majewski, 2003) and TOURIN II, 1995–1997. The latter included a project named "Tourism development in rural and forested areas." TOURIN I consisted of a series of six-month projects supported with ECU[19] 4.5 million allocated to Poland in 1992 (Augustyn, 1998; Majewski, 2003). The Commission of the European Communities supported the program with ECU 8.0 million, including ECU 3.3 million for rural tourism development (Augustyn, 1998; Kmita, 1997; Ryżak, 2000).

From the perspective of rural tourism and agritourism development, the second edition of TOURIN was much more important as it was designed to establish a national system for online tourism information and service booking, and prepare a tourism development plan for rural and forested areas, especially including the general development and promotion plan, the development plan for active and specialized forms of tourism, and the development plan for rural accommodation resources (Augustyn, 1998). In 1996, the project resulted in establishing the Polish Federation of Rural Tourism "Hospitable Farms" which joined the European Federation of Rural Tourism (EuroGites) in 1997 (Augustyn, 1998).

PHARE-TOURIN II included the formulation of five-year strategic plans for rural tourism development and promotion, the preparation and implementation of a strategy for the development of rural accommodation, and the identification of development opportunities for active rural tourism. Pilot action projects implemented in 1996 and 1997 were related to active tourism, qualified tourism, rural tourism, and branded products. Five forms of active rural tourism were identified as having potential for further development, namely: hiking, relaxation in nature, water sports, horse riding, and getting to know rural life (Augustyn, 1998; Borne & Doliński, 2000).

As an outcome of the implementation of PHARE–TOURIN II, "A Strategy for the Polish Tourism Product Development for 1994–2005" was a detailed extension and follow-up of the 1994 program entitled "Foundation for Tourism Development" (Augustyn, 1998; Borne & Doliński, 2000). The basic assumption of the Strategy was to specify the actions that allow a competitive tourism product to be optimally positioned in the domestic and international market, with particular emphasis being placed on its sustainable development. This resulted in the creation of the "National Tourist Product Development Program" in 1997. EU experts invited to TOURIN I and TOURIN II projects identified rural tourism and agritourism as two out of five potential unique Polish tourism products (Baum, 2011, p. 112). The tourism product was described as being competitive in the domestic and international markets, mainly due to national landscape values, the cultural wealth of the Polish countryside, the preservation of traditional forms of farming and the environment being less polluted than in European Union countries (Augustyn, 1998; Borne & Doliński, 2000; MRiRW, ARiMR, 2001).

PHARE funds were used to implement TOURIN III and to establish the Local Grants Fund which, however, covered only three voivodeships: Krakow, Nowy Sącz, and Tarnów. Examples of projects financed with this fund include the Piedmont Agritourism complex composed of six modernized model agritourism farms (Smoleń, 2000). Nevertheless, funds granted by the European Union under the PHARE Program were insufficient in relation to the number of applications filed, even though Firlej (2006) believes this measure can be viewed as a pilot project which therefore was of extreme importance to the development of Polish rural tourism.

Note that the development of rural tourism, mainly including tourism infrastructure, was also indirectly supported under other PHARE investment programs such as PHARE STRUDER, STRUDER II, RAPID, and INRED.

Between 1994 and 2007 (until April 30, 2007), as part of what is referred to as the basic investment axis, the Agency for Restructuring and Modernization of Agriculture offered subsidized loan schemes for non-farm agritourism projects (axis IP/05: on-farm agritourism) related to investments in agriculture, agri-food processing, and services for the agricultural sector. The loans could be used to finance the conversion and repair of residential farm buildings intended for agritourism uses, the modernization of residential buildings, or the construction of campgrounds, including sanitary and water supply facilities. The total amount of loans granted under this scheme was PLN 89.9 million, distributed between 1,457 projects (see Table 3.5).

Thanks to financial support granted under domestic aid schemes, a total of 1,434 agritourism accommodation facilities were created from the day Poland joined the European Union to April 30, 2007 (Nurzyńska, 2009). In the post-accession period (until July 31, 2005), funds derived from soft loans supported the creation of 586 agritourism accommodation facilities in Poland (MRiRW, ARiMR, 2005).

Until the end of April 2004, preferential financial support was also granted to non-agricultural investments intended to create new rural jobs. The beneficiaries

Table 3.5 Number and amount of soft loans for agritourism businesses granted in Poland between 1994 and 2007

Specification	1994–1999	2000–2006	Jan. 1 to Apr. 30, 2007	Total
Number of loans granted	735	699	23	1457
Amount of loans granted (PLN million)	42.2	45.5	2.2	89.9
Average amount of loans granted (PLN thousand)	57.4	65.0	97.1	61.7

Source: Nurzyńska (2009).

could use the funds to purchase, modernize, adjust, build, and extend non-agricultural establishments active in rural communes, urban-rural communes or small towns (with a population of up to 20,000) or to buy essential non-agricultural machinery and equipment, without limitation. During 1996–2004, the banks granted 4,086 loans amounting to a total of over PLN 500 million, resulting in the creation of 18,650 new jobs (MRiRW, ARiMR, 2005b). In addition to those listed above, until the end of 2003, the Agency for Restructuring and Modernization of Agriculture also granted loans under the "small entrepreneurship program" to drive the creation of agricultural jobs in rural areas.

In 2000, in consonance with the reforms proposed in the Agenda 2000, the EU introduced new pre-accession aid instruments for Central and Eastern European countries, including SAPARD, a basic instrument of direct support for rural tourism and agritourism development (Table 3.6). In Poland, funds offered under the agritourism farms program were available during 2002–2005. The program enabled financing for measures related to: farm investments (including in agritourism); support for undertakings which invest in rural tourism and recreation; small rural tourism infrastructure; promotion and marketing activities; and training projects. Of the seven measures identified in the program, Measure 4 was focused on the diversification of economic activity in rural areas, which also means development of agritourism projects.

From July 17, 2002 to April 20, 2004 a total of 31,100 SAPARD applications were filed with the regional offices of the Agency for Restructuring and Modernization of Agriculture (this includes applications for financing under all measures implemented by the Agency), of which 24,400 resulted in signing agreements amounting to a total of nearly PLN 4.8 billion (2005 Newsletter). Table 3.7 presents a summary of SAPARD implementation (as at November 30, 2006).

According to data of the Ministry of Agriculture and Rural Development, most (4,133) applications were submitted under Scheme 4.2, "Creation of rural jobs," followed by Scheme 4.1, "Creation of additional revenue streams for agricultural holdings" (2,693). The smallest number of applicants (678) opted for Scheme 4.3, "Public tourism infrastructure in rural areas." Data delivered by the Agency shows that while the applications were submitted in all

Table 3.6 Non-repayable financing for the development of rural tourism in Poland, 1989–2019

Source of financing	Implementation period	Action	Source of information
Pre-accession period (before 2004)			
PHARE (Poland and Hungary Assistance for Restructuring Economies: Development of the Tourism Industry)	(1) 1992–1995 (2) 1994–1997 (3) 1997	1 TOURIN I ("A Strategy for the Polish Product Development for 1994–2005") 2 TOURIN II ("Development of Tourism in Rural and Forested Areas") 3 TOURIN III;	Sikora, 1995; Majewski, 1997; Augustyn, 1998; Borne & Doliński, 2000; Majewski, 2000b; Smoleń, 2000; MRiRW, ARiMR, 2001; Niedziółka, 2017;
SAPARD (Special Accession Programme for Agriculture and Rural Development)	2002–2006	Measure 4 "Diversification of economic activity in rural areas," including the following Schemes: 4.1. Creation of additional revenue streams for agricultural holdings; 4.2. Creation of rural jobs; 4.3. Public tourism infrastructure in rural areas; Financial resources for farmers, local government units (at commune and district level) and entrepreneurs active in rural areas.	Lisztwan, 2000; Skórnicki, 2000; Smoleń, 2000; Kutkowska, 2003; MRiRW, ARiMR, 2005b; Agency for Restructuring and Modernization of Agriculture, 2006; Gradziuk, 2007; Sobiecki, 2008; Balińska, 2016; Niedziółka, 2017;
Post-accession period (after 2004)			
Sectoral Operational Program (SOP) "Restructuring and Modernization of the Food Sector and Rural Development'	2004–2006	Measures: 2.3 "Development of rural regions and preservation and protection of cultural heritage"; Financial resources available to commune-level local government units and cultural institutions; 2.4 "Diversification of rural activities and similar activities in order to ensure variety of activities or alternative sources of income."	Jankowska, Kierzkowski, & Knopik, 2004; Agency for Restructuring and Modernization of Agriculture, 2006; Sobiecki, 2008; Brelik, 2015; Balińska, 2016;

(Continued)

Table 3.6 (Cont.)

Source of financing	Implementation period	Action	Source of information
		Financial resources for farmers, local government units (at commune and district level) and entrepreneurs active in rural areas. Financial resources for farmers, farmer families and legal persons to support additional non-agricultural or agriculture-related activities which involve the use of farming or regional resources	
"Integrated Regional" Operational Program for 2004–2006	2004–2006	Priority 1: Developing infrastructure to enhance regional competitiveness (investments in tourism and culture); Priority 3: Local development (tourism and cultural heritage projects)	Niedziółka, 2017;
2007–2013 RDP (Rural Development Program) (financed from EAFRD, the European Agricultural Fund for Rural Development)	2007–2013	Axis 3: Quality of rural living and diversification of the rural economy Measures: 1 Diversification to non-agricultural activities; 2 Establishment and development of micro-enterprises; 3 Basic services for the economy and rural population; 313, 322, 323. Village renewal and development; Axis 4: LEADER	Siemiński & Poczta, 2014; Wiatrak, 2014; Brelik, 2015; Rudnicki & Biczkowski, 2015; Balińska, 2016;
16 Regional Operational Programs (financed from ERDF, the European Regional Development Fund)	2007–2013	Priority 4: Revitalization of problem areas; Priority 6: Tourism and cultural environment	Jęczmyk, Uglis, & Fedorczuk, 2018;

2014–2020 RDP (Rural Development Program), financed from EAFRD, the European Agricultural Fund for Rural Development	2014–2020	Priority: Social inclusion, reduction of poverty and promotion of economic development in rural areas; Measure 6: "Farm development and economic activity'; Sub-measure 3: "Support in setting up a non-agricultural business in rural areas (start-up bonus for non-agricultural undertakings)' Measure 7: "Basic services and village renewal in rural areas," Sub-measure 1: "Investments in the development, improvement or enhancement of essential local services for the rural population, including leisure, cultural and related infrastructure'; Measure 13: LEADER (sub-measure: implementation of operations under local development strategies)	Siemiński & Poczta, 2014; Idziak, 2015; Idziak et al., 2015; Przezbórska-Skobiej, 2015a; Balińska, 2016;
Regional operational programs for 2014–2020, financed from the European Regional Development Fund (ERDF), the European Social Fund (ESF), the Cohesion Fund (CF), the European Agricultural Fund for Rural Development (EAFRD) and the European Maritime and Fisheries Fund (EMFF)	2014–2020	Investment priority 6c. "Preservation, protection, promotion and development of natural and cultural heritage"; Investment priority 6d. "Protection and restoration of biodiversity, soil protection and reclamation, and support for ecosystem services, including through the Natura 2000 program and green infrastructure"	Grad & Ferensztajn-Galardos, 2015;

Source: Own study.

Table 3.7 Special Accession Program for Agriculture and Rural Development (SAPARD): implementation summary (as at November 30, 2006)

Specification	Number of			Amount of (PLN million)			Utilization of the aggregate funds limit (%)	
	applications filed	agreements concluded	payments disbursed	the financial limit	agreements concluded	payments disbursed	agreements concluded	payments disbursed
Measure 1: processing and marketing improvements	1,788	1,342	1,179	1,708.9	1,671.6	1,193.0	97.82	69.81
Measure 2: farm investments	15,586	13,742	12,432	661.4	637.2	566.4	96.34	85.63
Measure 3: development of and improvements to the rural infrastructure	6,230	4,493	4,582	2,034.0	2,023.8	1,973.4	99.50	97.02
Measure 4: diversification of economic activity in rural areas	7,504	4,854	3,763	438.4	437.0	292.0	99.6	66.2
Total	31,098	24,431	21,956	4,842.7	4,801.3	4,024.7	98.49	83.11

Source: SAPARD monitoring data, Management Information System of the Agency for Restructuring and Modernization of Agriculture, made on March 15, 2006. http://www.arimr.gov.pl/pliki/9/1/0/SAPARD.xls (accessed on October 31, 2007).

voivodeships, the largest numbers were recorded in the Małopolskie voivodeship (785, i.e. 10.5% of the total number), in the Wielkopolskie voivodeship (727, 9.7%), and in the Podkarpackie voivodeship (676, 9.0%). The largest number of payments were disbursed in the Małopolskie voivodeship (325, i.e. 13.1% of total number of payments disbursed in Poland), in the Podkarpackie voivodeship (323, 13.0%), and in the Lubelskie voivodeship (205, 8.2%). The largest number of payments (1,222) were disbursed under Scheme 4.2, followed by Scheme 4.1 (990) and Scheme 4.3 (277) (MRiRW, ARiMR, 2005b).

SAPARD was provided with limited financial resources allocated to strictly defined purposes, and was supposed to support—rather than replace—the national policy for rural and agricultural development.

In turn, according to Smoleń (2000), the way agritourism development was financed with European Union funds clearly shows that pre-accession funds played a complementary role. Indeed, SAPARD was to designed to "directly support the development of specific agritourism farms and undertakings whereas PHARE (Social and Economic Cohesion) was supposed to provide a general framework for agritourism development" (Smoleń, 2000, p. 50).

Of all the programs financed with EU structural funds, the Sectoral Operational Program (SOP) "Restructuring and modernization of the food sector and rural development (2004–2006)" played a major role in developing and supporting agritourism farms. Funds provided under Measure 2.4 "Diversification of rural activities and similar activities in order to ensure variety of activities or alternative sources of income" were allocated to projects which consisted in farms engaging in additional non-agricultural (but agriculture-related) activities. For instance, ca. 1,300 agritourism projects and 970 projects related to tourism and leisure were planned. A total of ca. 6,450 projects were implemented under Measure 2.4, including 5,160 in individual farms and 1,290 in farms held by legal persons (MRiRW, ARiMR, 2005b). The application submission period was from September 15, 2004 to April 14, 2006 (in late 2005, the call for applications was suspended in eight voivodeships due to exhaustion of funds in what is referred to as regional envelopes).

The maximum amount of co-financing available under this Measure was PLN 100,000, however, no more than 50% of eligible costs. Agritourism farms could allocate the financing granted under the SOP to (without limitation): extending, altering, reconstructing or repairing and upgrading the existing residential and farm buildings (and their equipment) to be used for agritourism purposes; preparing leisure facilities; purchasing tourism or leisure equipment to be used in their tourism activity; purchasing animals to be used for therapeutic, sports or recreational purposes; purchasing computer hardware and software; developing land for the purposes of activities planned under the project (MRiRW, ARiMR, 2005b). In the case of projects involving agritourism activities or services related to tourism and leisure which include the creation of rural tourism accommodation resources, aid could be granted only upon submitting a certificate of compliance with classification requirements issued by an accommodation classification authority. Therefore, the beneficiaries were required to join the Polish Federation of Rural Tourism "Hospitable Farms" and undergo the classification process. As a consequence, the Agency for Restructuring and Modernization of Agriculture entered into cooperation with the Federation on the opinion-making and classification of agricultural facilities and services related to tourism and leisure (PFTW 'Gospodarstwa Gościnne', 2005).

Also, rural residents could access funds available under other operational programs, such as the Sectoral Operational Program "Human Resources Development" (financed both from national resources and from the European Social Fund, ESF), the Sectoral Operational Program "Improvements in Enterprise Competitiveness" (co-financed from the European Regional Development Fund, ERDF) and the Integrated Regional Development Program (financed from two structural funds, i.e. the ERDF and the ESF).

The farmers could also use rural investment loans granted for agritourism projects under the European Fund for the Development of Polish Villages ("Counterpart Fund") established in 1990. According to Foundation reports, 328 agreements on agritourism investment loans were entered into during 2001–2004. The tasks implemented were worth a total of PLN 28.4 million (the amount of loans granted in support of these projects was PLN 22.7 million). The projects resulted in creating 3,620 new beds, 158 sports, recreation, and catering facilities, and 86 new jobs (EFRWP, 2005).

In the 2007–2013 programming period, rural tourism and agritourism activities were already supported from the new agricultural fund under the Operating Rural Development Program, i.e. from the European Agricultural Fund for Rural Development (EAFRD). While the 2007–2013 RDP continued to be a supporting tool for rural tourism development, it had a declining share among other non-agricultural rural activities. Between 2007 and 2013, tourism activity accounted for less than 10% of projects supported with funds allocated to the creation of non-agricultural revenue streams (Wiatrak, 2014). Support for rural tourism projects was mainly financed under the third axis of the RDP ("Improvement of the quality of rural living and diversification of the rural economy"). Note that agritourism received relatively greater support than rural tourism, as evidenced by the number of projects implemented (Siemiński & Poczta, 2014). Between June 5, 2008 and October 21, 2008, farmers (or their household members) launched 4,012 projects worth a total of PLN 334.5 million under Measure 311 "Diversification into non-agricultural activities" alone; the financial resources were capped at PLN 229.0 million.

During 2014–2020, RDP funds have continued to act as a development enabler for tourism services in rural areas. However, support was allocated differently than during 2007–2013 because legal entities engaged in a commercial activity were ineligible. It is also important that local governments and their organizational units were assigned a special role in forming conditions for the development of tourism services and were designated as authorized beneficiaries of EU support (Siemiński & Poczta, 2014).

It can be therefore concluded that both in the pre-accession period and after Poland joined the European Union, rural tourism and agritourism could be supported with domestic resources (e.g. soft loans subsidized by the Agency for Restructuring and Modernization of Agriculture) and foreign funds (PHARE pre-accession funds, SAPARD, SOP "Restructuring and modernization of the food sector and rural development 2004–2006," 2007–2013 and 2014–2020 Rural Development Programs). Rural tourism and agritourism development funds available in Poland enabled the creation of new agritourism farms and rural tourism operators, and allowed the financing of many investments made to diversify rural economic activities (whether non-agricultural or agriculture-related). However, funds other than national and EU resources need to be mentioned, too, as they played a significant role in supporting the development of rural tourism and agritourism in Poland.

3.6 Union policies vs. rural tourism development

Viewed as being part of multipurpose rural development, agritourism and rural tourism have long been taken into consideration in European Union's community policies, especially in the Common Agricultural Policy, the regional policy, and the tourism policy (Table 3.8). These are tools used in the pursuit of many strategic objectives of the Community policy. The broad range of its functions makes tourism a matter of interest to many policies, whether focused on tourism itself or on agriculture, rural areas, regional issues, environmental protection, small and medium enterprise development or employment. As a consequence, tourism is eligible for support under diverse Community funds. The Community adopted a clear attitude towards rural tourism and agritourism only in the 1990s; it boils down to the following statement: "Rural tourism is a fundamental

Table 3.8 Tourism in community policies: the cross-sectoral nature of tourism and European policies. Sources of financing

European policies	European funds used in supporting development policies
Common Agricultural and Rural Development Policy	European Agricultural Fund for Rural Development (EAFRD)
Regional policy	European Regional Development Fund (ERDF)
Environmental policy	Cohesion Fund
SME development policy	European Regional Development Fund (ERDF)
Employment policy	European Social Fund (ESF)
Tourism policy	European Agricultural Fund for Rural Development (EAFRD), European Regional Development Fund (ERDF), European Social Fund (ESF), Cohesion Fund

Source: Own study.

element of a true rural policy [...] and has a special role in spatial planning as an important and complementary element of a rural development strategy" (CdR 19/95, 1995). Conversely, the European Union's tourism policy is viewed as a combination of comprehensive tasks covered by different domains of the Union's policy rather than being considered at sector level. First of all, it provides support for the regional policy which is intended to reduce the internal disparities between Community regions and countries through a harmonious economic development assisted by structural funds. The European Union uses its structural funds to finance the development of tourism while also helping aspiring member countries by providing them with pre-accession funds.

The formation of the tourism policy and of policies supported by it, the Union's experience in that respect, and the number of documents developed and adopted are multidimensional, ample, and complicated issues which cannot be easily presented in synthetic terms; all the more since the EU's tourism policy is viewed as a combination of comprehensive tasks covered by different domains of the Union's policy rather than being considered at sector level. It provides support for the regional policy whose objectives include reducing the internal disparities between Community regions and countries through a harmonious economic development assisted by structural funds. The European Union uses its structural funds to finance the development of tourism while also helping aspiring member countries by providing them with pre-accession funds.

Although tourism develops rapidly and the tourism industry is of tremendous importance to the European economy, the European Communities have not paid much attention to it for many years. No separate tourism policy existed and the European Commission claimed that tourism was a matter to be dealt with by member country governments rather than by the Community. However, the European Commission's policy had an indirect impact on tourism through legal acts adopted and measures taken in other areas.

The special Council Resolution of April 10, 1984 on a Community policy on tourism (together with the European Commission's "Community policy on tourism. Initial guidelines" attached thereto) is believed to be one of the EU's first legal acts focused on tourism. While these documents confirmed the emergence of the Union's tourism policy as a separate domain, they also provided the first definition of priority areas for Community action, including: preserving the architectural heritage of less developed areas, and promoting social, cultural and rural tourism. According to Alejziak (2008), the two documents were so important mostly because they indicated the role and importance of tourism as a supporting tool for measures taken by the European Communities to pursue a number of key strategic goals, such as promoting sustainable economic development or small and medium enterprise development. The EU Council pointed out that tourism needs must be addressed in the Community's decision-making process and that tourism issues must be discussed between member states and the European Commission.

However, the actual development of the EU's tourism policy started only in the 1990s. Named "the Year of the Tourism," 1990 marked the beginning of proposals for a common tourism policy. The purpose of this initiative was to reinforce the concept of the European Single Market and emphasize the integrative role of tourism. The Action Plan for Tourism, developed the next year, was adopted under the Council Decision of July 13, 1992. The provisions of the Plan included taking action to diversify the forms of tourism into cultural, rural or ecological tourism, for instance. However, only the reform of the Treaty of Rome (initiated by the Single European Act in 1987) and the 1992 Treaty of Maastricht resulted in supplementing the Community's earlier initiatives with provisions regarding tourism, even though tourism and natural environment were beyond the competencies of the European Economic Community, as defined in the Treaty of Maastricht. The Treaty includes a provision that in the future, the Community will also take measures related to tourism (Article 3/k). However, according to Bąk (2003), this provision suggests that tourism is part of different areas of Community action rather than being covered by its common tourism policy. Importantly, the Treaty of Maastricht also defined the relationships between tourism and the environment by identifying the following priorities and lines of action: protection of the natural environment; and protection of the cultural and historical heritage. Also, an indirect reference was made to environmentally-friendly tourism which includes rural tourism and agritourism.

In 1992, the Commission also submitted the 5th Environment Action Program (European Community Program regarding policy and action for environmental protection and sustainable development), supported by the Report on the condition of the environment in the Community. According to Zangari (2003), due to bad experience from the rapid development of mass tourism from the 1960s to the 1980s, the 5th Program indicated the need for developing alternative, sustainable forms of tourism "to achieve a balance between the development of the tourism industry and the protection of the environment with the use of defined tools and methods." An action plan was also drawn up for tourism development, including such forms of tourism as cultural, social, and rural tourism (and the relationships between tourism and the environment). The implementation of the programs was planned to involve the organization of diverse trainings; improvements in information flow and in access to assistance systems for people employed in a rural setting; and incentives for improving the quality of tourism services and for supporting projects that make it easier to access the rural tourism market. Only since that moment has tourism been considered by the European Union as one of major areas of its policy. This is reflected in areas and action lines such as: the freedom to move across EU countries; protection and security of tourists; protection of tourism values; a system of tourism statistics; vocational education; and support for certain kinds of tourism, especially including social tourism, youth tourism, and agritourism.

During 1993–1995, due to new challenges facing the tourism sector, such as the need for improving its competitiveness, enhance the quality of tourism products, and address the requirements of the tourism industry under the EU's general economic policy, the European Union implemented an action plan to assist the tourism sector, although it did not have much success. In turn, in 1995, the European Commission presented the Green Book which described the measures taken by the Communities in the tourism sector and outlined the prospects for further growth. It also stated that formulating a tourism policy of the European Union would enable boosting that sector's efficiency. The next year, the European Commission presented "Philoxenia," a draft multinational assistance program for the tourism industry for 1997–2000 which laid strong emphasis on the need for making tourism part of the Community policy. According to Walasek and Embacher (2005), although "Philoxenia" addressed the challenges and problems related to European tourism development, it started with a delay and proved to be poorly effective due to having a small budget and being met with opposition from some member countries.

The new tourism policy proposed by the Commission in 2006 was designed to support the sector in a context of new challenges and to promote its competitiveness. The Commission communication entitled "A renewed EU Tourism Policy—Towards a stronger partnership for European Tourism" was an important document designed to establish a Community tourism policy. It states that

> The Commission intends to put in place a renewed European tourism policy, based on the experiences gained so far and responding to the challenges of today. The main aim of this policy will be to improve the competitiveness of the European tourism industry and create more and better jobs through the sustainable growth of tourism [...] In implementing this policy, the Commission will develop a close partnership with Member States' authorities and the stakeholders in the tourism industry.

It also specifies the main areas of focus for the policy, including the promotion of and support for sustainable tourism. In November 2007, having considered the documents, initiatives, and actions of the Commission and other Community institutions, the European Parliament adopted a resolution on a renewed EU tourism policy entitled "Towards a stronger partnership for European Tourism," which clearly highlights the environmental and social dimension of tourism development and the role it plays in regional development. It also states that due to the multidimensionality of tourism itself, the development, promotion, and implementation of an equitable, efficient, and sustainable tourism policy requires a broad dialog between social partners, stakeholders, public authorities of all (European, national, regional, local) levels, and public and private operators. Hence, the European tourism policy should be complementary to policies in place in member states.

The growing interest in the creation of a common tourism policy was also reflected by the appointment of the Advisory Committee on Tourism in 1986. Its role boils down to providing information and advice and cooperating in all tourism measures taken in the EU. However, the European Commission—which supervises the execution of treaty provisions and decisions made by Community authorities—has the greatest impact on the formation of the European Union's tourism policy through its right of initiative and its right to issue implementing acts for the European Council's regulations. Within the Commission's 24 Directorates-General in charge of specific domains, tourism comes under the supervision of DG Enterprise (XXIII) which also includes the Tourism Department. Within the Commission, it has an important coordination function related to taking account of tourism in draft legislation, programs, and other EU actions. In DG XXIII, an important role is also played by working groups responsible for a detailed analysis of problems related to the functioning and development of tourism (tourism quality, staff training, sustainable development, etc.). The Tourism Department cooperates closely with the Council of Ministers, the Economic and Social Committee, the Committee of the Regions, and the European Parliament. Members of the EU advisory committee in Directorate-General XXIII include the European Federation of Rural Tourism (EuroGites) which delivers opinions on matters regarding rural tourism.

The creation and formation of the European Union's common policy for tourism is still ongoing. Debates and consultations are in progress; conferences have been organized to discuss the development of the tourism industry. The development of the tourism policy is also impacted by the fact that it is viewed as an important instrument of regional planning, and especially as a driver of economic mobilization and a stimulus for the development of remote regions. Alejziak (2008) also finds it important that tourism has a positive effect on the preservation of rural life. In a context of the depopulation of many European rural areas, this is an aspect of major importance for the formation of the common tourism policy.

One of the first reports on the development of rural tourism in European Union countries was created in the late 1980s. Already then Grolleau (1987), the author of the report, noted the problem of how to define the terms related to rural areas and rural tourism. He stated that rural tourism includes "all forms of tourism which develop in a rural setting," although the rural setting can have different meanings in Italy, Germany, the Netherlands, Belgium Luxembourg, and France (Grolleau, 1987, p. 34). He also noted that rural accommodation facilities differed between European Union member countries, and identified a separate group of agritourism accommodation facilities related to farms. However, he found that even that group was heterogeneous and although farms exist in all EU member countries, they strongly differ between Belgium, Denmark, Portugal, Italy or France (Grolleau, 1987, p. 32). As late as in 2003, Kutkowska (2003, p. 59) wrote that in Central and Eastern European countries other than Poland, agritourism was "virtually unheard of," and that CEE rural areas witnessed the development of other forms of tourism. For instance, there were "some forms of rural accommodation rental" in Hungary, just like in former Yugoslavia and Estonia (Kutkowska, 2003, p. 59, after: Drzewiecki, 2001). Nevertheless, McMahon analyzed the development of rural tourism and agritourism in Central and Eastern European countries already in 1996 (McMahon, 1996, pp. 175–182), and found that these forms of tourism offered relatively cheap tourism products because they usually do not require huge investments as they rely on the existing resources of farms and other rural buildings. However, he emphasized that due to growing tourist requirements for accommodation facilities, that group, too, will be forced to invest and improve its marketing efforts, especially including promotion activities (McMahon, 1996, p. 182). The development of rural tourism in that period was driven by tourism development trends which suggested that mass tourism slowly started to lose its role to individual tourism. Hence, tourism preferences and expectations began to change, too (Pothoff, 1993). Rural tourism, especially agritourism, started to be a topic addressed in diverse strategic documents and rural development plans as a component of multipurpose rural development (Kłodziński, 2006, pp. 63–77), as a basis for sustainable development of small farms, as a factor in rural and farm diversification (Grolleau, 1987, pp. 35–37; OECD, 2009), as an additional revenue stream for farmers and other rural dwellers (OECD, 2009), as a way to mitigate the operational risks faced by farms, and as a way to counteract the urbanization or excessive depopulation of rural areas (Table 3.9). It also is a factor in the economic mobilization of rural areas; indeed, engaging in a tourism-related economic activity could create additional jobs for the farming and non-farming rural population. Over time, it could become their main activity. This kind of business can also create jobs for rural residents. In the early 2000s, Drzewiecki (2001), based on European Commission data, claimed that 23% of European tourists visit a rural destination each year; the share of rural tourists was particularly high in the following nations: Germany (43%), Netherlands (39%), Denmark (35%), France (29%), UK and Portugal (28% in each country), Ireland and Spain (27% in each country), and Belgium (25%) (Świetlikowska, 1998). As a consequence, the greatest number of accommodation facilities were established in Western Europe, especially in Germany, France, UK, and Ireland (Drzewiecki, 1995).

"Rural Tourism in Europe. An Explanation of Success and Failure Factors," a paper by Veer and Tuunter (2005), analyzed the development of rural tourism in Europe and identified its development drivers (factors of success or failure) at country level. The authors paid particular attention to: tourism development policies for rural areas; development strategies; the entrepreneurial spirit of the rural population; organizational structures (e.g. associations); agritourism trainings; range of products offered; the agritourism and rural tourism market (with emphasis being placed on the importance of the domestic market); marketing; and legal regulations which have a considerable effect on the development capacity of both forms of tourism.

Table 3.9 Tourism in the context of the transformation of the Common Agricultural Policy into the EU Rural Development Policy: development milestones

CAP evolution period	Highlights	Years	References to rural tourism
Preparatory period	• 1957: signing of the Treaties of Rome, Italy • 1958: the Stresa Conference, Italy	1957[a]–1961	– No references to tourism
Implementation of CAP rules	–	1962–1968	
Functioning of the CAP with remedial actions	• 1968: the Mansholt Plan • 1972 and 1975: "Social and Structural" Directives • 1973: first enlargement of the European Communities (Denmark, Ireland, United Kingdom) • 1980: second enlargement (Greece)	1968–1984	• No direct references to tourism; • strengthening the structural nature of the CAP (shifting towards a rural development policy)
Reforms to the CAP	• 1985: Green Paper (Evolution and future of the CAP) • 1986: Single European Act (SEA); • 1986: third enlargement (Portugal, Spain) • 1992: MacSharry CAP reform • 1997: Agenda 2000 • 1999: Council Regulation (EC) No. 1257/1999 of May 17, 1999 on support for rural development from the EAGGF[c] and amending and repealing certain Regulations • 1995: fourth enlargement (Austria, Finland, Sweden)	1985–2002	• Encouraging farmers to engage in non-agricultural activities (including tourism); • reallocation of agricultural land to other uses (e.g. recreation and leisure); • diversification: multipurpose rural development concepts (creation of the CRDP[b])

(*Continued*)

Table 3.9 (Cont.)

CAP evolution period	Highlights	Years	References to rural tourism
New CAP	• 2003: the Luxembourg CAP reform agreement • 2005: Council Regulation (EC) No. 1698/2005 of September 20, 2005 on support for rural development by the EAFRD[d] • 2004: fifth enlargement (Cyprus, Czech Republic, Estonia, Hungary, Latvia, Lithuania, Malta, Poland, Slovakia, Slovenia) • 2007: sixth enlargement (Bulgaria and Romania) • 2007–2008: Health Check (CAP Health Check) • 2013: seventh enlargement (Croatia)	2003–2013	– Providing support for activities taken to diversify the rural economy, including encouraging tourism-related business and investment (small infrastructure, recreational infrastructure, development and marketing of rural tourism services, preservation and improvement of rural heritage)
"New opening" in the CAP	– New financial perspective for 2014–2020 2020: Brexit, the withdrawal of the UK from the EU	2014–2020	– Strengthening the rural development policy (transferring funds from the 1st to the 2nd pillar)

Source: Own study based on: Czyżewski and Stępień (2012); Marzewski (2012); Przezbórska (2009); Przezbórska-Skobiej (2015a); Stańko (2009).

a Although Stańko (2009, p. 54) identified 1958–1961 as a preparatory period, the Treaty establishing the European Economic Community was signed on March 25, 1957. However, the content, objectives and nature of the CAP result from the provisions of the Treaty of Rome and from the objectives set for the establishment of European Communities.
b Common Rural Development Policy.
c European Agricultural Guidance and Guarantee Fund.
d European Agricultural Fund for Rural Development.

In 2006, the European Federation of Rural Tourism[20] referred to European rural tourism as a "sleeping giant" (EuroGites, 2006) and stated that European rural areas are home to ca. 180,000 rural accommodation facilities (including agritourism farms) registered as EuroGites members. Together, they offered a total of over 2 million beds (one-third being located in Spain, one of the most popular European tourism destinations). According to EuroGites, rural tourism directly and indirectly created ca. 500,000 jobs. Moreover, the development of rural tourism provides rural areas with an income of ca. EUR 37 billion (EuroGites, 2006). European rural areas are home to a total of ca. 350,000 (affiliated and non-affiliated) accommodation facilities[21] which offer ca. 4 million beds. Their operations generate an income of over EUR 150 billion; an important aspect is that nearly all of this amount is earned directly by the rural population.

Three years earlier (in 2003), the 1st European Congress of Rural Tourism was held in Spain. Its summary included similar conclusions regarding the development of rural tourism, and noted also that agritourism incomes had a share of ca. 15% in the entire tourism market of what was at the time the EU (Konkluzje... [Conclusions...], 2003). In 2001, that share varied in the range of 10% to 20%, according to Roberts and Hall (2001, p. 1). At that time, the expenses of one-day visitors and other expenditure related to rural tourism and recreation in European countries were estimated at over EUR 65 billion (Roberts & Hall, 2001).

In recent years, more attention has been paid to the decreasing development pace of European agritourism and rural tourism. Furthermore, some countries have witnessed a decline in the number of agritourism farms for reasons which include the tourists being less interested in the agritourism offer.

The purpose of this section is to determine the position of tourism in the Common Agricultural Policy, which gradually shifts to a Common Rural Development Policy. An analysis of Community documents and legal acts and a review of the relevant literature were carried out in the pursuit of that goal.

Tourism, and especially rural tourism, has been a matter of interest to the Common Agricultural Policy almost since it was established. However, the Community started to adopt a clearly defined attitude towards rural tourism and agritourism only in the 1990s when the Common Agricultural Policy turned into the Common Rural Development Policy (Przezbórska, 2009). At that time, the position and responsibilities of rural tourism evolved under the impact of subsequent reforms and events that marked the history of the Communities, especially including the adoption of relevant documents and legal acts, the seven subsequent enlargements of the European Union, and Brexit (Table 3.9). As a consequence of these transformations, tourism saw its importance grow considerably and started to be viewed as an enabler of many goals set under the rural development policy, the structural policy, and the regional policy (Marzewski, 2012). The increase in the importance of tourism in the European policy was expressed by the following statement: "rural tourism is a fundamental part of the rural development policy and a crucial and complementary component of the rural development strategy while also playing a special role in land use planning" (CdR 19/95, 1995). Table 3.9 presents the development milestones of the Common Agricultural Policy and the way it gradually evolved into the Common Rural Development Policy in a context of changing attitudes towards rural tourism and factors which affected the increase in importance of tourism as an enabler of many strategic goals in rural areas.

In its actions, the European Union relies on a great deal of measures with an indirect effect on tourism development, including funds allocated to rural development. The multipurpose rural development is an extremely extensive term defined in a number of different ways. Generally, it is believed to be a concept of rural mobilization and economic diversification, in which the future of the rural population is associated not only with agriculture but also with alternative sectors, including tourism. According to Kłodziński (2006), the strategy for a multidirectional rural development consists in a greater diversification of the whole rural economy and, thus, means moving away from having a single function (mainly the production of agricultural raw materials). Kamiński (1995) lists the following among the basic elements of multipurpose rural development:

1   agricultural production activity,
2   non-agricultural activity directly related to agriculture (supply of materials and productive inputs, production services, buying-in, storage, transport, trade in agricultural produce, etc.);
3   non-agricultural activity not directly related or totally unrelated to agriculture (including agritourism, i.e. organizing leisure and holidays, catering, hospitality, landscape maintenance, environmental protection, and any other production and service activity without any link to agriculture).

Europe saw the first serious scenarios for rural development when the European Economic Community was established, although a unified rural development concept was created only in the mid-1980s. At that time, the Common Agricultural Policy came up with a proposal to shift away from the philosophy of maximizing agricultural output to a policy of rural development through extensification of production, introduction of new non-agricultural functions to rural areas, promotion of employment forms alternative to farming, and support for what is referred to as the structural policy. According to Adamowicz (1997) and Wilkin, the MacSharry reforms marked the beginning of a gradual transformation of the Common Agricultural Policy into a Common Rural Development Policy. This is best reflected by the fact that the share of funds allocated to that policy grew from one budget period to another. Borkowski (2001) noted that "the European Union's shift away from a traditional agricultural policy towards a rural policy should be a beneficial move to Poland. Indeed, rural areas can be viewed as having an untapped potential for development."

The Future of Rural Society, a document published by the EEC Commission (CEC, 1988), can be considered as a next step towards the formation of a modern development model for European rural areas. It emphasizes the decreasing role of agriculture as an employer and booster of regional product, while also paying more attention to rural areas viewed as a whole which, depending on their geographic diversity, were divided into: rapidly developing rural areas adjacent to urban and tourism centers; declining areas whose development depends on the diversification and stimulation of the local economy; and depopulated or depopulating areas which require particular support to keep at least a minimum population.

Another milestone in the formation of the EU's rural development policy was the European Charter for Rural Areas. Drawn up and adopted in 1996, it set out a new scope of tasks

> enabling a sustainable and harmonious rural development in Europe [...] which lays down grounds for sustainable resource management by initiating new functions in the agriculture, forestry, pisciculture and fisheries sectors, including nature and landscape preservation [...] and participation in agritourism and recreation.

The assumption behind the Charter was that rural natural resources should be managed in a diversified way; that economic, socio-cultural and environmental functions should be balanced; and that the rural cultural and historical heritage should be protected. Importantly, the Charter also emphasizes that

> the parties should take essential legal, fiscal and administrative measures to promote the development or rural tourism, with particular focus on agritourism. This should be done by supporting different forms of rural holidays and by taking initiatives that could encourage farmers to host tourists in their farms.

At the end of 1996, the Cork Declaration ("A living countryside") was adopted at a conference held in Cork, Ireland. It included a program for the establishment of a new rural support policy in EU member countries through measures such as: a comprehensive, integrated, and sustainable approach to an interdisciplinary rural development policy (with respect to agriculture, a diversified economic activity, natural resource management, environmental protection, and assistance for culture, tourism, and recreation development); diversification or rural economic and social activities; and sustainable development which means maintaining the unique nature of rural landscape.

Subsequent amendments to the rural development policy were the consequence of the "Agenda 2000 — for a stronger and wider Europe," a set of documents published in 1997 by the European Commission which specified the Union's financial and budgetary measures for 2000–2006. Ensuring a fair standard of living and stable agricultural incomes for the rural community; diversification of incomes; creation of supplementary or alternative jobs and revenue streams for farmers and farming families; and development of sustainable agriculture continued to be the key objectives of the Common Agricultural Policy. Hence, agriculture was no longer considered to be of key importance, and was transforming into a multipurpose sector; rural tourism and agritourism were cited among the most popular options for the diversification of the rural economy.

Major changes to the rural development policy were brought by the 2003 reform agreed upon between the EU ministers of agriculture in Luxembourg. The main amendments consisted in the reallocation of agricultural funds to rural development, a procedure which was implemented during 2004–2006. In November 2003, a conference on rural development was held in Salzburg. The attendees concluded that "diversification is needed both inside and outside the agricultural sector in order to guarantee adequate levels of income to preserve sustainable rural communities," and that public support must be allocated to the European rural development policy to attain that goal.

The rural development policy for 2007–2013 was formulated as a result of measures taken during 2000–2003, including at the European Council meeting in Lisbon in 2000, at the conferences held in Gothenburg (2001) and Salzburg (2003), and during the 2004–2006 budget negotiations. The growing importance of multipurpose, multidimensional rural areas was reflected by the establishment of the European Agricultural Fund for Rural Development and, at the same time, by much more funds being allocated to rural diversification and environmental protection.

The analysis of documents and programs published by the European Union, and of other relevant literature, suggests that the Community started to adopt a clearly defined attitude towards rural tourism and agritourism only in the 1990s. This is expressed by the following statement: "rural tourism is a fundamental part of the rural development policy and a crucial and complementary component of the rural development strategy while also playing a special role in land use planning." In many countries, e.g. Austria, Germany, Ireland or the UK, rural tourism is believed to be a major driver of local economy; to emphasize its importance, tourism is referred to as an "industry of the future." However, the rural development policy and, as a consequence, the development policy for rural tourism and agritourism, must comply with the rules established by the European Union. This is because multipurpose development also includes active measures taken to protect the environment and climate and to preserve the local culture and traditions. Otherwise, rural areas will lose their attractiveness and, thus, their potential for development. Hence, a durable, sustainable, and environmentally friendly rural tourism is a way to solve some rural problems and bridge the gaps between regions. The broad range of its forms and functions makes tourism a matter of interest to many policies and therefore makes it eligible for support under different Community funds.

Currently, tourism plays a major role in rural development. Especially, it is believed to have a mobilizing effect on the rural community, and is emphasized to be of great importance to the creation of new rural jobs and to growth and stabilization of rural incomes (especially including farming incomes). As another important aspect, tourism has a multiplier effect on the development of local infrastructure and more. The ability to access financial support from different Community funds (in particular, the European Agricultural Fund for Rural Development) is crucial for the development of rural tourism. In the 2014–2020 financial perspective, the rural development policy is reinforced by the reallocation of funds from the 1st to the 2nd pillar,[22] which also is an essential measure.

In summary, it needs to be emphasized that its interdisciplinary nature and the broad range of its functions make tourism a matter of interest to many Union policies, including the common agricultural and rural policy, the regional policy, the environmental protection policy, the policy for small and medium enterprise development, and the employment policy. The tourism policy developed by the EU provides support for the regional policy which is intended to reduce the internal disparities between Community regions and countries through a harmonious economic development assisted by structural funds. The formation of the EU's tourism policy is impacted by a number of factors. In Poland, the priorities of multipurpose rural and agricultural development include diversifying the activity to ensure alternative revenue streams. Hence, it is essential to support "the creation of all forms of small enterprise in rural areas, services for the economy and rural residents, local initiatives taken to renew and develop rural areas [...], and activities related to agritourism and rural tourism" (*Strategia rozwoju obszarów wiejskich i rolnictwa na lata 2007–2013*).

Notes

1 Gromada, the Tourism and Holiday Cooperative, was established in 1937. Kazimierz Wyszomirski, a Polish activist of the people's movement, educator and member of the Management Board of the Union of Rural Youth, was its co-creator and first president in 1937–1939. The creation and functioning of Gromada is related to the beginnings of organized rural tourism in Poland. Already before World War II, the cooperative organized what is referred to as summer holiday stays for children and poor families from Łódź, an important industrial center.

2 The Jabłoński family are the characters of *W Jezioranach* (In Jeziorany), a radio series broadcasted by the 1st Program of the Polish Radio which started in 1960 and was supposed to be a rural equivalent of *Matysiakowie* (The Matysiak family), a highly popular series. The main characters are members of the Jabłoński family who live in Jeziorany, a fictional village. In turn, *Matysiakowie* is the oldest and longest-running Polish radio series, and the longest story ever in the history of global radio broadcasting. It was started by the Polish Radio in 1956. The series presents a story of a fictional eponymous family living in Dobra Street in the Powiśle district of uptown Warsaw. In its early years of broadcasting, it was followed by 12 million listeners (based on: https://www.polskieradio.pl/241/4717/Artykul/1598248,Matysiakowie-o-audycji and https://www2.polskieradio.pl/jeziorany/, accessed on July 12, 2020).

3 Agritourism often includes ecological tourism (eco-tourism) which is an option for the urban population to have a true rural experience and a relaxing time in a natural environment (and is chosen by environmentally focused people) (Majewski, 2000a; Zaręba, 2000); and eco-agritourism in organic farms or farms converting to organic production methods (Łopata, 2000; Łabaj, 1998, pp. 52–53; Jalinik, 2005a; Wojciechowska, 2009).

4 Similar terms exist in many languages, e.g.: *Landtourismus, Landurlaub, Ferien auf dem Lande* in German; *tourisme rural* in French; *turismo rural* or *turismo en espacio rural* in Spanish; *cel'skij turizm* (сельский туризм), *turizm v sel'skoj mestnosti* (*туризм в сельскёй местнёсти*) or *derevenskij turizm* (деревенский туризм) in Russian (after: Wojciechowska, 2009, p. 35 and Durydiwka, 2012, p. 51, a supplemented compilation).

5 Durydiwka (2012, p. 48) adds two more reasons, namely: tourism can be an urbanization driver for the territories where it develops; this means it can entail certain cultural and economic changes or changes to land development patterns in areas considered rural. Note also that rural tourism consists of many different forms and kinds of tourism activity which developed in different parts of the world.

6 Detailed summaries of how agritourism is defined in both Polish and international literature can be found in papers by, e.g. Bott-Alama (2005, p. 17); Wojciechowska (2009); Durydiwka (2012, p. 50).

7 Majewski (2005, p. 106) additionally listed *turystyka farmerska, turystyka zagrodowa*, and *turystyka w gospodarstwie rolnym*, which are Polish equivalents of *farm tourism* or *rural homestead tourism* and are derived directly from the English and German languages (cf. Wojciechowska, 2009); Kłodziński (2005, p. 10) also added the following terms: *green tourism* and *soft tourism*. However, these names are not widespread in Poland and "are even opposed by some tourism experts" (Majewski, 2005, p. 106; cf. Wojciechowska, 2009, p. 34). In turn, Wojciechowska (2009, p. 22) makes a distinction between the Polish terms "*agroturystyka* and "*agroturyzm*"; she considers the latter to be an umbrella term for agritourism and rural tourism. Based on "*turyzm*" a Polish term defined by Leszczycki, Wojciechowska defined *agroturyzm* as "all theoretical concepts which refer in time and space to tourism related to rural areas and agriculture" (Wojciechowska, 2009, p. 22, after: Leszczycki, 1937). Based on a broad review of the relevant literature, Philip, Hunter, and Blackstock (2010, p. 755) developed a synthetic summary of definitions of *agritourism/agrotourism* and other similar terms (*farm tourism, farm-based tourism*). Their analysis resulted in developing a typology of terms related to agritourism, with five types of agritourism being identified: (1) agritourism in a non-active farm; (2) agritourism in an active farm which, however, offers only a passive experience to the tourists; (3) agritourism in an active farm which offers an intermediate experience to the tourists; (4) agritourism in an active farm where the tourists have a direct contact with a "staged" farm; (5) agritourism in an active farm where the tourists have a direct contact with an authentic farm (Philip, Hunter, & Blackstock, 2010, p. 756).

8 Similar terms can be found in other languages: *agri-tourisme* in French, *agriturismo* in Italian, *agroturizm* (агрётуризм) in Russian (for a broader description of the terminology applicable to tourism activities which develop in rural areas, see Wojciechowska, 2007a, 2009, pp. 34–44; Philip, Hunter, & Blackstock, 2010, pp. 754–758; and Durydiwka, 2012, p. 48).

9 Pursuant to the Personal Income Tax Act of July 26, 1991 (Article 21.1, Item 43), farm owners can be exempted from income tax on agritourism activities if they meet the following conditions: (1) residential buildings where rooms are rented out are owned by the farm (as per the agricultural tax act); (2) rooms are rented out to holidaymakers (the exemption is not applicable if rooms are rented out on a permanent basis); (3) rooms rented out are located in residential buildings; (4) accommodation facilities are located in rural areas; (5) up to five rooms are rented out (excluding the dining room, living room, porch, and other accommodation accessible to the public). In order to be exempt from personal income tax, the operator of the agritourism farm must meet all conditions provided for in the article quoted above. However, pursuant to the Assumptions for the draft act on personal income tax on agricultural activities and on amendments to other acts (draft from June 4, 2013), incomes from sale of

agritourism services are also subject to taxation (if the number of rooms does not exceed 5, the agritourism activity is included in, and taxed together with, the agricultural activity) (Cholewa & Nachtman, 2014, p. 116).

10 According to providers of agritourism services, the term "agritourism" includes different forms of accommodation, catering, and recreation services as well as leisure, sports or even healthcare or rehabilitation (Sznajder & Przezbórska, 2006).

11 For the tourists, agritourism means tourism and recreation activities related to the exploration of agricultural production and/or having a holiday in a rural and agricultural environment (Sznajder & Przezbórska, 2006; cf. Zawadka, 2010a).

12 Więckowski (2014, p. 23, after: Urry, 2000) suggests that the forms and kinds of tourism which currently are on a development path will require many tourism-related terms to be redefined, including tourism or tourism space. Indeed, it would be difficult to use the existing terminology in studying, for instance, the four currently evolving types of tourism travel: corporeal travel, physical movement of objects, virtual travel, and imaginative travel.

13 Jaworski and Lawson (2005) and Kozak (2013, p. 2) found this to be caused by the fact agritourism is increasingly often a business run by urban outmigrants ("townies"). According to the most liberal definitions, agritourism also includes businesses run by people who previously had nothing to do with farming and operate their tourism undertakings in former farm buildings adjusted to the needs and requirements of tourists. However, a question arises whether they provide an authentic experience.

14 The logo of the "Hospitable Farms" Polish Federation of Rural Tourism is a stork in its nest on the roof of a house.

15 The following taxation forms are available in Poland: flat tax; tax on registered income without deductible costs (17% of revenue; the taxpayer is required to keep a register of revenues); constant amount tax (revenue from renting out up to 12 guest rooms and revenue from serving meals; however, the farm owner cannot employ more than two persons or two adult family members); and taxation of revenue under the general rules, i.e. keeping a revenue and expense ledger (Personal Income Tax Act of July 26, 1991, as amended, Journal of Laws [Dz. U.] of 1991, No. 80, Item 350).

16 The tax can be introduced by communes holding the status of health resort communes (a total of 42 of them exist in Poland). Pursuant to Article 48 of the Act of July 28, 2005 on spa treatment, health resorts, health-resort protection areas and health resort communes (Journal of Laws [Dz. U.] 2017.0.1056), "in order to preserve the healthcare functions of a health resort, health resort communes shall have the right to charge the resort tax as provided for in separate regulations."

17 Construction Law Act of July 7, 1994 (Journal of Laws [Dz.U.] of 1994, No. 89, Item 414).

18 Fire Protection Act of August 24, 1991, as amended (Journal of Laws [Dz. U.] of 1991, No. 81, Item 351).

19 European Currency Unit (ECU): precursor to the euro; it was a basket of the currencies of the European Community member states, used as the unit of account of the European Community (Gandolfo, 1987) before being replaced by the euro on January 1, 1999, at parity (UBC Sauder School of Business, n. d.). The ECU itself replaced the European Unit of Account, also at parity, on March 13, 1979. The European Exchange Rate Mechanism attempted to minimize fluctuations between member state currencies and the ECU. The ECU was also used in some international financial transactions, where its advantage was that securities denominated in ECUs provided investors with the opportunity for foreign diversification without reliance on the currency of a single country (Scott, 2003).

20 Also known as EuroGites.

21 According to EuroGites, rural tourism accommodation includes: agritourism farms, private rural accommodation, bed and breakfast services, and self-contained housing units. These operators also offer traditional rural cuisine and tourism services related to the stay of guests in accommodation facilities.
22 The CAP consists of two pillars: the first covers direct payments and measures for market support, whereas the second includes measures for multi-annual rural development. The two pillars are complementary to each other. The rural development policy falls within their scopes (Czyżewski & Stępień, 2012).

References

Adamowicz, M. (1997). Wspólna Polityka Rolna UE. Skutki reformy i perspektywy zmian [Common Agricultural Policy of the EU. Effects of the reform and prospects for changes]. *Zagadnienia Ekonomiki Rolnej* no. 16/97, IERiGŻ i SER PTE, Warszawa, pp. 43–58.

Agency for Restructuring and Modernisation of Agriculture (2006). ARMA – year after accession, Warsaw. Retrieved from: https://www.arimr.gov.pl/fileadmin/pliki/zdjecia_strony/224/2182_06lamanie_w-ang.pdf (access: 15. 07. 2020).

Agenda 2000, For a Stronger and Wider Union (1997). Bulletin of the EU Commission, Brussels–Luxembourg.

Airey, D. (1994). Education for tourism in Poland: the PHARE programme. *Tourism Management*, 15(6), 467–471. https://doi.org/10.1016/0261-5177(94)90068-X.

Alejziak, W. (2008). Międzynarodowa polityka turystyczna, Serwis Internetowy ITiR, AWF w Krakowie. Retrieved from: http://itir.awf.krakow.pl/eot/mptur_wa.doc (access: 10. 11. 2008), pp. 1–18.

Andrychowicz, A. (2000). Źródła finansowania działalności gospodarczej na obszarach wiejskich [Sources of financing economic activity in rural areas]. In: Majewski, J. (ed.), *Agroturystyka w programach pomocowych dla wsi. Materiały konferencyjne. POLAGRA FARM 2000*, pp. 21–40. Międzynarodowe Targi Poznańskie i Akademia Rolnicza w Poznaniu, Poznań.

Andrzejewska, O. (1999). *Wczasy pod gruszą – agroturystyka w modzie* [*Holidays under the pear tree – agritourism in fashion*]. Report Rolnictwo. BOSS-Inf. Ekon., no. 38(505).

Augustyn, M. (1998). National strategies for rural tourism development and sustainability: the Polish experience. *Journal of Sustainable Tourism*, 6 (3), 191–209. Short Run Pre, Exeter, UK.

Bajcar, A. (1969). Regiony turystyczne Polski [Tourist regions of Poland]. *Geografia w Szkole*, 3 (4). Wydawnictwa Szkolne i Pedagogiczne, Warszawa.

Bąk, S. A. (2003). *Procesy integracji z UE w sektorach turystyki i kultury w świetle Traktatu Amsterdam. a nauczanie turyzmu w Polsce. Turystyka czynnikiem integracji międzynarodowej*. Wydawnictwo WSIiZ w Rzeszowie, Rzeszów, p. 26.

Balińska, A. (2009). Kierunki rozwoju agroturystyki w Polsce i wybranych krajach europejskich [Directions of development of agritourism in Poland and selected European countries]. Witryna Wiejska – wsparcie dla aktywnych społeczności lokalnych. Retrieved from: http://www.witrynawiejska.org.pl/strona.php?p=1891&c=5248 (access:5. 02. 2020).

Balińska, A. (2016). *Znaczenie turystyki w rozwoju gmin wiejskich na przykładzie obszarów peryferyjnych wschodniego pogranicza Polski* [*The importance of tourism in the development of rural communes on the example of peripheral areas of the eastern border of Poland*]. Wydawnictwo Szkoły Głównej Gospodarstw Wiejskiego w Warszawie, Warszawa.

Bański, J. (2006). *Geografia polskiej wsi* [*The geography of the Polish countryside*]. Wydawnictwo Naukowe PWN, Warszawa.

Bański, J., Rudolf, A., Przybył, C., Bednarek-Szczepańska, M., Czapiewski, K., Mazur, M., & Pieniążek, W. (2012). *Turystyka wiejska, w tym agroturystyka, jako element zrównoważonego i wielofunkcyjnego rozwoju obszarów wiejskich* [*Rural tourism, including agritourism, as an element of sustainable and multifunctional development of rural areas*]. Raport końcowy z Projektu finansowanego ze środków z Funduszy Counterpart Funds (CPF) Sektorowych Agrolinia (PL9005). Agrotec Polska Sp. z o.o., & IGiPZ PAN, Warszawa. Retrieved from: http://www.bip.minrol. gov.pl/DesktopDefault.aspx?TabOrgId=1683&LangId=0 (access: 21. 01. 2020).

Baum, S. (2011). The tourist potential of rural areas in Poland. *CEJSH*, 11, pp. 107–135. Retrieved from: http://cejsh.icm.edu.pl/cejsh

Bednarek-Szczepańska, M. (2017). Rural Tourism – "An apple of the eye" of rural policy in Poland. *EUROPA XXI*, 32, "32 Environmental and Demographic Challenges for Territorial Development," pp. 37–50. Institute of Geography and Spatial Organization Polish Academy of Sciences. http://doi.org/10.7163/Eu21.2017).32.3. Retrieved from: https://rcin.org.pl/igipz/dlibra/publication/83147 (access: 13. 07. 2020).

Biderman, E. (1981). Struktura i wykorzystanie bazy noclegowej w obiektach turystycznych i wczasowo-wypoczynkowych województwa poznańskiego [Structure and use of accommodation facilities in tourist and holiday-recreational facilities of the Poznań province]. *Kronika Wielkopolski*, no. 1, Państwowe Wydawnictwo Naukowe, pp. 76–94.

Bieńkowski, A. (2001). Turystyka jako forma rewitalizacji terenów wiejskich [Tourism as a form of revitalization of rural areas]. *Zeszyty Naukowe Ostrołęckiego Towarzystwa Naukowego*, no. 15, pp. 83–94. Retrieved from: http://mazowsze.hist.pl/28/ Zeszyty_Naukowe_Ostroleckiego_Towarzystwa_Naukowego/644/2001/22879/ (access: 31. 07. 2020).

Borkowski, J. (2001). Obszary wiejskie – niewykorzystany potencjał rozwojowy. In: Kolarska-Bobińska, L., Rosner, A., & Wilkin, J. (eds.), *Przyszłość wsi polskiej. Wizje, strategie, koncepcje*. ISP, Warszawa, pp. 33–41.

Borne, H., & Doliński, A. (2000). *Organizacja turystyki*, Wydawnictwa Szkolne i Pedagogiczne, S.A.,Warszawa.

Bott-Alama, A. (2005). Uwarunkowania rozwoju turystyki wiejskiej w województwie zachodniopomorskim [Conditions for the development of rural tourism in the West Pomeranian Voivodeship]. *Rozprawy i Studia*, DLXXV, p. 501. Wydawnictwo Naukowe Uniwersytetu Szczecińskiego, Szczecin.

Brelik, A. (2015). *Dobra publiczne na obszarach wiejskich jako czynnik rozwoju działalności agroturystycznej na Pomorzu Zachodnim* [*Public goods in rural areas as a factor in the development of agritourism in Western Pomerania*]. Wydawnictwo Naukowe PWN, Warszawa.

Budzich-Szukała, U. (2000). Kredyty preferencyjne i inne formy wspierania agroturystyki (na przykładzie działalności programu "Agrolinia 2000') [Preferential loans and other forms of supporting agritourism (on the example of the operation of the "Agrolinia 2000" program)]. In: Majewski, J. (ed.), *Agroturystyka w programach pomocowych dla wsi. Materiały konferencyjne. POLAGRA FARM 2000*, pp. 51–55. Międzynarodowe Targi Poznańskie i Akademia Rolnicza w Poznaniu. Poznań.

Budzinowski, R. (2003). Prawne pojęcie działalności rolniczej [Legal concept of agricultural activity]. *Prawo i Administracja, 2*, pp. 167–178. Piła.

CdR 19/95 (1995). Opinion of the committee of the Regions on a policy for the development of rural tourism in the regions of the European Union. *Official Journal*C 210, August 14, P. 0099, Brussels. Retrieved from: https://eur-lex.europa.eu/legal-content/EN/TXT/?uri=CELEX%3A51995IR0019 (access: 20. 06. 2020).

CEC (1988). *The Future of Rural Society*, COM(88) 501 final. Commission of the European Communities, Brussels.

CEC (1990). *Community Action to Promote Rural Tourism. Communication from the Commission*, COM(1990) 438 final. Commission of the European Communities, Brussels. Retrieved from: http://aei.pitt.edu/3700/1/3700.pdf (access: 12. 12. 2019).

Center for Profitable Agriculture. University of Tennessee – Farm Bureau Partnership (2005). Agritourism in focus. A guide for Tennessee farmers. Retrieved from: https://trace.tennessee.edu/cgi/viewcontent.cgi?article=1020&context=utk_agexmkt (access: 20. 05. 2020).

Chądrzyński, M. (2009). Wybrane przesłanki funkcjonowania i rozwoju małych i średnich przedsiębiorstw na obszarach wiejskich w Polsce [Selected premises for the functioning and development of small and medium-sized enterprises in rural areas in Poland]. *Zeszyty Naukowe SGGW. Ekonomika i Organizacja Gospodarki Żywnościowej*, no. 79, pp. 125–135.

Chojnicki, Z. (1966). Region w ujęciu geograficzno-systemowym [Region in terms of geography and systems]. In: Czyż, T. (ed.), *Podstwy regionalizacji geograficznej*. Wydawnictwo Naukowe Bogucki, pp. 7–43.

Cholewa, I., & Nachtman, G. (2014). Analiza przewidywanych skutków wprowadzenia reformy podatkowej w polskim rolnictwie na tle rozwiązań niemieckich [Analysis of the anticipated effects of introducing the tax reform in Polish agriculture in comparison with German solutions]. *Zagadnienia Ekonomiki Rolnej*, 2 (339), 104–126.

Cichowska, J., & Klimek, A. (2010). Rola i znaczenie instytucji wspierających rozwój usług agroturystycznych na przykładzie województwa kujawsko-pomorskiego [Role and siginificance of institutions supporting development of agritourism services on an example of kujawsko-pomorskie voideship]. *Infrastruktura i ekologia terenów wiejskich*, no. 13, pp. 63–73. Polska Akademia Nauk, Oddział w Krakowie, Komisja Technicznej Infrastruktury Wsi, Kraków.

Czerwiński, J. (2011). *Podstawy turystyki [The basics of tourism]*. Wrocław–Poznań: Wydawnictwo Wyższej Szkoły Bankowej w Poznaniu.

Czyżewski, A., & Stępień, S. (2012). Wspólna Polityka Rolna – doświadczenia i przyszłość. *Przegląd Prawa Rolnego*, 2 (11), Wydawnictwo Naukowe UAM, pp. 161–185.

Dębniewska, M. (2000). Ekonomiczne aspekty rozwoju turystyki [Economic aspects of tourism development]. *Roczniki Naukowe Stowarzyszenia Ekonomistów Rolnictwa i Agrobiznesu*, 2 (2), pp. 16–20.

Dębniewska, M., & Tkaczuk, M. (1997). *Agroturystyka: koszty, ceny, efekty [Agritourism: costs, prices, effects]*. Poltext, Warszawa.

Derek, M. (2008). Funkcja turystyczna jako czynnik rozwoju lokalnego w Polsce [Tourist function as a factor of local development in Poland]. Uniwersytet Warszawski, Wydział Geografii i Studiów Regionalnych, Warszawa, p. 186. Retrieved from: http://wgsr.uw.edu.pl/wgsr/wp-content/uploads/2018/11/1_pdfsam_doktora t-w-pdf.pdf (access: 11. 04. 2020).

Dionysopoulou, P., Katsoni, V., & Argyropoulou, A. (2014). Agritourism marketing strategy and typology investigation. *Journal of Tourism Research*, 9, pp. 12–27. Retrieved from: http://jotr.eu/pdf_files/V9.pdf (access:2. 03. 2014).

Długokręcka, M. (1998). Agroturystyka jako forma przedsiębiorczości kobiet [Agritourism as a form of female entrepreneurship]. In: *Zrównoważony rozwój turystyki wiejskiej – idee, działania, efekty. Materiały konferencyjne VI Ogólnopolskiego Sympozjum Agroturystycznego*. Centrum Doradztwa i Edukacji w Rolnictwie. Kraków.

Domański, R. (2000). Kredyty preferencyjne i inne formy wspierania agroturystyki w programach Fundacji Wspomagania Wsi [Preferential loans and other forms of supporting agritourism in the programs of the Rural Development Foundation]. In: Majewski, J. (ed.), *Agroturystyka w programach pomocowych dla wsi. Materiały konferencyjne. POLAGRA FARM 2000*, pp. 56–62. Międzynarodowe Targi Poznańskie i Akademia Rolnicza w Poznaniu. Poznań.

Dorocki, S., Rachwał, T., Szymańska, A. I., & Zdon-Korzeniowska, M. (2012). Spatial conditions for agritourism development on the example of Poland and France. *Current Issues of Tourism Research, 2*, pp. 20–29. Retrieved from: https://www.resea rchgate.net/publication/235874225_Spatial_Conditions_for_Agritourism_Developm ent_on_the_Example_of_Poland_and_France (access: 28. 05. 2020).

Drzewiecki, M. (1980). *Rola turystyki w rozwoju ekonomicznym wsi pomorskich [The role of tourism in the economic development of Pomeranian villages]*. Instytut Turystyki, Warszawa.

Drzewiecki, M. (1983). *Status wsi letniskowej [Status of a holiday resort]*. Wieś współczesna, no. 11/321. Naczelny Komitet Zjednoczonego Stronnictwa Ludowego (NK ZSL). Warszawa.

Drzewiecki, M. (1985). Rola przestrzeni wiejskiej w rekreacji [The role of rural space in recreation]. *Problemy Turystyki*, no. 1, Instytut Turystyki, Warszawa, pp. 41–56.

Drzewiecki, M. (1992). *Wiejska przestrzeń rekreacyjna [Rural Recreational Space]*. Instytut Turystyki. Warszawa.

Drzewiecki, M. (1993). Problemy rozwoju agroturystyki w Polsce w latach dziewięćdziesiątych [Problems of the development of agritourism in Poland in the nineties]. *Problemy Turystyki* 4(62), pp. 25–31.

Drzewiecki, M. (1995). *Agroturystyka. Założenia – uwarunkowania – działania*. Instytut Wydawniczy Świadectwo, Bydgoszcz.

Drzewiecki, M. (1998). Pojęcie turystyki wiejskiej [The concept of rural tourism]. *Turyzm*, 8 (1), pp. 21–27.

Drzewiecki, M. (2001). *Podstawy agroturystyki [The basics of agritourism]*. Oficyna Wydwanicza OPO, Bydgoszcz.

Drzewiecki, M. (2002). *Podstawy agroturystyki [The basics of agritourism]*. 2nd edition, corrected and supplemented. Oficyna Wydwanicza OPO, Bydgoszcz.

Drzewiecki, M. (2005). Agroturystyka w Polsce – stan obecny i tendencje rozwojowe [Agritourism in Poland – current state and development trends]. In: Sawicki, B., & Bergier, J. (eds.), *Uwarunkowania rozwoju turystyki związanej z obszarami wiejskim*. Wydawnictwo Państwowej Wyższej Szkoły Zawodowej im. Papieża Jana Pawła II, Biała Podlaska, pp. 46–51.

Drzewiecki, M. (2009). *Agroturystyka współczesna w Polsce [Contemporary agritourism in Poland]*. Wyższa Szkoła Turystki i Hotelarstwa w Gdańsku, Gdańsk, p. 88.

Durydiwka, M. (2005). Zróżnicowanie rozwoju funkcji turystycznej na obszarach wiejskich w Polsce [Diversification of the development of the tourist function in rural areas in Poland]. In: Sawicki, B., & Bergier, J. (eds.), *Uwarunkowania rozwoju turystyki związanej z obszarami wiejskimi*. Wyd. PWSZ im. Papieża Jana Pawła II w Białej Podlasce, Biała Podlaska, pp. 17–19.

Durydiwka, M. (2012). *Czynniki rozwoju i zróżnicowanie funkcji turystycznej na obszarach wiejskich w Polsce* [*Factors of development and diversification of the tourist function in rural areas in Poland*]. Wydawnictwo Uniwersytetu Warszawskiego, Wydział Geografii i Studiów Regionalnych, Warszawa, p. 376.

Encyklopedia PWN. Wydawnictwo Naukowe PWN. Warszawa. Retrieved from: http://encyklopedia.pwn.pl/.

EuroGites (2006). *Turystyka wiejska w Europie – Drzemiący Gigant* [*Rural tourism in Europe – The Sleeping Giant*]. EuroGîtes Presentation. Europejska Federacja Turystyki Wiejskiej.

Europejski Fundusz Rozwoju Wsi Polskiej [The European Fund for Rural Development of Poland] (2005). Efekty finansowo-rzeczowe działalności Funduszu [Financial and material effects of the Fund's operations]. Retrieved from: http://www.efrwp.com.pl/efekty.htm (access: 25. 11. 2005).

Falkowski, J. (2016). Koncepcja typologii i regionalizacji turystyczno-rekreacyjnej w ujęciu krajowym (Polska) i globalnym (Świat) [The concept of typology and regionalization of tourism and recreation in national (Polish) and global (world) context]. *Geography and Tourism*, 4 (1), Kazimierz Wielki University Press, pp. 7–21. Retrieved from: http://www.geography.and.tourism.ukw.edu.pl/artykuly/vol4.no1_2016/gat-vol4-2016-1_01-falkowski.pdf (access: 27. 08. 2019).

Faracik, R. (2006). *Turystyka w strefie podmiejskiej Krakowa* [*Tourism in the suburban area of Krakow*]. Instytut Geografii i Gospodarki Przestrzennej, Uniwersytet Jagielloński, Kraków.

Faracik, R. (2011). Turystyka w strefie podmiejskiej Krakowskiego Obszaru Metropolitalnego [Tourism in suburban zone of Kraków Metropolitan Area]. In: Durydiwka, M., & Duda-Gromada, K. (eds.), *Przestrzeń turystyczna. Czynniki, różnorodność, zmiany*. Uniwersytet Warszawski, Wydział Geografii i Studiów Regionalnych, Warszawa. Retrieved from: https://www.researchgate.net/publication/328858068_Turystyka_w_strefie_podmiejskiej_Krakowskiego_Obszaru_Metropolitalnego_TOURISM_IN_SUBURBAN_ZONE_OF_KRAKOW_METROPOLITAN_AREA (access: 14. 04. 2020).

Faracik, R., Kubal, M., Kurek, W., & Pawlusiński, R. (2014). The Transformation of tourism model in the Polish Carpathians – reporting on the last 20 years of experiences. *Zeszyty Naukowe Uniwersytetu Szczecińskiego. Ekonomiczne Problemy Turystyki*, 4 (28), pp. 285–306. Retrieved from: http://yadda.icm.edu.pl/yadda/element/bwmeta1.element.ekon-element-000171364147 (access: 13. 07. 2020).

Filar, W. (1996). Finansowanie gospodarstw agroturystycznych [Financing of agritourism farms]. In: Jarosz, A. (ed.), *Agroturystyka jako szansa aktywizacji gospodarczej wiejskich regionów turystycznych Małopolski Wschodniej*, pp. 113–116. Oficyna Wydawnicza Politechniki Rzeszowskiej, Rzeszów – Boguchwała.

Firlej, K. (2006). The Polish sector of agri-tourism and rural tourism development. *Roczniki Naukowe Stowarzyszenia Ekonomistów Rolnictwa i Agrobiznesu*, 8 (6), pp. 43–49. Retrieved from: https://depot.ceon.pl/bitstream/handle/123456789/6837/The%20polish%20sector%20of%20agri-tourism%20and%20rural%20development.pdf?sequence=1&isAllowed=y (access: 16. 06. 2020).

Firlej, K., & Niedziółka, A. (2007). Agritourism as a factor of local development in the Malopolska Region. *Roczniki Naukowe Stowarzyszenia Ekonomistów Rolnictwa i Agrobiznesu*, IX (2), pp. 92–96.

Foryś, G. (2016). *Gospodarstwa i stowarzyszenia agroturystyczne w Polsce.* W poszukiwaniu ruchu społecznego, Wydawnictwo Naukowe Scholar, Warszawa.

Foryś, G. (2017). Stowarzyszenia agroturystyczne w Polsce jako składnik nowego ruchu turystycznego [Agritourist associations in Poland as an element of a new social movement. In: Wojciechowska, J. (ed.), *Sieci współpracy w turystyce wiejskiej.* Stan obecny i nowe wyzwania. Kraków–Łódź, pp. 99–113.

Froń, R. (2019). Świadczenie usług enoturystycznych [Provision of enotourism services]. Retrieved from: http://paragrafwkieliszku.pl/swiadczenie_uslug_enotur ystycznych/ (access: 24. 06. 2020).

Gandolfo, G. (1987). *International Economics II: International Monetary Theory and Open-Economy Macroeconomics.* Berlin: Springer Verlag, pp. 380–393, 404–411.

Gaworecki, W. W. (2003a). *Turystyka [Tourism].* PWE, Warszawa.

Gaworecki, W. W. (2003b). Wybrane kierunki badań naukowych w sferze turystyki polskiej [Selected directions of scientific research in the field of Polish tourism]. In: Gołembski, G. (ed.), *Kierunki badań naukowych w turystyce.* Wydawnictwo Naukowe PWN, Warszawa, pp. 213–224.

Gaworecki, W. W. (2010). *Turystyka [Tourism].* Polskie Wydawnictwo Ekonomiczne, Warszawa, p. 438.

Getz, D., & Page, S. J. (1997). Conclusions and implications for rural business development. In: Page, S. J., & Getz, D. eds. *The Business of Rural Tourism.* International Perspective, London.

Gołembski, G. (ed.), (1999). *Regionalne aspekty rozwoju turystyki [Regional aspects of tourism development].* Wydawnicto Naukowe PWN. Warszawa–Poznań, p. 206.

Gołembski, G. (ed.), (2002). *Kompendium wiedzy o turystyce [A compendium of knowledge about tourism].* Wydawnictwo Naukowe PWN, Warszawa–Poznań.

Grad, B., & Ferensztajn-Galardos, E. (2015). Możliwości wsparcia finansowego rozwoju turystyki i agroturystyki na wsi w perspektywie finansowej Unii Europejskiej na lata 2014–2020 [Financial support of tourism and agritourism development on the country in the financing perspective of the European Union for the period 2014–2020]. *Studia Komitetu Przestrzennego Zagospodarowania Kraju,* no. 162, pp. 27–45. Polska Akademia, Nauk Komitet Przestrzennego Zagospodarowania Kraju. Warszawa. Retrieved from: https://journals.pan.pl/Content/97550/mainfile.pdf (access: 20. 06. 2020).

Gradziuk, K. (2007). Realizacja programu SAPARD w Polsce – zróżnicowanie regionalne [Implementation of the SAPARD program in Poland – regional differentiation]. Program Wieloletni 2005–2009. Ekonomiczne i społeczne uwarunkowania rozwoju polskiej gospodarki żywnościowej po wstąpieniu Polski do Unii Europejskiej. Instytut Ekonomiki Rolnictwa i Gospodarki Żywnościowej, Państwowy Instytut Badawczy, Warszawa. Retrieved from: https://www.ierigz.waw.pl/publikacje/raporty-program u-wieloletniego-2005-2009/757,19,3,0,1314185806.html (access: 12. 06. 2020).

Grolleau, H. (1987). *Rural Tourism in the 12 Member States of the European Economic Community Countries,* Commission of the European Communities, Directorate-General for Transport (Tourism Service).

Gurgul, E. (ed.), (2005). *Agroturystyka jako element rozwoju i promocji region [Agritourism as an element of the development and promotion of the region].* Wydawnictwo Politechniki Częstochowskiej, Częstochowa, p. 90.

Hayden, G. (1996). Środki Agencji Restrukturyzacji i Modernizacji Rolnictwa jako sposób na obniżenie kosztów inwestycji w agroturystyce [Funds of the Agency for Restructuring and Modernization of Agriculture as a way to reduce investment costs in agritourism]. In: Jarosz, A. (ed.), *Agroturystyka jako szansa* aktywizacji gospodarczej wiejskich regionów turystycznych Małopolski Wschod-niej*, pp. 129–136. Oficyna Wydawnicza Politechniki Rzeszowskiej, Rzeszów – Boguchwała.

Iakovidou, O., Partalidou, M., & Manos, B. (2000). *Rural tourism. Agritourism: A challenge for the development of the Greek countryside.* International Seminar "Agritourism and Rural tourism. A Key Option for the Rural Integrated and Sustainable Development Strategy." IAERT, Perugia, Italy, pp. 65–70.

Idziak, W. (2015). *Turystyka wiejska i agroturystyka w perspektywie finansowej 2014–2020 dla MRiRW*, broszura informacyjna, Zespół Projektowy Konsorcjum Bluehill Sp. z o. o. oraz Quality Watch Sp. z o.o. pod przewodnictwem dr Wacława Idziaka, Warszawa. https://www.gov.pl/web/rolnictwo/wsparcie-dla-rozwoju-turystyki-wiejskiej-i-agrotur ystyki-w-ramach-programu-rozwoju-obszarow-wiejskich-na-lata-2014-2020-oraz-z-kra jowych-i-regionalnych-programow-operacyjnych (access: 17. 05. 2020).

Idziak, W., Idziak, P., & Kamiński, R. (2015). *Wsparcie dla rozwoju turystyki wiejskiej i agroturystyki w ramach Programu Rozwoju Obszarów Wiejskich na lata 2014–2020 oraz z krajowych i regionalnych programów operacyjnych. Ekspertyza dla Ministerstwa Rolnictwa i Rozwoju Wsi [Support for the development of rural tourism and agritourism under the Rural Development Program for 2014–2020 and from national and regional operational programs. Expertise for the Ministry of Agriculture and Rural Development]*, Warsaw. Retrieved from: https://www.gov.pl/web/rolnictwo/wsparcie-dla-rozwo ju-turystyki-wiejskiej-i-agroturystyki-w-ramach-programu-rozwoju-obszarow-wiejs kich-na-lata-2014-2020-oraz-z-krajowych-i-regionalnych-programow-operacyjnych (access: 20 June 2020.

Iwicki, S. (2001). *Rola turystyki w zrównoważonym i wielofunkcyjnym rozwoju obszarów wiejskich* [The role of tourism in sustainable and multifunctional rural development]. *Zeszyty Naukowe Wyższej Pomorskiej Szkoły Turystyki i Hotelarstwa*, no. 1, pp. 27–41.

Izydorczyk, A. (1975). *Organizacja turystyki w Polsce [Organization of Tourism in Poland]*. Instytut Turystyki, Warszawa.

Jackowski, A. (1981). *Typologia funkcjonalna miejscowości turystycznych na przykładzie województwa nowosądeckiego [Functional typology of tourist destinations: (on the example of the Nowy Sącz Province]*. Uniwersytet Jagielloński, Kraków, p. 173.

Jalinik, M. (2005a). Rozważania na temat pojęć i definicji w agroturystyce [Considerations on concepts and definitions in agritourism]. In: Sawicki, B., & Bergier, J. (eds.), *Uwarunkowania rozwoju turystyki związanej z obszarami wiejskimi*. Wyd. Panstwowej Wyższej Szkoły Zawodowej im. Papieża Jana Pawła II w Białej Podlasce, Biała Podlaska, pp. 97–101.

Jalinik, M. (2005b). Kontrowersje wokół wybranych pojęć, definicji i kategoryzacji w agroturystyce [Controversies around selected concepts, definitions and categorization in agritourism]. *Zagadnienia Ekonomiki Rolnej*, 3 (304), pp. 83–89.

Jankowska, A., Kierzkowski, T., & Knopik, R. (2004). Fundusze pomocowe dla Polski po akcesji – fundusze strukturalne i Fundusz Spójności [Aid funds for Poland after accession – structural funds and the Cohesion Fund]. Polska Agencja Rozwoju Przedsiębiorczości. Warszawa. Retrieved from: https://poig.parp.gov.pl/files/74/81/ 104/fundusze_pomocowe.pdf (access 20. 07. 2020).

Jaworski, A., & Lawson, S. (2005). Discourses of Polish agritourism: global, local, pragmatic. In: Jaworski, A., & Pritchard, A. (eds.), *Discourse, Communication, and Tourism*. Channel View Publications, Clevedon, pp. 123–149.

Jęczmyk, A., Uglis, J., & Fedorczuk, A. (2018). *Wykorzystanie środków unijnych w gospodarce turystycznej – studium przypadku sektora hotelarskiego w województwie wielkopolskim* [The use of EU funds in the tourism economy – a case study of the hotel sector in the Greater Poland VProvince]. *Intercathedra*, 35 (2), 135–142.

Jędrzejczyk, I. (1996). Finansowanie rozwoju biznesu agroturystycznego [Financing the development of agritourism business]. In: Jarosz, A. (ed.), *Agroturystyka jako szansa aktywizacji gospodarczej wiejskich regionów turystycznych Małopolski Wschodniej*, pp. 117–128. Oficyna Wydawnicza Politechniki Rzeszowskiej, Rzeszów – Boguchwała.

Kachniewska, M. (2011). Funkcja turystyczna jako determinanta jakości życia na wsi [Tourist function as a determinant of the quality of life in the countryside]. *Folia Pomeranae Universitatis Technologiae Stetinensi Oeconomica* 288 (64), pp. 53–72. Retrieved from: https://www.zut.edu.pl/fileadmin/pliki/wydawnictwo/Folia/Oeconomica/288/Kachniewska.pdf (access: 15. 07. 2020).

Kamiński, W. (1995). *Warianty wielofunkcyjnego rozwoju wsi – uwarunkowania przestrzenne* [Warianty wielofunkcyjnego rozwoju wsi – uwarunkowania przestrzenne]. *Zeszyty Naukowe Akademii Rolniczej im. H. Kołłątaja w Krakowie*, 295 (43), pp. 19–25.

Kapała, A. (2008). Prawne pojęcie agroturystyki [Legal concept of farm tourism]. *Przegląd Prawa Rolnego*, 1 (3), pp. 99–115. Retrieved from: https://repozytorium.amu.edu.pl/bitstream/10593/9092/1/005_Anna_Kapa%C5%82a_Prawne_poj%C4%99cie_agroturystyki_100_115.pdf (access: 15. 04. 2020).

Kapała, A. (2010). Regulacja prawna agroturystyki – uwagi de lege ferenda [Legal regulation of agritourism "de lege ferenda" remarks]. *Przegląd Prawa Rolnego*, 2 (7), pp. 101–120. Retrieved from: https://www.researchgate.net/publication/277187084_Regulacja_prawna_agroturystyki_-_uwagi_de_lege_ferenda (access: 15. 04. 2020).

Karolczak, M., Rzeńca, P., & Wojciechowska, J. (2002). Analiza porównawcza rozwoju agroturystyki w Śladkowie Małym i Wojciechowie [A comparative analysis of the development of agritourism in Śladków Mały and Wojciechów]. In: *Turystyka wiejska w perspektywie europejskiej. VIII Ogólnopolskie Sympozjum Agroturystyczne w Wysowej*. Wydawnictwo Krajowego Centrum Doradztwa Rozwoju Rolnictwa i Obszarów Wiejskich w Brwinowie, Brwinów.

Kielesińska, A. (2014). Wybrane aspekty synergii agroturystyki i rozwoju lokalnego [Selected aspects of synergy of agritourism and local development]. *Zeszyty Naukowe Politechniki Śląskiej. Seria Organizacja i Zarządzanie*, no. 68, pp. 41–50. Retrieved from: http://www.woiz.polsl.pl/znwoiz/z68/Kielesinska%20A%20%20po%20recenzjach%20%20i%20poprawie.pdf (access: 20. 07. 2020).

Kiper, T., & Özdemir, G. (2012). Tourism planning in rural areas and organisation possibilities. *Landsc. Plan.*, Ozyavuz, M. (ed.), June 13. Retrieved from: https://www.intechopen.com/books/landscape-planning/tourism-planning-in-rural-areas-and-organization-possibilities (access: 12. 02. 2020). doi:10.5772/39072..

Kłodziński, M. (2005). Turystyka wiejska w procesie zrównoważonego rozwoju obszarów. Wiejskich [Rural tourism in the process of sustainable development of rural areas]. *Woda Środowisko Obszary Wiejskie*, 5, issue 1 (13), pp. 9–22.

Kłodziński, M. (2006). Strategia wielofunkcyjnego rozwoju gmin w Polsce. *Zagadnienia Doradztwa Rolniczego*, no. 2/96, CDiEwR, Poznań.

Kmita, E. (1994). *Agroturystyka jako szansa aktywizacji społeczno-gospodarczej środowisk wiejskich* [Agritourism as an opportunity for socio-economic activation of rural environments]. *Zagadnienia Doradztwa Rolniczego*, no. 2/94, pp. 14–20, Poznań.

Kmita, E. (1997). System kategoryzacji wiejskiej bazy noclegowej w Polsce [System of categorization of rural accommodation base in Poland]. *Zagadnienia Doradztwa Rolniczego*, no. 4/97, CDiEwR, Poznań, pp. 86–102.

Kmita, E. (1999). Polsko-niemiecka współpraca w dziedzinie agroturystyki w latach 1992–1998 [Polish-German cooperation in the field of agritourism in 1992–1998]. *Zagadnienia Doradztwa Rolniczego*, 1 (1), 111–112.

Kmita-Dziasek, E. (2004). Turystyka wiejska w Polsce – od rozproszonych działań do kompleksowej strategii [Rural tourism in Poland – from dispersed activities to a comprehensive strategy]. *Zagadnienia Doradztwa Rolniczego*, 1 (37), pp. 128–137.

Knecht, D. (2009). *Agroturystyka w agrobiznesie* [*Agritourism in agribusiness*]. Wydawnictwo, C.H. Beck, Warszawa.

Kobyłecki, W. (2003). Uwarunkowania prawne organizowania agroturystyki [Legal conditions for organizing agritourism]. In: Mirończuk, A. (ed.), *Turystyka wiejska i agroturystyka. Stan i perspektywy rozwoju*, pp. 117–138. Wydawnictwo Akademii Podlaskiej, Siedlce.

Konkluzje z I Europejskiego Kongresu Turystyki Wiejskiej [*Conclusions from the 1st European Congress of Rural Tourism*] (2003). Spain. Retrieved from: http://www. europeanrtcongress.org/en/conclusions.php (access: 21. 03. 2006).

Kosmaczewska, J. (2007). *Wpływ agroturystyki na rozwój ekonomiczno-społeczny gminy* [*The impact of agritourism on the economic and social development of the commune*]. Bogucki Wydawnictwo Naukowe, Poznań.

Kozak, M. W. (2013). *Barriers to the Development of Agritourism*, Conference Proceedings "Active Countryside Tourism," 23–25 Jan. International Center for Research in Events, Tourism and Hospitality (ICRETH), Leeds Metropolitan University, Leeds, UK.

Kożuchowska, B. (2000). Podstawowe pojęcia, cechy, składniki agroturystyki oraz formy samoorganizacji usługodawców [Basic concepts, features, components of agritourism and forms of self-organization of service providers]. In: Świetlikowska, U. (ed.), *Agroturystyka*, [*Agritourism*]. Fundacja Programów Pomocy dla Rolnictwa (FAPA), Warszawa, pp. 22–26.

Kruczek, Z., & Sacha, S. (1994). *Geografia atrakcji turystycznych Polski* [*Geography of Polish tourist attractions*]. Wydawnictwo Ostoja, Kraków.

Kruczek, Z., & Zmyślony, P. (2010). *Regiony turystyczne* [*Tourist regions*]. Wydawnictwo Proksenia, Kraków, p. 222.

Kruczek, Z., & Zmyślony, P. (2014). *Regiony turystyczne. Podstawy teoretyczne. Studium przypadków* [*Tourist regions. Theoretical basics. Case studies*]. Wydawnictwo Proksenia, Kraków.

Kubal, M., & Mika, K. (2012). Agritourism in Poland – the legal model and the realities of the market. *Current Issues of Tourism Research*, 2 (1), pp. 4–11. Retrieved from: http://bazekon.icm.edu.pl/bazekon/element/bwmeta1.element.ekon-element-00017130 4855 (access: 14. 06. 2020).

Kurtyka, I. (2000). *Gospodarstwa agroturystyczne jako element wielofunkcyjnego rozwoju obszarów wiejskich Sudetów* [*Agritourism Farms as an Element of Multifunctional Development of Rural Areas of the Sudetes*]. Praca doktorska opracowana w Katedrze Ekonomiki i Organizacji Rolnictwa Akademii Rolniczej we Wrocławiu. Wrocław.

Kutkowska, B. (2003). *Podstawy rozwoju agroturystyki ze szczególnym uwzględnieniem agroturystyki na Dolnym Śląsku* [*The Bases of Agrotourism Development with Particular Consideration of Agrotourism in Lower Silesia*]. Wydawnictwo Akademii Rolniczej we Wrocławiu, Wrocław.

Łabaj, M. (1998). *Priorytet strategiczny – rozwój turystyki w gminie Zawoja*[*Strategic priority – development of tourism in the Zawoja commune*]. VI Ogólnopolskie Sympozjum Agroturystyczne "Zrównoważony rozwój turystyki wiejskiej – idee, działania, efekty," Lubniewice, 14–17.09. Centrum Doradztwa i Edukacji w Rolnictwie, Oddział w Krakowie, Kraków, pp. 51–60.

Lane, B. (1994). What is rural tourism? *Journal of Sustainable Tourism*, pp. 7–21.

Łazarek, R. (1972). *Ekonomika i organizacja turystyki* [*Economics and organization of tourism*]. Polskie Wydawnictwo Ekonomiczne, Warszawa.

Leszczycki, S. (1937). *Podhale jako region uzdrowiskowy: rozważania z geografii turyzmu* [*Podhale as a spa region: considerations on the geography of tourism*]. Prace Studium Turyzmu UJ, vol. 1. Wydawnictwo Uniwersytetu Jagiellońskiego w Krakowie, Kraków.

Lijewski, T., Mikułowski, B., & Wyrzykowski, J. (2002). *Geografia turystyki Polski* [*The geography of Polish tourism*]. Polskie Wydawnictwo Ekonomiczne, Warszawa.

Liszewski, S. (2009). Przestrzeń turystyczna Polski. Koncepcja regionalizacji turystycznej [The tourist space of Poland. Tourist regionalization concept]. *Folia Turistica*, no. 21, "Regiony turystyczne," Akademia Wychowania Fizycznego im. Bronisława Czecha w Krakowie, Kraków, pp. 17–30. Retrieved from: http://www.folia-turistica.pl/attachm ents/article/402/FT_21_2009.pdf (access: 15. 07. 2020).

Lisztwan, I. (2000). Możliwości uzyskania wsparcia z programu SAPARD [Possibilities of obtaining support from the SAPARD program]. In: Majewski, J. (ed.), *Agroturystyka w programach pomocowych dla wsi. Materiały konferencyjne. POLAGRA FARM 2000*, pp. 11–20. Międzynarodowe Targi Poznańskie i Akademia Rolnicza w Poznaniu. Poznań.

Łopata, J. (2000). *Ekoturystyka. Urlop u EKOrolników* [*Ecotourism. Vacation with ECO farmers*]. ECEAT-Poland.

Maciąg, J. (1996). Źródła i perspektywy turystyki wiejskiej. Od wywczasów do agroturystyki [Sources and perspectives of rural tourism. From vacation to agritourism]. *Wieś i Rolnictwo*, no. 3, 92.

Majewski, J. (1995). *Turystyka wiejska* [*Rural tourism*]. Wczasy pod gruszą, Poradnik dla rolników rozpoczynających działalność agroturystyczną. Zachodnie Centrum Organizacji, Zielona Góra, pp. 9–40.

Majewski, J. (1997). Turystyka wiejska – przejściowa moda czy rzeczywista szansa wsi? [Rural tourism – a temporary fashion or a real rural opportunity?]. In: Goryńska-Bittner, B., & Kaczmarek, Z. (eds.), *Wielofunkcyjność wsi w warunkach bezrobocia. Zakład Filozofii i Myśli Społecznej, Akademia Rolnicza im. A. Cieszkowskiego*, Poznań, pp. 45–60.

Majewski, J. (2000a). *Agroturystyka to też biznes* [*Agritourism is also a business*]. Fundacja Wspomagania Wsi, Warszawa.

Majewski, J. (2000b). Koncepcja produktu turystycznego jako kryterium wspierania projektów turystyki wiejskiej [The concept of a tourism product as a criterion for supporting rural tourism projects]. In: Majewski, J. (ed.), *Agroturystyka w programach pomocowych dla wsi. Materiały konferencyjne. POLAGRA FARM 2000*, pp. 2–10. Międzynarodowe Targi Poznańskie i Akademia Rolnicza w Poznaniu. Poznań.

Majewski, J. (2003). Turystyka wiejska w programie PHARE TOURIN – kierunki strategiczne i ich realizacja [Rural Tourism in PHARE TOURIN programmes: Strategic Options and their Implementation]. In: *Zeszyty Naukowe Akademii Rolniczej im. H. Kołłątaja w Krakowie. Turystyka wiejska w Polsce – od rozproszonych działań do kompleksowej strategii*. Kraków, pp. 33–48.

Majewski, J. (2005). *Definiowanie terminu agroturystyka – pojęcia wąskie i szerokie* [*Defining the term agritourism – narrow and broad concepts*].

Majewski, J. (2010). *Wiejskość jako rdzeń produktu turystycznego – użyteczność podejść geograficznego i ekonomicznego* [The countryside as the core of the tourist product – utility of geographic and economic approaches]. *Acta Scientiarum Polonorum. Oeconomia,* 9 (4), pp. 287–294.

Majewski, J., & Lane, B. (2001). *Turystyka wiejska i rozwój lokalny* [*Rural tourism and local development*]. Fundacja Fundusz Współpracy, Poznań.

Majewski, J., & Zawistowska, H. (2015). *Konferencja Międzynarodowa – Turystyka Wiejska i Agroturystyka – Nowe Paradygmaty dla XXI w. – sprawozdanie* [*International Conference – Rural Tourism and Agritourism – New Paradigms for the 21st century – report*]. Studia KPZK, no. 162. Polska Akademia Nauk Komitet Przestrzennego Zagospodarowania Kraju, Warszawa. Retrieved from: http://journals.pan. pl/dlibra/publication/112285/edition/97552/content (access: 20. 07. 2020).

Majewski, J., & Zmyślony, P. (2014). Wiejski charakter – podmiejska lokalizacja. Turystyka na obszarze metropolitalnym Poznania [Rural character – suburban location. Rural tourism in Poznań metropolitan area]. *Turystyka i Rekreacja,* 11 (1). Akademia Wychowania Fizycznego i. Józefa Piłsudskiego w Warszawie, Warszawa. Retrieved from: https://www.awf.edu.pl/__data/assets/pdf_file/0005/20696/TiR_11-1_ 2014.pdf (access: 20. 05. 2020).

Marcinkiewicz, C. (2013). Rozwój i stan polskiej agroturystyki [The development and condition of Polish agritourism]. Wyższa Szkoła Humanitas w Sosnowcu, Oficyna Wydawnicza "Humanitas', Sosnowiec. Retrieved from: http://www.sbc.org.pl/dlibra/ docmetadata?id=80828&from=publication (access: 12. 01. 2020).

Marzejon-Frycz, I. (2012). Specyfika agroturystyki Polski na tle wybranych krajów Unii Europejskiej [Specific of Polish agritourism in the context of chosen EU countries]. *Ekonomia. Rynek, Gospodarka, Społeczeństwo,* no. 29, pp. 66–85. Retrieved from: http://ekonomia.wne.uw.edu.pl/ekonomia/getFile/337 (access: 12. 06. 2020).

Marzewski, T. (2012). Turystyka we Wspólnej Polityce Rolnej UE, Studia Periegetica. *Zeszyty Naukowe Wielkopolskiej Wyższej Szkoły Turystyki i Zarządzania w Poznaniu,* no. 7/2012, pp. 9–29.

Matczak, A. (2015). *Ewolucja turystyki na obszarach wiejskich* [*Evolution of Tourism in Rural Areas*]. Studia KPZK, no. 162. Polska Akademia Nauk Komitet Przestrzennego Zagospodarowania Kraju, Warszawa. Retrieved from: http://journals.pan. pl/dlibra/publication/112285/edition/97552/content (access: 20. 07. 2020).

Matczak, A., & Suliborski, A. (1984). Funkcja turystyczna regionu Zbiornika Otmuchowskiego [Tourist function of the Otmuchów Reservoir region]. *Acta Universitatis Lodziensis. Folia Geographica,* no. 3, pp. 99–117.

Matias, A., Nijkamp, P., & Sarmento, M. (eds.), (2011). *Tourism Economics. Impact Analysis.* Physica-Verlag. Berlin–Heidelberg. Retrieved from: https:// books.google.pl/books?id=vzcufw6qcCoC&printsec=frontcover&hl=pl#v=onepag e&q&f=false access: 12. 04. 2020).

Mazurski, K. R. (2009). Region turystyczny jako pojęcie [Tourist region as a concept]. *Folia Turistica,* no. 21, "Regiony turystyczne," pp. 7–15. Akademia Wychowania Fizycznego im. B. Czecha w Krakowie, Kraków. Retrieved from: http://www.folia -turistica.pl/attachments/article/402/FT_21_2009.pdf (access: 21. 06. 2020).

McMahon, F. (1996). Rural and agri-tourism in Central and Eastern Europe. In: Richards, G. (ed.), *Educating for Quality,* Tilburg University Press, pp. 175–182.

Mikuta, B., & Żelazna, K. (2004). *Organizacja ruchu turystycznego na wsi* [*Organization of tourist movement in the countryside*]. Wydawnictwo Format-AB, warszawa, p. 299.

Mileska, M. I. (1963). *Regiony turystyczne Polski. Stan obecny i potencjalne warunki rozwoju* [*Tourist regions of Poland. Current state and potential development conditions*]. Prace Geograficzne no. 43, Instytut Geografii PAN. Państwowe Wydawnictwo Naukowe, Warszawa. Retrieved from: https://rcin.org.pl/Content/16859/WA51_21988_r1963_nr43_Prace-Geogr.pdf (access: 12. 06. 2020).

MRiRW, ARiMR (2001). Działanie w zakresie agroturystyki i turystyki wiejskiej [Action in the field of agritourism and rural tourism]. *Biuletyn Informacyjny* no. 5/2001, Warszawa, pp. 6–12.

MRiRW, ARiMR (2005a). Efekty pomocy kredytowej dla gospodarstw rolnych 2005 [Outcomes of farm loan schemes in 2005]. *Biuletyn Informacyjny* no. 10/2005 (100), Warszawa, pp. 23–25.

MRiRW, ARiMR (2005b). Rozwój agroturystyki na obszarach wiejskich [Agritourism development in rural areas]. *Biuletyn Informacyjny* no. 7–8 (98), Warszawa, pp. 11–18.

Niedziółka, A. (2007). Znaczenie kategoryzacji obiektów noclegowych w działalności agroturystycznej na przykładzie wybranych gmin województwa małopolskiego [The importance of the categorization of accommodation facilities in agritourism activities on the example of selected communes of the Małopolskie Voivodeship]. In: Kurek, W., & Palusiński, R.(eds.). *Studia nad turystyką. Prace ekonomiczne i społeczne*, pp. 123–149. IGiGP UJ, Kraków.

Niedziółka, A. (2017). *Stan i czynniki rozwoju usług agroturystycznych w powiecie nowotarskim w perspektywie wsparcia agroturystyki z programów Unii Europejskiej w latach 2014–2020* [*The condition and development factors of agritourism services in the Nowy Targ county in the perspective of supporting agritourism from the European Union programs in 2014–2020*]. Wyd. Stowarzyszenie Naukowe Instytut Gospodarki i Rynku, Kraków.

Nurzyńska, I. (2009). Wsparcie agroturystyki w ramach instrumentów wdrażanych przez Agencję Restrukturyzacji i Modernizacji Rolnictwa, w: *Perspektywy rozwoju i promocji turystyki wiejskiej i agroturystyki w Polsce*. AGROTRAVEL. Kielce, 17–18 kwietnia.

OECD (1994). *Tourism Strategies and Rural Development*, report. Paris. OECD/GD (94)49.

OECD (2009). *The Role of Agriculture and Farm Household Diversification in the Rural Economy of Poland*. OECD, Trade and Agriculture Directorate. Retrieved from: http://www.oecd.org/tad/agricultural-policies/43245592.pdf (access: 10. 11. 2019).

Pająk, W. (2002). *Portugalski model rozwoju turystyki wiejskiej* [*Portuguese model of rural tourism development*]. Wrocław: U. Wroc. Maszyn., p. 16.

Pałka-Łebek, E. (2016). Współpraca gospodarstw agroturystycznych z lokalnymi stowarzyszeniami wspierającymi rozwój turystyki na przykładzie województwa świętokrzyskiego [Cooperation between agritourism farms and institutions supporting development of tourism in the świętokrzyskie voivodship]. *Studia obszarów wiejskich*, 42, pp. 55–64, http://dx.doi.org/10.7163/SOW.42.4.

Paluszek, J. (2008). Dywersyfikacja wykorzystania zasobów ziemi w gospodarstwach rolnych w Polsce [Diversification of the use of land resources on farms in Poland]. *Roczniki Naukowe Stowarzyszenia Ekonomistów Rolnictwa i Agrobiznesu*, 10(3), pp. 424–430.

Panansiuk, A. (2019). Tourism management by public administration institutions. *Scientific Journal of the Military University of Land Forces*, 51, no. 2 (192), pp. 364–376,

doi:10.5604/01.3001.0013.2610. Retrieved from: http://yadda.icm.edu.pl/baztech/elem
ent/bwmetal.element.baztech-19b030ab-d7d9-42c1-8142-d409cd1c7fb8 (access: 15. 07.
2020).

Passaris, S., Sokólska, J., & Vinaver, K. (2002). *Rozwój obszarów wiejskich i turystyka* [*Rural development and tourism*]. Wydawnictwo Narodowej Fundacji
Ochrony Środowiska, Białystok-Paryż.

PFTW 'Gospodarstwa Gościnne' (2005). Agroturystyka, Wieś polska zaprasza, Fundusze UE. Polska Federacja Turystyki Wiejskiej "Gospodarstwa Gościnne".
Retrieved from: http://www.agroturystyka.pl/index.php?d=0&i=182&inc=cms (access:
21. 11. 2005).

Philip, S., Hunter, C., & Blackstock, K. (2010). A typology for defining agritourism.
Tourism Management, 31 (6), pp. 754–758. Retrieved from: http://www.macaulay.ac.
uk/LADSS/papers/AgritourismTypologyPaper.pdf (access: 12. 03. 2014).

Pieńkos, K., & Tymińska, U. (2013). Baza zakwaterowania w gospodarstwach agroturystycznych jako czynnik rozwoju turystyki na obszarach wiejskich [Accommodation
base in agritourism farms as a factor of tourism development in rural areas]. In:
Sikorska-Wolak, I. (ed.), *Ekonomiczne i społeczne aspekty rozwoju turystyki wiejskiej.*
Wydawnictwo SGGW, pp. 125–134. Retrieved from: http://keekid.wne.sggw.pl/wp-con
tent/uploads/2013/01/Ekonomiczne-i-spo%C5%82eczne-aspekty-rozwoju-turystyki-wie
jskiej.pdf (access: 31. 07. 2020).

Płocka, J. (2009). *Wybrane zagadnienia z zagospodarowania turystycznego* [*Selected
issues of tourism development*]. Centrum Kształcenia Ustawiczego, Toruń.

Pomajda, W. (1996). Działania pomocowe programu PHARE Fundacji Programów
Pomocy dla Rolnictwa (FAPA) na rzecz modernizacji i restrukturyzacji polskiego
rolnictwa [Aid activities of the PHARE program of the Foundation of Assistance
Programs for Agriculture (FAPA) for the modernization and restructuring of Polish
agriculture]. In: Jarosz, A. (ed.), *Agroturystyka jako szansa aktywizacji gospodarczej
wiejskich regionów turystycznych Małopolski Wschodniej*, pp. 151–157. Oficyna
Wydawnicza Politechniki Rzeszowskiej, Rzeszów – Boguchwała.

Pothoff, H. (1993). Trendy w europejskiej turystyce wiejskiej. Gannon, A., Fox, C.,
Nejez, M., Pichler, G., & Pothoff, H. (eds.), *Agroturystyka a rozwój wsi*. CDiEwR,
Kraków.

Przeorek-Smyka, R. (2009). Źródła finansowania aktywności turystycznej na obszarach wiejskich [Sources of financing tourist activity in rural areas]. *Prace Naukowe
Uniwersytetu Ekonomicznego we Wrocławiu* no. 50. Gospodarka turystyczna w
regionie. Przedsiębiorstwo. Samorząd. Współpraca, pp. 203–211.

Przezbórska, L. (1998). Turystyka wiejska czy agroturystyka. Terminologia, formy
turystyki alternatywnej i ich rozwój na przestrzeni wieków [Rural tourism or agritourism. terminology, forms of alternative tourism and their development over the
centuries]. *Zagadnienia Doradztwa Rolniczego*, 1 (16), pp. 65–73.

Przezbórska, L. (2002). Analiza ekonomiczna funkcjonowania gospodarstw zajmujących się turystyką wiejską i agroturystyką w Wielkopolsce oraz perspektywy ich
rozwoju [Economic analysis of the functioning of farms dealing with rural tourism
and agritourism in Greater Poland and the prospects for their development]. Praca
doktorska, tekst niepublikowany.

Przezbórska, L. (2006). Finansowanie agroturystyki i turystyki wiejskiej w okresie przed
i po akcesji Polski do Unii Europejskiej [Financing agritourism and rural tourism in
the period before and after Poland's accession to the European Union]. In: Czternasty
& Sapa (eds.), *Wsparcie finansowe sektora rolno-żywnościowego w Polsce i*

Wielkopolsce z krajowych i unijnych środków budżetowych, Zeszyty Naukowe no. 74, Wydawnictwo Akademii Ekonomicznej w Poznaniu, Poznań, pp. 152–164.

Przezbórska, L. (2007). Determinanty rozwoju agroturystyki w Polsce (na przykładzie wybranych regionów) [Determinants of Agri-tourism Development in Poland (On the Example of Chosen Regions)]. *Acta Scientiarum Polonorum, Oeconomia*, 6(2), pp. 113–121.

Przezbórska, L. (2009). *Rola polityk wspólnotowych w rozwoju turystyki wiejskiej i agroturystyki w Polsce* [The role of the European Union's policies in development of rural tourism and agritourism in Poland]. *Zeszyty Naukowe SGGW w Warszawie Polityki Europejskie, Finanse i Marketing [European Policies, Finance and Marketing]*, no. 2 (51-vol. I), Wydawnictwo SGGW, Warszawa, pp. 191–207.

Przezbórska, L., & Hegarty, C. (2004). Rozwój turystyki wiejskiej i agroturystyki w starych i nowych krajach członkowskich UE – na przykładzie Irlandii i Polski [Development of rural tourism and agritourism in old and new EU member states – on the example of Ireland and Poland]. In: *Szanse i zagrożenia rozwoju polskich obszarów wiejskich w rozszerzonej Unii Europejskiej*, Akademia Rolnicza w Szczecinie, Politechnika Koszalińska, Szczecin, pp. 411–426.

Przezbórska, L., & Lira, J. (2011). Walory środowiska przyrodniczego jako czynnik rozwoju turystyki wiejskiej w Polsce [Natural environment values as a determinant of rural tourism development in Poland]. *Prace Naukowe Uniwersytetu Ekonomicznego we Wrocławiu* no. 166 (Polityka ekonomiczna), pp. 568–579.

Przezbórska-Skobiej, L. (2015a). Polityka rozwoju obszarów wiejskich Unii Europejskiej a rozwój turystyki wiejskiej [European Union rural development policy vs. rural tourism development]. *Studia Komitetu Przestrzennego Zagospodarowania Kraju*, no. 162, pp. 9–25. Polska Akademia, Nauk Komitet Przestrzennego Zagospodarowania Kraju. Warszawa. Retrieved from: http://journals.pan.pl/dlibra/publication/112280/edition/97547/content (access: 20. 06. 2020).

Przezbórska-Skobiej, L. (2015b). *Uwarunkowania rozwoju turystyki wiejskiej w Polsce. Analiza regionalna, subregionalna i lokalna [Determinants of the Development of Rural Tourism in Poland – a Regional, Sub-regional and Local Analysis]* Wydawnictwo Uniwersytetu Przyrodniczego w Poznaniu, Poznań.

Przezbórska-Skobiej, L., & Lira, J. (2012). Przestrzeń agroturystyczna Polski i ocena jej atrakcyjności [Agritourism space of Poland and its valuation]. *Prace Naukowe Uniwersytetu Ekonomicznego we Wrocławiu* 242 (19), pp. 637–645.

Przezbórska-Skobiej, L., & Sobotka, S. (2016). Propozycja delimitacji regionów agroturystycznych w Polsce [Proposal of delimitation of agritourism regions in Poland]. *Wieś i Rolnictwo*, no. 2 (171), Instytut Rozwoju Wsi i Rolnictwa Polskiej Akademii Nauk. Warszawa, pp. 173–197. Retrieved from: http://kwartalnik.irwirpan.waw.pl/dir_upload/photo/3cdb89565df0f771326ac26cbc68.pdf (access: 12. 03. 2020).

Raciborski, J. (2000). Wymagania prawne w turystyce wiejskiej. In: Świetlikowska, U. (ed.), *Agroturystyka [Agritourism]*. Fundacja Programów Pomocy dla Rolnictwa (FAPA), Warszawa, pp. 447–465.

Raciborski, J. (2003). Prawo finansowe – uwarunkowania rozwoju turystyki wiejskiej. In: Burzyński, T., & Łabaj, M.(eds.), *Turystyka rekreacyjna oraz turystyka specjalistyczna*, pp. 99–112. ARiMR, MRiRW, FAPA, Warszawa.

Raciborski, J. (2005). Świadczenie usług turystycznych na wsi w świetle przepisów prawnych Polski i Unii Europejskiej. Kmita-Dziasek, E. (ed.), *Prawno-finansowe uwarunkowania prowadzenia usług turystycznych na polskiej wsi po akcesji do Unii*

Europejskiej. Poradnik praktyczny, pp. 9–22. Centrum Doradztwa Rolniczego w Brwinowie – Oddział w Krakowie, Kraków.

Raciborski, J., Turlejska, H., Zielińska, R., Kmita-Dziasek, E., Śliz, J., Rzutki, M., & Jasiński, J. (2005). *Prawno-finansowe uwarunkowania prowadzenia usług turystycznych na polskiej wsi po akcesji do UE. Poradnik praktyczny [Legal and financial conditions for running tourist services in the Polish countryside after accession to the EU. Practical Guide]*. Centrum Doradztwa Rolniczego w Brwinowie, Oddział w Krakowie, Kraków.

Roberts, L., & Hall, D. (2001). *Rural Tourism and Recreation: Principles to Practice.* CABI Publishing, Wallingford, Oxon, UK.

Roman, M. (2014). The organisational and legal aspects of agritourism activities in Poland. *Problemy Drobnych Gospodarstw Rolnych [Problems of Small Agricultural Holdings]*, no. 1/2014, pp. 69–77. Retrieved from: http://yadda.icm.edu.pl/yadda/element/bwmeta1.element.ekon-element-000171461326 (access: 27. 06. 2020).

Roman, M. (2018). *Innowacyjność agroturystyki jako czynnik poprawy konkurencyjności turystycznej makroregionu Polski Wschodniej [Innovativeness of agritourism as a factor of improving the tourist competitiveness of the macroregion of Eastern Poland]*. Wydawnictwo SGGW, Warszawa.

Roman, M., & Niedziółka, A. (2017). *Agroturystyka jako forma przedsiębiorczości na obszarach wiejskich [Agritourism as a form of entrepreneurship in rural areas]*. Wydawnictwo Szkoły Głównej Gospodarstw Wiejskiego w Warszawie, Warszawa.

Rudnicki, R., & Biczkowski, M. (2015). Agroturystyka jako forma aktywizacji pozarolniczej działalności gospodarstw rolnych w Polsce – stan, zróżnicowanie przestrzenne oraz wpływ środków PROW 2007–2013 [Agritourism as a form of non-agricultural business activity stimulation in agricultural holdings in Poland – quantitative approach, spatial diversification analysis and significance of the European Agricultural Fund for Rural Development 2007–2013]. *Studia Komitetu Przestrzennego Zagospodarowania Kraju*, no. 162, pp. 83–108. Polska Akademia, Nauk Komitet Przestrzennego Zagospodarowania Kraju. Warszawa. Retrieved from: http://journals.pan.pl/dlibra/publication/112287/edition/97556/content (access: 20. 06. 2020).

Ryżak, K. (2000). *Polska Federacja Turystyki Wiejskiej "Gospodarstwa Gościnne' – Efekt instytucjonalizacji ruchu agroturystycznego [the Polish Federation of Rural Tourism "Hospitable Farms' – The effect of institutionalization of the agritourism movement]*. Praca magisterska napisana pod kierunkiem prof. dr hab. Piotra Hübnera, Uniwersytet. M. Kopernika w Toruniu, Instytut Socjologii, Toruń.

Saja, A. (2009). *Rozwój bazy noclegowej jako warunek rozwoju turystycznego regionów, Inwestycje w turystyce[Development of accommodation facilities as a condition for the tourist development of regions, Investments in tourism]*. Konferencja naukowa, 23.10.Polska Organizacja Turystyczna, Poznań. Retrieved from: http://pot.gov.pl. pot.potsite.pl/do-pobrania/materialy-szkoleniowe/ (access: 12. 03. 2020).

Scott, D. L. (2003). *Wall Street Words.* Boston, MA: Houghton Mifflin Co. p. 130.

Sharpley, R., & Sharpley, J. (1997). *Rural Tourism. An Introduction.* International Thompson Business Press, London.

Siemiński, P., & Poczta, W. (2014). *Możliwości rozwoju agroturystyki i turystyki wiejskiej w ramach PROW 2007–2013 i 2014–2020* [Possibilities of the development of Agro-Tourism and rural tourism under the RDP 2007–2013 and 2014–2020]. *Annals of the Polish Association of Agricultural and Agribusiness Economists (PAAAE)* 16 (6): 432–437.

Sikora, J. (1995). *Agroturystyczny biznes [Agritourism business]*. Poradnik organizatora turystyki. Oficyna Wydawnicza Ośrodka Postępu Organizacyjnego, Bydgoszcz.

Sikora, J. (1999). *Organizacja ruchu turystycznego na wsi* [*Organization of tourist traffic in the countryside*]. Wydawnictwa Szkolne i Pedagogiczne, S. A., Warszawa.

Sikora, J. (2012). *Agroturystyka. Przedsiębiorczość na obszarach wiejskich* [*Agritourism. Entrepreneurship in rural areas*].Wydawnictwo, C. H. Beck, Warszawa.

Sikora, J., & Jęczmyk, A. (2005). Czynniki wspierające i bariery ograniczające rozwój agroturystyki [Supporting factors and barriers limiting the development of agritourism]. In: Sawicki, B., & Bergier, J. (eds.), *Uwarunkowania rozwoju turystyki związanej z obszarami wiejskimi*. Wydawnictwo Państwowej Wyższej Szkoły Zawodowej im. Papieża Jana Pawła II, Biała Podlaska, pp. 37–45.

Skórnicki, H. (2000). *Fundusze przedakcesyjne SAPARD* [*SAPARD pre-accession funds*]. Prawo rolne Unii Europejskiej a polski sektor rolny, FAPA, Warszawa.

Słownik Języka Polskiego PWN [*PWN Polish Language Dictionary*]. Wydawnictwo Naukowe PWN, Warszawa. Retrieved from: http://sjp.pwn.pl/.

Smoleń, D. (2000). Jak uzyskać pomoc na rozwój agroturystyki z funduszy przedakcesyjnych Unii Europejskiej? (PHARE) [How to get aid for the development of agritourism from the pre-accession funds of the European Union? (PHARE)]. In: Majewski, J. (ed.), *Agroturystyka w programach pomocowych dla wsi. Materiały konferencyjne. POLAGRA FARM 2000*, pp. 41–50. Międzynarodowe Targi Poznańskie i Akademia Rolnicza w Poznaniu. Poznań.

Sobiecki, M. (2008). Wdrażanie SPO ROL na przykładzie działania 2.3 "Odnowa wsi oraz zachowanie i ochrona dziedzictwa kulturowego" [Implementation of SPO ROL on the Example of the 2.3 Measure "Development of Rural Regions and Preservation and Protection of Cultural Heritage']. *Prace Naukowe Uniwersytetu Ekonomicznego we Wrocławiu, Spójność społeczna, gospodarcza i terytorialna w polityce Unii Europejskiej*, no. 21, pp. 302–315.

Stańko, S. (2009). Wspólna Polityka Rolna, jej reformy i perspektywy rozwoju podstawowych rynków rolnych w Polsce. In: Adamowicz, M. (ed.), *Wspólna Polityka Rolna Unii Europejskiej. Uwarunkowania, mechanizmy, efekty*, Wydawnictwo SGGW, pp. 51–67.

Strategia rozwoju obszarów wiejskich i rolnictwa na lata 2007–2013 (z elementami prognozy do roku 2020) [*Strategy for the development of rural areas and agriculture for the years 2007–2013 (with elements of forecasts until 2020)*] (2004). Ministerstwo Rolnictwa i Rozwoju Wsi, Warszawa.

Strzelczak, L., & Cieślak, J. (1995). *Organizacja usług agroturystycznych w wybranych gospodarstwach na terenie Bieszczadów* [*Organization of agritourism services in selected farms in the Bieszczady Mountains*]. Ekologiczne i ekonomiczne uwarunkowania rozwoju gospodarczego Karpat Południowo-Wschodnich. Materiały Międzynarodowej Konferencji Naukowej "Bieszczady', 25–29 maja 1995, pp. 322–326. Narodowy Fundusz Ochrony Środowiska i Gospodarki Wodnej, Akademia Rolnicza im. H. Kołłątaja w Krakowie, Urząd Wojewódzki w Krakowie, Kraków.

Strzembicki, L. (1997). Uwarunkowania rozwoju agroturystyki w Polsce [Conditions for the development of agritourism in Poland]. In: Harasymowicz, S., & Nałęcka, D. (eds.), *Agroturystyka w społeczno-ekonomicznym rozwoju środowiska wiejskiego. Materiały z ogólnopolskiej konferencji naukowo-metodycznej w Białej Podlaskiej, 23–24 maja 1997*, pp. 12–15. Polskie Towarzystwo Naukowe Kultury Fizycznej, Biała Podlaska.

Strzembicki, L. (2003). Instytucjonalne aspekty zarządzania turystyką wiejską w Polsce. Podstawowe problemy i tendencje [Institutional aspects of rural tourism management in Poland. Basic problems and trends]. *Zeszyty Naukowe Akademii Rolniczej w Krakowie*, z. 90. Kraków, pp. 19–31.

Strzembicki, L. (2009). Uwarunkowania rozwoju agroturystyki i usług towarzyszących na obszarach wiejskich [Conditions for the development of agritourism and accompanying services in rural areas]. Retrieved from: www.rcie.lodz.pl (access: 26. 05. 2020).

Strzembicki, L., & Kmita, E. (1994). Agroturystyka formą przedsiębiorczości ludności rolniczej w Polsce. Alternatywne źródła dochodów ludności wiejskiej [Agritourism as a form of entrepreneurship of the agricultural population in Poland. Alternative sources of income for the rural population.]. *Biuletyn Regionalny Zakładu Doradztwa Rolniczego*, no. 309. Akademia Rolnicza w Krakowie, Kraków, p. 14.

Surdacka, E. (2017). Pojęcie i geneza rozwoju agroturystyki [Concept and development genesis of agritourism]. *Autobusy: technika, eksploatacja, systemy transportowe*, 18, no. 6, pp. 1778–1783, CD.

Świetlikowska, U. (ed.), (1998). *Agroturystyka [Agritourism]*. Wydawnictwo SGGW, Warszawa.

Sznajder, M., & Przezbórska, L. (2006). *Agroturystyka [Agritourism]*. Polskie Wydawnictwo Ekonomiczne. Warszawa.

Sznajder, M., Przezbórska, L., & Scrimgeour, F. (2009). *Agritourism*. CAB Publishing, Wallingford, UK.

Szwichtenberg, A. (1998). Pojmowanie turystyki wiejskiej w Polsce i na świecie [Understanding rural tourism in Poland and in the world].*Turyzm* 8, issue 1, pp. 29–37.

Szymla, Z. (2000). *Determinanty rozwoju regionalnego [Determinants of regional development]*. Zakład Narodowy im. Ossolińskich, Wrocław.

Talarowska, M. (2005). *Rozwój agroturystyki w Wielkopolsce w latach 1990–2003 (na przykładzie wybranych gospodarstw) [Development of agritourism in Wielkopolska in 1990–2003 (on the example of selected farms]*. Praca magisterska napisana pod kierunkiem naukowym dr, L. Przezbórskiej. Katedra Ekonomiki Gospodarki Żywnościowej. Akademia Rolnicza w Poznaniu. Poznań.

Trzpioła, K. (2008). Agroturystyka czy działalność gospodarcza na terenie wiejskim – dylematy przedsiębiorcy ubiegającego się o wsparcie unijne na przykładzie uczestniczek projektu Nowe Kwalifi kacje Kobiet [Agritourism or registered business in rural areas: dilemma of entrepreneur applying for EU funds on the example of participants of the project 'Women's New Qualifications']. *Zeszyty Naukowe SGGW. Ekonomika i Organizacja Gospodarki Żywnościowej* no. 65, pp. 219–230.

Tyran, E. (2005). Uwarunkowania działalności agroturystycznej [Determinants of agritourism activity]. In: Musiał, W., Kania, J., & Leśniak, L.(eds.), *Agroturystyka i usługi towarzyszące*, pp. 44–48. Małopolskie Stowarzyszenie Doradztwa Rolniczego z s. W Akademii Rolniczej w Krakowie, Kraków.

Tyran, E. (2010). Dywersyfikacja jako proces dostosowawczy gospodarstw województwa małopolskiego [Diversification as a process of adapting farms in the Małopolskie Province]. *Roczniki Nauk Rolniczych Seria G Ekonomika Rolnictwa*, 97 (4), pp. 200–209.

UBC Sauder School of Business(n.d.). A brief history of the ECU, the predecessor of the euro. Retrieved from: http://fx.sauder.ubc.ca/ECU.html (access: 25. 04. 2020).

Urry, J. (2000). *Sociology beyond Societies, Mobilities for the Twenty First Century*. Routledge, London.

Ustawa z dnia 23 kwietnia 1964 r. – Kodeks cywilny. *Dz.U.* 1964 no. 16 poz. 93., Retrieved from: https://isap.sejm.gov.pl/isap.nsf/DocDetails.xsp?id=WDU19640160093 (access: 25. 06. 2020).

Ustawa z dnia 15 listopada 1984 r. o podatku rolnym. *Dz. U.* 1984 no. 52 poz. 268., Retrieved from: https://isap.sejm.gov.pl/isap.nsf/DocDetails.xsp?id=WDU19840520268 (access: 25. 06. 2020).

Ustawa z dnia 7 kwietnia 1989 r. Prawo o stowarzyszeniach. *Dz. U.* 1989 no. 20 poz. 104.

Veer, M., & Tuunter, E. (2005). Rural tourism in Europe. An Explanation of success and failure factors. Stichting Recreatie Expert and Innovation Centre, Hague. Retrieved from: www.stichtingrecreatie.agro.nl (access: 22. 02. 2020).

Walasek, J., & Embacher, H. (2005). Agroturystyka – doświadczenia europejskie. Polityka turystyczna Unii Europejskiej. In: Musiał, W., Kania, J., & Leśniak, L. (eds.), *Agroturystyka i usługi towarzyszące*, MSDR z s. w AR w Krakowie, Kraków, pp. 79–106.

Warszyńska, J. (1985). Funkcja turystyczna Karpat Polskich [Tourist function of the Polish Carpathians]. *Folia Geographica, Ser. Geogr.-Oecon.*, 18, pp. 79–104.

Wiatrak, A. P. (1995). Czynniki określające rozwój turystyki na obszarach wiejskich [Factors Determining the Development of Tourism in Rural Areas]. *Problemy Turystyki*, 1/2 (67/68), pp. 15–24.

Wiatrak, A. P. (1996). Wpływ agroturystyki na zagospodarowanie obszarów wiejskich [The Impact of Agritourism on the Development of Rural Areas]. *Zagadnienia Ekonomiki Rolnej*, 1 (252), pp. 35–45.

Wiatrak, A. P. (2009). Wspomaganie instytucjonalne rozwoju turystyki wiejskiej [Institutional suport of development of rural tourism]. *Roczniki Naukowe Stowarzyszenia Ekonomistów Rolnictwa i Agrobiznesu*, 11 (4), pp. 347–351. Retrieved from: http://agro.icm.edu.pl/agro/element/bwmeta1.element.dl-catalog-9f0acbb5-7d7c-42fd-a90c-5fee5a93435c (access: 10. 05. 2020).

Wiatrak, T. (2014). Wspieranie rozwoju turystyki wiejskiej przez PROW 2007–2013 [Supporting the Development of Rural Tourism by RDP 2007–2013]. In: Jastrzębski, C. (ed.), *Infrastruktura okołoturystyczna jako element wzbogacający ofertę obszarów wiejskich*. Wydawnictwo Wyższa Szkoła Ekonomii, Prawa i Nauk Medycznych im. prof. Edwarda Lipińskiego w Kielcach, Kielce, pp. 19–31.

Wicks, B.E. & Merrett, C.D. (2003). Agritourism: An economic opportunity for Illinois. *Rural Research Report*, 14(9), pp. 1–8.

Więckowski, M. (2014). Tourism Space: An Attempt at a Fresh Look. *Turyzm*, 24(1), pp. 17–24. DOI: doi:10.2478/tour-2014-0002.

Wilkin, J. (2002). *Budowa instytucji wspierających rozwój wsi i rolnictwa w kontekście integracji Polski z Unią Europejską* [*Building institutions supporting the development of rural areas and agriculture in the context of Poland's integration with the European Union*]. Wieú i rolnictwo – perspektywy rozwoju. Instytut Ekonomiki Rolnictwa i Gospodarki Żywnościowej, Warszawa, pp. 199–205.

Wojciechowska, J. (2005). Efekty rozwoju agroturystyki – koncepcja modelu [The Effects of Agritourism Development – Model Concept]. In: Sawicki, B., & Bergier, J. (eds.), *Uwarunkowania rozwoju turystyki związanej z obszarami wiejskimi*. Wydawnictwo Państwowej Wyższej Szkoły Zawodowej w Białej Podlasce, Biała Podlaska, pp. 93–96.

Wojciechowska, J. (2006). Geneza oraz ewolucja turystyki na obszarach wiejskich w Polsce [Tourism in rural areas in Poland – the origins and evolution]. *Folia Turistica*, no. 17, pp. 99–119.

Wojciechowska, J. (2007a). Systematyka i wyróżniki pojęć dotyczących turystyki na obszarach wiejskich [Systematics and distinguishing features of terms related to tourism in rural areas]. In: Kurek, W., & Mika, M. (eds.), *Studia nad turystyką. Tradycje, stan obecny i perspektywy, badawcze. Geograficzne, społeczne i ekonomiczne aspekty turystyki.* Instytut Geografii i Gospodarki Przestrzennej Uniwersytetu Jagiellońskiego, Kraków, pp. 259–308.

Wojciechowska, J. (2007b). Formy i mechanizmy działań podmiotów wspierających rozwój agroturystyki [Forms and mechanisms of activities of entities supporting the development of agritourism]. In: Sikora, J. (ed.), *Turystyka wiejska a edukacja – różne poziomy, różne wymiary.* Wydawnictwo Akademii Rolniczej im. Augusta Cieszkowskiegow Poznaniu, pp. 30–38.

Wojciechowska, J. (2009). *Procesy i uwarunkowania rozwoju agroturystyki w Polsce [Processes and conditions for the development of agritourism in Poland].* Rozprawy habilitacyjne Uniwersytetu Łódzkiego. Wydawnictwo Uniwersytetu Łódzkiego, Łódź, p. 167.

Wojciechowska, J. (2012). Ścieżki rozwoju organizacyjnego turystyki w Polsce – od rewolucyjnego po ewolucyjny system [The paths of the organizational development of tourism in Poland – from revolutionary to evolutionary]. In: Rapacz, A. (ed.), *Wyzwania współczesnej polityki turystycznej. Problemy funkcjonowania rynku turystycznego.* Wydawnictwo Uniwersytetu Ekonomicznego we Wrocławiu, Wrocław, pp. 102–111. Retrieved from: https://www.dbc.wroc.pl/Content/18620/download/ (access: 10. 07. 2020).

Wyrwicz, E. (1996). *Ekonomiczne funkcje turystyki [Economic functions of tourism].* Turystyka na wsi. Dodatek do tygodnika Chłopska Droga.

Wysocka, I. (1975). Funkcje turystyczne gmin w świetle wielkości i rodzaju bazy noclegowej [Tourist functions of communes in the light of the size and type of accommodation base]. *Biuletyn Informacyjny Instytutu Turystyki,* no. 4/5, Warszawa.

Yeoman, J. (2000). The Importance of Rural Tourism and Agritourism in Rural Development. *Roczniki Naukowe Stowarzyszenia Ekonomistów Rolnych i Agrobiznesu,* II (1), pp. 41–51.

Zangari, G. (2003). *Dokumenty UE dotyczące turystyki wiejskiej [EU documents on rural tourism].* Turystyka rekreacyjna oraz turystyka specjalistyczna. ARiMR, MRiRW, FAPA – Projekt współfinansowany ze środków UE w ramach programu SAPARD, przeznaczonych na działanie 6 obejmujące Szkolenia zawodowe, wdrażane przez FAPA w imieniu i na rzecz MRiRW, Warszawa, pp. 21–33.

Zaręba, D. (2000). *Ekoturystyka [Ecotourism].* Wyzwania i nadzieje. Wydawnictwo Naukowe PWN, Warszawa.

Zawadka, J. (2010a). *Ekonomiczno-społeczne determinanty rozwoju agroturystyki na Lubelszczyźnie (na przykładzie wybranych gmin wiejskich) [Economic and Social determinants of agritourism development in the Lublin region (on the example of selected rural communes)].* Wyd. SGGW, Warszawa. Retrieved from: http://keekid. wne.sggw.pl/wp-content/uploads/2013/01/Zawadka-J.-Ekonomiczno-spo%C5%82ecz ne-determinanty-rozwoju-agroturystyki-na-Lubelszczy%C5%BAnie-na-przyk%C5% 82adzie-wybranych-gmin-wiejskich.pdf (access: 15. 06. 2020).

Zawadka, J. (2010b). Ewolucja działalności agroturystycznej w Polsce i typologia wiejskich gospodarstw turystycznych [Evolution of agritouristic activity in Poland and typology of rural tourism farms]. *Acta Scientiarium Polonorum. Oeconomia,* 9 (4), pp. 627–638. Retrieved from: https://js.wne.sggw.pl/index.php/aspe/article/view/ 3988/3514 (access: 23. 06. 2020).

Żelazna, K. (2004). Podstawy prowadzenia działalności turystycznej na obszarach wiejskich. Turystyka wiejska a agroturystyka [Basics of running a tourist activity in rural areas. Rural tourism and agritourism]. In: Mikuta, B. (ed.), *Organizacja ruchu turystycznego,* pp. 108–110. Format AB, Warszawa.

4 Evolution of rural tourism in Poland, 1989–2019

Joanna Kosmaczewska

This chapter presents the changes in tourism service delivery in Polish rural areas between 1989 and 2019. The shift from a centrally planned economy to free market mechanisms, the liquidation of state-owned agricultural holdings, the agrarian fragmentation, and capital deficiencies facing rural areas are the reasons why the development of the tourism function (in consonance with the multipurpose development concept) is viewed as one possible remedy to the progressive underdevelopment and marginalization of rural areas. This chapter includes a review of evaluation of research projects addressing the development of the rural tourism function and the residents' attitude to tourists. Also, it describes changes in how rural tourism/agritourism staff are trained, and in how tourism products are developed based on regional and traditional food. Moreover, the author used the Hellwig's development pattern method to evaluate—and present the territorial differences in—the development of the tourism function in Polish rural areas.

4.1 Tourism function of Polish rural areas: research overview

The shift away from a centrally planned economy towards free market mechanisms triggered a series of socioeconomic transformations in Poland. On the one hand, there was development of entrepreneurship and empowerment of local government and local communities. On the other hand, there was inflation, corporate bankruptcies, and unemployment as the adverse consequences of the transformation. The liquidation of state-owned agricultural holdings, agrarian fragmentation, and capital deficiencies resulted in rural pauperization and backwardness. Therefore, in accordance with the multi-purpose development concept, the development of the tourism function was viewed as one possible antidote to the progressing economic underdevelopment and marginalization of rural areas. Local government units at all levels (voivodeship, district, commune) were provided with legal and financial instruments, enabling them to have a considerable impact on the structure, pace, and targets of changes taking place in their respective territories. Usually, their activities are centered around similar goals, such as a long-term increase in global incomes of residents and communal budgets, and the development of enterprise based on a rational use of existing resources. However, the effects of these measures are determined not only by the efficiency of local government units themselves, but also by differences in available resources which largely result from the territories being part of different states in the era of partitions. The differences which date back to those times had an impact on the type and extent of consequences of the clash of specific social groups and economic operators with the realities of a liberal market economy, and seem to be of a persistent nature (Bański & Stola, 2002). Therefore, the differences in the development of the tourism function in Polish rural areas do not depend solely on natural resources, and continue to be an interesting subject of scientific research to representatives of different fields of science.

The change in measures taken to implement the tourism function in Polish rural areas, as witnessed during 1989–2019, will be presented using bibliometric analysis and content analysis methods. The first step consisted in a bibliometric analysis based on Google Scholar (GS) and Web of Science (WoS), the most widely known scholarly search engines. The preliminary analysis of keywords found a considerable number of papers not related to the tourism function in Polish rural areas. Hence, the results were narrowed down to papers which included the relevant keywords in their title. Also, when it comes to Google Scholar, separate analyses were carried out for papers written in English and Polish.

The quantitative analysis of literature shows a flagrant disproportion between the number of available papers written entirely in Polish and that of papers written entirely in English. Another noticeable aspect (especially in the first two decades after Poland entered the realities of a free market economy) is the disproportion between the number of agritourism papers and that of rural tourism papers. It seems quite unusual because rural tourism has a broader

Table 4.1 Bibliometric analysis of available literature resources from 1989 to 2019[a]

Key words	GS 1989–1999		WoS 1989–1999	GS 2000–2009		WoS 2000–2009	GS 2010–2019		WoS 2010–2019
	P	E	E	P	E	E	P	E	E
[rural tourism] +Poland	12	3	0	68	10	3	116	26	16
[agritourism/ agroturism] +Poland	27	1	0	136	9	1	151	49	26
tourism+Poland	3	21	13	39	66	17	53	310	94
tourism	42	11 200	2773	300	43 700	6144	2708	49 460	25469

Source: Own elaboration based on Google Scholar and Web of Science [April 9, 2020].

a When searching for papers written in Polish, the declension of keywords was taken into account and the word "Poland" was skipped because it resulted in the results being excessively narrowed down. Erroneous records, if any, were removed manually.

meaning. The reasons for this state of affairs should be sought in the socialist era when organized forms of rural holidays were referred to as a "summer holiday stay" or "camping holidays" (promoted by the Central Physical Culture Committee and the National Paid-Leave Fund). Following the political and economic transformation, these activities turned into individual trips referred to as agritourism. Note also that at the early development stage of the tourism function of Polish rural areas, "rural tourism" and "agritourism" were often used as synonyms (Majewski, 1994; Dębniewska & Suchta, 1996).

In the first post-transformation decade (1989–1999), papers on the development of tourism functions in Polish rural areas emphasized the role played in that period by agricultural consultancy centers (Boczar-Rozewicz, 1997; Woźniak et al., 1998; Jaźwińska, 1999; Wiśniewska, 1998) and government institutions which established a preferential legal and financial framework (Patura, 1995; Raciborski, 1997; Gąsiorowska, 1996; Wiatrak, 1997) making it easier to pursue an agritourism activity (including dedicated credit facilities and the exemption from income tax for owners who rent up to five rooms).[1] That period was also marked by numerous attempts to define rural tourism and agritourism in the Polish realities using terms adapted from papers written in English (Davies & Gilbert, 1992; Lane, 1992) or new nomenclature developed by the authors, mainly based on the territorial and functional scope of rural areas (Drzewiecki, 1995, 1998; Szwichtenberg, 1998) and on the environmental impacts of tourism (Bogucki et al., 1995). "Deep countryside," a term that emerged somehow in contrast to suburban, urbanized, and industrialized rural areas, means rural space particularly suitable for the development of the tourism function (Grolleau & Ramus, 1986; Drzewiecki, 1998). While having a good environmental status and an attractive, harmonious cultural landscape with some regional and folkloristic features, it is visited by a limited number of tourists. Another term was also introduced, namely the rural recreational space (Drzewiecki, 1992) which is dominated by agricultural economy and offers land-use patterns and natural values that are conducive to leisure activities. The author evaluates the quality of rural recreational space using metrics such as:

- population density;
- share of individual farms in the area of agricultural land;
- share of grassland and pasture in the area of agricultural land;
- share of woods and water bodies in the total area of a commune;
- types of rural settlement;
- share of people who earn their living from non-agricultural activities (Drzewiecki, 1992).

The valuation of attractiveness of all Polish communes, as carried out by Drzewiecki (1992), found that 66% of the national territory (1369 communes) had adequate tourism assets to locally develop rural tourism and agritourism. The score method he used became a basis for assessing the attractiveness of rural areas at regional and local levels as part of studies carried out by many authors. Sikora (1999) used it to identify seven regions of Polish rural tourism (Pomerania, Masuria, Greater Poland, Central Poland, Eastern Poland, Sudetes, and Carpathians). Other papers published at that time analyzed the potential of rural areas to develop the tourism function based on the available infrastructure (Niedzwiecka-Filipiak & Potyrala, 1996; Jagusiewicz, 1999) or natural and cultural assets (Rodacka, 1977; Stepaniuk, 1999; Matusek-Psiuk, 1999). These issues were usually addressed at a regional or local level. Considering the nature of tourism activities and their impact on geographic and social features, Liszewski (1995) proposed five separate types of tourism space (the exploration space; the penetration space; the assimilation space; the colonization space; and the urbanization space). He believes that the implementation of tourism functions in the form of agritourism and rural tourism matched the "tourism assimilation" pattern in the first post-transformation decade in Poland. This means that the tourism function was delivered in consonance with the local natural and cultural resources while often enhancing them with new value (Liszewski, 1995). However, that period was also marked by progressing rural colonization. This is especially true for coastal areas (Szwichtenberg, 1991; Dziegięć, 1995) and upland areas (Kurek, 1990, 1995, 1996; Dziegięć, 1995) which became home to large accommodation facilities, such as guest houses or holiday youth centers. The second model of rural colonization by tourists is found in suburban areas and around water bodies where people build what are referred to as second houses (Liszewski, 1995; Kowalczyk, 1994). Rural transformation processes driven by tourism urbanization are also analyzed in terms of changes in settlement patterns, population density, demographic and professional structure, land use, technical infrastructure, and economic situation of rural residents (Dębniewska & Suchta, 1995; Kurek, 1995; Dąbrowska, 1996; Szymańska, 1996; Sikora, 1998; Wiatrak, 1996; Dziegięć, 1995). There are calls for making the tourism function development only a part of multi-purpose rural development, and making it compliant with sustainable development guidelines (Szwichtenberg, 1996; Suchta & Koczowski, 1992). Some papers written in Polish are available which present the development of the tourism function in Western European rural areas; this is viewed as a kind of good practice and a point of reference (Targosz, 1997; Pastuszek, 1994; Snopek, 1996; Kraszewska, 1993; Maton, 1999). Also, the first studies were published addressing the behavior of buyers of rural tourism services in Poland (Strzembicki, 1999) and the way rural tourism is viewed by the population of host locations (Jalinik, 1998).

English-language literature published at that time includes analyses of changes driven by rural tourism development in the context of multi-purpose development and of sustainable development (Augustyn, 1998; Kaleta, 1994; Kowalczyk, 1996), as well as some comparisons between processes taking place in different Central and Eastern European (CEE) countries (McMahon, 1996). Also, case studies are available that illustrate the transformation triggered by tourism development in selected regions: Polish Carpathians (Kurek, 1996) and Lower Silesia (Grykień, 1998).

Table 4.2 Content analysis of publications available in the first post-transformation decade (1989–1999)

Topic[a]	Papers in English with a Polish context	Papers in Polish
1989–1999		
Defining rural tourism and agritourism (and its forms) in Poland	Davies & Gilbert, 1992; Lane, 1992; Lane, 1994; Wolańska & Lisowska, 1998	Drzewiecki, 1995, 1998; Szwichtenberg, 1991, 1993, 1996, 1998; Bogucki et al., 1995; Szwichtenberg, 1993, 1996
Sustainability / multi-purpose rural development	Augustyn, 1998; Kaleta, 1994; Kowalczyk, 1996	Suchta & Koczowski, 1992
Comparisons in the Central and Eastern European context	McMahon, 1996	–
Role of consultancy in rural tourism/agritourism	–	Boczar-Rozewicz, 1997; Woźniak et al., 1998; Jaźwińska, 1999; Wiśniewska, 1998; Lewczuk et al., 1995
Legal and financial aspects of agritourism activities	–	Patura, 1995; Raciborski, 1997; Gąsiorowska, 1996; Wiatrak, 1997
Tourism vs. land-use and socio-economic transformation of rural areas	–	Ginda, 1995; Dębniewska & Suchta, 1996; Dąbrowska, 1996; Dziegięć, 1990, 1992, 1995; Drzewiecki, 1994
Good practices and examples from European countries (UK, Denmark, France, Belgium)	–	Targosz, 1997; Pastuszek, 1994; Snopek, 1996; Kraszewska, 1993; Maton, 1999

Source: Own elaboration.

a Selected topics and selected examples of papers.

In this literature review, the pre-accession period is somehow a bridge between the first (1989–1999) and the second (2000–2009) decade of changes that followed the transformation. At that time, some publications emerged addressing the financing for rural tourism activities (Wiatrak, 1997; Tyran, 2005; Niedziółka, 2010; Przezbórska, 2002; Majewski, 2003). Also, branded rural tourism products were created (Sawicki, 2009; Czapiewska, 2009; Niedziółka, 2010; Tyran, 2009; Perepeczko, 2009) and improvements could be seen in the quality of rural tourism services thanks to support available under programs such as PHARE and SAPARD. This was manifested in the introduction of a voluntary categorization scheme, in improved equipment of agritourism farms

(Kraszewska, 1994; Sokolowicz & Krawczyk, 2010; Sammel & Dańczak, 2002; Sokół, 2009), and in the accommodation providers having a greater understanding of tourist expectations (Kosmaczewska & Górka, 2009; Brodziński, 2006; Strzembicki, 2002, 2005; Żelazna & Woźniczko, 2005). Continued efforts were taken to delimit rural areas according to how much they are suitable for the development of various forms of tourism (Wiatrak, 2008; Wojciechowska, 2009; Uglis & Jęczmyk, 2009; Balińska & Sikorska-Wolak, 2009; Karbowiak & Lejmel, 2000; Jalinik, 2009). An evolution could be observed in the method for a quantitative evaluation of development drivers of tourism functions, proposed by Drzewiecki (1992). Natural and landscape values of the geographic environment are usually considered to be the basic determinants of the establishment and development of agritourism farms; the differences between them and their volatility are viewed as important aspects (Drzewiecki, 2005; Przezbórska, 2007). Based on her own definition of agritourism space, Wojciechowska (2009) proposed that Polish voivodeships be divided into groups and sub-groups in accordance with pre-established development levels of agritourism accommodation resources. This allowed her to identify 12 regions with an outstanding agritourism offering. Moreover, the assessment and regionalization of rural areas according to the potential and level of development of the tourism function in Poland during 2000–2009 also relied on multidimensional benchmarking methods, including taxonomic methods such as Hellwig's or Perkal's aggregate indicator (Gołembski, 2002; Milewski, 2005; Rapacz, 2000; Świeca et al., 2007; Przezbórska, 2006; Kosmaczewska, 2009a). In most cases, this was done with the synthetic indicator of tourism attractiveness, expressed with characteristics that differ between the studies (e.g. characteristics of natural and cultural values and of the availability of a tourism destination, Tucki, 2009), or with the indicator of functional development of tourism which usually provides a picture of the tourism infrastructure and tourist traffic in the territory concerned. Some other papers emerged which indicated the role of social capital as an endogenous asset of local rural communities in the context of rural tourism development (Sikora & Wartecka-Ważyńska, 2009; Kosmaczewska, 2009a). Kosmaczewska (2009a) used Hellwig's development pattern to rank rural communes of the Wielkopolskie voivodeship according to their social capital. As a next step, she analyzed the development of the tourism function in selected communes at the highest and lowest levels of social capital using the Baretje–Defert index of tourism function. Her conclusion was that no general pattern (such that the higher the social capital level, the higher the development level of the tourism function in the region) could be identified in the Wielkopolskie voivodeship (Kosmaczewska, 2009a). In this case, the absence of such a relationship can be explained by the fact that the commune, depending on the level of other types of capital, can choose any function which plays a dominant role in its development (which must not necessarily be tourism development); and that high levels of social capital do not necessarily mean the local community has leaders capable of accelerating development processes (Herbst, 2007: Kłodziński, 2006; Sikora & Wartecka-Ważyńska, 2007). Furthermore, if capital is viewed not only in a context of social resources but also in its political and cultural dimension (SPCC), then the community can be thought of as a strategic resource which enables sustainable tourism development (Macbeth et al., 2004). The second post-transformation decade saw the continuation of research on the role of local government in forming and managing the tourism function at commune level. The development of rural tourism and agritourism became a major part of local development strategies for many communes (Nawrocka, 2001; Mickiewicz, 2009; Grzebyk & Lechwar, 2007; Czaplicka-Kozłowska, 2009; Herbst, 2007, Kłodziński, 2006; Sikora & Wartecka-Ważyńska, 2007; Sikora & Wartecka-Ważyńska, 2009; Tucki & Świeca, 2008; Majewska, 2008; Wiatrak, 2008). Globalization and knowledge economy bring a new challenge which is general competition, including between regions and communes. In the economic theory, competitiveness was initially considered in a context of economic operators or tourism products. However, as time went by, it started to affect the tourist domain, too (Kozak, 2006; Grabowski, 2008; Woźniak, 2008; Tyran, 2006; Tyran, 2008). This resulted in the emergence of previously undiagnosed issues, such as reputation, identity or brand of a territorial unit which, however, were much more often ascribed to cities than to rural areas at that stage (Zrobek, 2007). Competitive edge is also built for the tourist product itself, and therefore a number of studies emphasize the unique components of products which are based on tangible and intangible aspects of local cultural heritage (Zbikowski & Kasprzyk, 2010; Midura, 2008; Kosmaczewska, 2009b; Krupa, 2010; Kuźniar, 2010, Domagalska-Gredys, 2008; Ziernicka-Wojtaszek & Zawora, 2010; Bogucka, 2010). On the other hand, using the Internet as a promotion tool seems to be of importance, as reflected in numerous publications addressing this issue (Stepaniuk, 2009a, 2009b; Król, 2007; Pisarek, 2007). Moreover, agritourism farms shift towards organic production methods, which are also viewed as a specialization measure and a business advantage (Wos, 2009; Bogucka, 2010; Ziółkowski, 2006; Niedziółka, 2009; Popławski, 2009). The differences in location, available resources and products offered became a basis for the typology of agritourism farms (Wojciechowska, 2007b; Jalinik, 2007). Also, the literature published in that decade emphasizes the role of cooperation in agritourism associations (Maćkowiak, 2010a; Bodak, Jagoda, & Pietroń-Pyszczek, 2006; Niedziółka, 2008; Krzyżanowska, 2009), territorial partnerships, and cluster initiatives (Basińska & Poczta, 2009; Roman, 2009; Kosmaczewska, 2009a; Furmankiewicz & Foryś, 2006; Furmankiewicz, 2006). Another important topic of discussion was the process of education in rural tourism and agritourism (Sikora & Wartecka-Ważyńska, 2007; Sikorska-Wolak, 2007; Kurczewski, 2007; Sawicki, 2007; Orłowski & Woźniczko, 2007; Maciejewska, 2003; Prochorowicz, 2007).

In the second post-transformation decade, English-language literature on the development of the tourism function in Polish rural areas continued to be very scarce, thus preventing any broader analyses from being performed. Most of the available papers address certain topics and territories on a piecemeal basis, except for a book by Sznajder, Przezborska-Skobiej and

Scrimgeour (2009) which focuses on agricultural economics, rural development, marketing, rural policy, and different products and services described in a broad context. Another example of research performed that period is the survey carried out in 2005 with farmers based in the Podlaskie voivodeship to check their knowledge of the agritourism concept and their awareness of its environmental impacts (Zarski et al., 2005). Other authors considered the possible development trends of rural tourism and agritourism (Baum & Gramzow, 2009; Hall, 2004; Bański, 2004a, 2004b; Pawlusiński & Piziak, 2009) and of broadly defined entrepreneurship (Przezbórska, 2005a; Bott-Alama, 2004; Durydiwka, 2003; Marks et al., 2009; Gołębiewska, 2008). For instance, Przezbórska (2005b) identified three types of SMEs based on interviews with owners of small and medium rural tourism and agritourism enterprises in western Poland (*n* = 183):

- active, resilient, and rapidly developing agritourism or rural tourism enterprises making their living mainly from tourism activity;
- enterprises with a mixed structure of income, with tourism activities as an important income stream;
- enterprises with a mixed structure of income, with tourism as a less important or an unimportant income stream.

Moreover, a survey was carried out by Kosmaczewska (2008) with agritourism farm owners, local entrepreneurs, and people who did not provide any agritourism services in selected communes (*n* = 765). The study found that agritourism fosters an entrepreneurial mindset among rural residents, especially in the group with a tertiary education. Some papers are also available which emphasize the role of associations (Drag & Czerniec, 2004), of collaboration in agritourism clusters (Szymoniuk, 2003), and of the use of local cultural assets in developing an agritourism product (Tyran, 2007; Niedziółka & Lis, 2008).

In this literature review, the period of Poland's EU membership is somehow a bridge between the second (2000–2009) and the third (2010–2019) decade of changes that followed the transformation. New papers continue to tackle previously examined issues, namely the territorial differences in the development potential of rural tourism and agritourism, the changing conditions for pursuing a rural tourism or agritourism activity, and the resulting entrepreneurial mindsets. While that period saw a considerable increase in the number of publications, most discussions and analyses were carried at local (commune) or regional (voivodeship) level or as a case study. Furthermore, most authors followed an evolutionary approach by developing previously presented research problems. Table 4.4 shows a selection of topics addressed in the literature.

What needs to be particularly emphasized is that the third post-transformation decade witnessed an evolution in rural tourism and agritourism products in Poland. New papers emerged, highlighting the importance of suburban

Table 4.3 Content analysis of publications available in the second post-transformation decade (2000–2009)

Topic[a]	Papers in English with a Polish context	Papers in Polish
Agritourism and rural tourism brands / marketing activities	–	Sawicki, 2009; Czapiewska, 2009; Tyran, 2009; Perepeczko, 2009; Brelik, 2005
Financing for rural tourism / agritourism development	–	Tyran, 2005; Przezbórska, 2002; Majewski, 2003;
Resources and equipment of agritourism farms / enotourism	Zarski et al., 2005;	Sammel & Dańczak, 2002; Sokół, 2009
Agritourism and rural tourism development viewed by tourists / viewed by residents / viewed by local action groups (consumer preferences)	–	Kosmaczewska & Górka, 2009; Brodziński, 2006; Strzembicki, 2002, 2005, 2006; Želazna & Woźniczko, 2005
Conditions for the development of rural tourism / valuation	–	Perepeczko, 2003; Siekierski, 2009; Wojciechowska, 2006, 2009; Hełdak, 2002; Gołembski, 2002; Milewski, 2005; Kosmaczewska, 2009a; Chudy-Hyski, 2009; Przezbórska, 2000, 2006, 2007
Rural tourism / agritourism development trends (including in borderland / protected zones)	Baum & Gramzow, 2009; Bański, 2004a; Pawlusiński & Piziak, 2009; Sznajder et al., 2009	Wiatrak, 2008; Wojciechowska, 2006; Uglis & Jęczmyk, 2009; Balińska & Sikorska-Wolak, 2009; Karbowiak & Lejmel, 2000
The role of local units in the development of rural tourism / potential / role of the leader	–	Nawrocka, 2001; Mickiewicz, 2009; Grzebyk & Lechwar, 2007; Czaplicka-Kozłowska, 2009; Herbst, 2007, Kłodziński, 2006; Sikora & Wartecka-Ważyńska, 2007, 2009; Tucki & Świeca, 2008; Majewska, 2008; Wiatrak, 2008; Wojciechowska, 2007a
Tourism competitiveness / territorial marketing	Tyran, 2006, 2008	Kozak, 2006; Grabowski, 2008; Woźniak, 2008; Zrobek, 2007
The importance of the Internet	–	Stepaniuk, 2009a, 2009b; Król, 2007; Pisarek, 2007
Rural tourism / agritourism vs. entrepreneurship	Przezbórska, 2005b; Bott-Alama, 2004; Durydiwka, 2003; Marks et al., 2009; Gołębiewska, 2008; Kosmaczewska, 2008	Kurtyka, 2005, 2008; Sikora & Wartecka-Ważyńska, 2009, Balińska, 2001; Golik & Firlej, 2002; Jeleń, 2003; Siekierski & Popławski, 2009; Firlej, 2002; Otłowska et al., 2006; Kurtyka, 2008; Hasiński et al., 2004

(Continued)

Table 4.3 (Cont.)

Topic[a]	Papers in English with a Polish context	Papers in Polish
Rural tourism / agritourism vs. education	–	Sikorska-Wolak, 2007; Kurczewski, 2007; Sawicki, 2007; Orłowski & Woźniczko, 2007; Maciejewska, 2003; Prochorowicz, 2007
Eco-tourism / eco-agritourism	–	Wos, 2009; Bogucka, 2010; Musiał, 2007; Ziółkowski, 2006; Niedziółka, 2009; Popławski, 2009
Cultural heritage, traditional and regional products (including medicinal and handicraft products)	Tyran, 2007; Niedziółka & Lis, 2008	Midura, 2008; Kosmaczewska, 2009b; Domagalska-Gredys, 2008
Typology of rural tourism farms	Sznajder & Przezbórska, 2004	Wojciechowska, 2007b; Jalinik, 2007
Agricultural associations / agricultural consultancy	Drag & Czerniec, 2004; Szymoniuk, 2003	Bodak, Jagoda, & Pietroń-Pyszek, 2006; Niedziółka, 2009; Krzyżanowska, 2009
Role of cooperation: territorial partnership / cluster initiatives	Szymoniuk, 2003	Basińska & Poczta, 2009; Roman, 2009; Kosmaczewska, 2009c; Furmankiewicz & Foryś, 2006; Furmankiewicz, 2006
References to other countries (Ireland, Finland, Italy, CEE)	Hall, 2000a, 2000b, 2004; Sznajder et al., 2009	Hegarty & Przezbórska, 2005

Source: Own elaboration.

a Selected topics and selected examples of papers.

Table 4.4 Content analysis of publications available in the third post-transformation decade (2010–2019)

Topic[a]	Papers in English with a Polish context	Papers in Polish
Sustainability / multi-purpose rural development	Strzelecka & Wicks, 2010; Adamowicz, 2010; Wojciechowska, 2011	Bański et al., 2012
Entrepreneurship	Kosmaczewska, 2011; Szyguła, 2014; Dorocki et al., 2015; Marks-Bielska et al., 2014	Niedziółka, 2017; Bajgier-Kowalska et al., 2016, 2019; Dorocki et al., 2012; Kurtyka, 2010; Woźniak, 2014; Karbowiak, 2013; Roman & Niedziółka, 2017

(Continued)

Table 4.4 (Cont.)

Topic[a]	Papers in English with a Polish context	Papers in Polish
Rural tourism in metro-politan areas	Sznajder, 2017; Pawlikowska-Piechotka et al., 2016	Majewski & Zmyślony, 2014
Product / innovation / quality / co-creation by tourists	Miczyńska-Kowalska, 2017; Kurtyka-Marcak & Kutkowska, 2017; Poczta-Wajda & Poczta, 2016	Sikorska-Wolak & Zawadka, 2016; Kuźniar & Surmacz, 2015; Maćkowiak & Jęczmyk, 2013; Pałka, 2015; Kosmac-zewska, 2012; Duda, 2019
Rural tourism vs. dis-abled people	Niedziółka, 2017	Czachara & Krupa, 2011; Popiel, 2016; Roman & Roman, 2019
Specialization in rural tourism / agritourism (theme villages, leisure, educative farms, care facilities, api-tourism, astro-tourism)	Kubal-Czerwińska & Piziak, 2010; Kuźniar & Witek, 2016; Wos, 2014; Tobiasz-Lis et al., 2019; Mitura et al., 2017; Idziak et al., 2015	Sala, K., 2016; Cichowska, 2014; Sieczko, 2011b; Woź-niczko, 2014; Smoleńska & Machnik, 2013; Jankun et al., 2016; Kosmaczewska, 2009d; Bogusz & Wojcieszak, 2018; Kmita-Dziasek & Bogusz, 2017; Kmita-Dziasek, 2015; Oleśnie-wicz et al., 2016; Wojcieszak & Wojcieszak, 2018; Chmie-lewska, 2018a, 2018b; Stępnik, 2017; Roman & Wojcieszak, 2018; Poczta & Malchrowicz-Mośko, 2016; Oleśniewicz, Widawski, & Markiewicz-Pat-kowska, 2016
Collaboration / net-working / consultancy	Borkowska-Niszczota, 2015; Abrham, 2014; Zajda et al., 2017	Bednarek-Szczepańska, 2010; Maćkowiak & Graja-Zwo-lińska, 2017; Sieczko, 2011a; Zawadka, 2017; Pisarek et al., 2017; Balińska, 2017; Brelik & Bogusz, 2013; Maćkowiak, 2010a
Consumer behavior / feedback / expectations	Kamińska & Mularczyk, 2015; Spychała et al., 2017	Maćkowiak, 2010b; Maćkowiak & Budych-Tomkowiak, 2012; Zawadka, 2016; Sikora & War-tecka-Ważyńska, 2013; Feczko, 2015a, 2015b; Wilk & Keck-Wilk, 2013; Balińska, 2014a, 2014b; Nowogródzka & Pie-niak-Lendzion, 2014; Wojcies-zak, 2017; Zawadka, 2012; Mika, 2013; Feczko, 2015a, 2015b; Balińska, 2010; Krysa & Basaj, 2010; Roman, 2010; Sikorska-Wolak, 2010

(Continued)

Table 4.4 (Cont.)

Topic[a]	Papers in English with a Polish context	Papers in Polish
Local community attitudes	Kosmaczewska et al., 2016b	Sikorska-Wolak & Zawadka, 2011; Kuźniar, 2015; Maćkowiak, 2010b; Mika, 2013
Service providers' profile	Dorocki et al., 2015;	Golian, 2016; Kurtyka, 2011
Internet	Król & Halva, 2017; Król, 2019	Król, 2016, 2017, 2018; Stepaniuk, 2010a, 2010b; Sammel, 2010; Kosmaczewska, 2010a; Pisarek et al., 2013; Dziechciarz, 2013; Gralak, 2016
Marketing activities / social media in rural tourism/agritourism	Werenowska, 2014; Uglis & Kosmaczewska, 2017; Dorocki et al., 2016	Krzyżanowska, 2014; Kuźniar, 2013; Matuszewska et al., 2016
Rural tourism / agritourism financing (including with EU funds)	Chirițescu, 2011;	Zawadka, 2015; Siemiński & Poczta, 2014; Korczak & Tomaszewski, 2015; Sammel & Prochorowicz, 2013; Przeorek, 2014; Patoła, 2016; Kozłowska-Burdziak, 2012; Marks-Bielska & Zielińska-Szczepkowska, 2011
References to other countries (Czech Republic, Germany, Slovenia, Spain, Ukraine, Hungary, France, Romania, Slovakia)	Jaszczak & Žukovskis, 2010; Wos, 2014; Potočnik-Slavič & Schmitz, 2013; Roman et al., 2018; Strzelecka et al., 2017; Abrham, 2014; Dorocki et al., 2013; Spychała et al., 2017; Król & Halva, 2017	Łukasiewicz, 2017

Source: Own elaboration.

a Selected topics and selected examples of papers.

(metropolitan) areas in this context (Majewski & Zmyślony, 2014; Sznajder et al., 2017) and the fact that they offer a different, location-specific product (Poczta & Malchrowicz-Mośko, 2016; Majewski, 2017). Analyses and studies were carried out of components of agritourism and tourism products offered in rural areas, pointing out their innovativeness, the growing role of specialization (Sikorska-Wolak & Zawadka, 2016), and collaborative networking. Some papers were published that addressed agritourism farms highly specialized in education, referred to as educational farms[2] (Kmita-Dziasek, 2015; Oleśniewicz et al., 2016; Kmita-Dziasek & Bogusz, 2017; Bogusz &

Wojcieszak, 2018), in apiculture and apitherapy (api-tourism, Wos, 2014) or in care services for elderly and disabled people (Stępnik, 2017; Chmielewska, 2018a, 2018b; Czachara & Krupa, 2011; Popiel, 2016; Wojcieszak & Wojcieszak, 2018; Roman & Wojcieszak, 2018; Roman & Roman, 2019). Thus, the social farming concept was introduced to the Polish literature (Matysiak & Michalska, 2016; Roman & Wojcieszak, 2018). Some started to point to the fact that the development level of the rural tourism function is not only largely decisive for the importance of tourism to socioeconomic development of a territorial unit but also considerably determines its tourism attractiveness. Also, Kosmaczewska (2013) made an attempt to identify the hierarchical position of the tourism function among the many economic functions of rural communes. Her studies demonstrated that the tourism function is not always an efficient driver of rural development, and that local development fueled by tourism can have different grounds and driving forces. Based on research tools such as structural modeling (value of the standardized regression coefficient), Data Envelopment Analysis (DEA), the Kruskal–Wallis ANOVA test, and interview surveys,[3]Kosmaczewska (2013) developed a theoretical cause-and-effect model that explains the development of rural tourism under the assumption that the relevant relationship is determined by the characteristics of commune types defined as summative indicators of: socioeconomic resources at commune level (development level indicator, DLI); management efficiency at commune level (financial efficiency indicator, FEI); and tourism attractiveness (tourism development indicator, TDI). Note that the typology of communes grouped by their potential for the development of the tourism function takes account not only of the stimulating or inhibiting nature of each indicator but also of their accurate value. Thus, it was possible to create four empirical typological groups of rural communes (n = 348). The grouping criterion was the relationship between the values of particular summative indicators; the following groups were identified on that basis:

- well-developed communes with a non-tourism function as the dominant function (FEI > DLI > TDI): integrated urban communes referred to as satellite communes;
- medium-developed communes with the tourism function as the dominant function (FEI > DLI ≤ TDI): intermediate rural areas with a developed tourism function, referred to as "star communes";
- poorly developed communes with a tourism potential (FEI < DLI ≤ TDI): intermediate rural areas which, while having an underdeveloped tourism function, demonstrate potential for tourism development; they are referred to as aspiring star communes;
- extremely poorly developed non-tourism communes (FEI < DLI < TDI): remote rural areas referred to as stones or peripheral communes.

The comparison of relationships between particular summative indicators of the three conditions for tourism development, i.e. the socioeconomic situation (DLI), the efficiency of financial management at commune level (FEI), and the commune's attractiveness to tourists (TDI), served as a basis for determining the relevance of particular conditions for tourism development at commune level. As a next step, the theoretical model was empirically verified in different types of communes. This is how Kosmaczewska (2013) found that weak communes (WPR < 0) are mostly located in what is referred to as the Reintegrated Territories which joined Poland in 1945 and were populated with immigrants from the Eastern Borderlands. The reason behind the socioeconomic underdevelopment of these communes is the nature of the "post-migrant" community which failed to fully develop regional awareness, and the prevailing agrarian structure. Moreover, from 1949 to 1991, the Reintegrated Territories had the country's largest concentration of state-owned agricultural holdings, which resulted in the elimination of individual farming. For the local population, the liquidation of their sole revenue stream meant high unemployment rates, a lack of future prospects, and social exclusion. Building a reality which, in addition to jobs, provided the employees with accommodation and social and cultural benefits actually resulted in making people helpless, indifferent to changes in their environment and unable to develop an entrepreneurial mindset. This continues to be a major development obstacle in these territories to the present day (Kosmaczewska, 2013). Also, the territorial distribution of supporting communes[4] (WZF > 0) is not identical to that of strong communes, as defined with the development level indicator (DLI). This means that communes which enjoy an advantageous socioeconomic situation are not necessarily efficient in their financial management. Thus, analyzing tourism development with the commune's socioeconomic situation as the sole criterion (without analyzing the managerial efficiency at local government level) should be viewed as insufficient (Kosmaczewska, 2013). Additionally, the territorial distribution of the indicator of attractiveness to tourists (TDI) largely overlaps with the distribution of communes with a positive indicator of financial management efficiency (FEI). This suggests that the resourcefulness of local authorities plays an important role in tourism development, and debunks the myth that every commune with valuable tourism resources can effectively develop the tourism function in its territory. It was also demonstrated that most communes surveyed can be referred to as medium communes and do not exhibit outstanding levels of development or financial management efficiency. This provides a practical explanation of what can be observed in Poland, namely a situation where a rural commune is unable to become a recognized tourist destination despite having considerable natural values; this indicates the key importance of local authorities and of residents themselves in the tourism development process. As pointed out by Kosmaczewska (2013), the planning and implementation of tourism development processes in a selected commune type should be based first on measures focused on the commune's leading

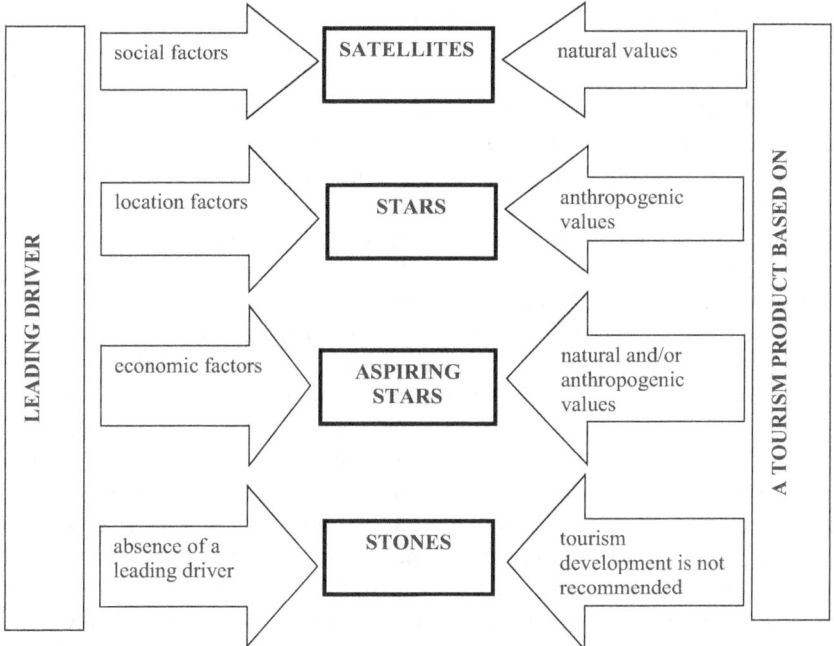

Figure 4.1 Tourism development in different commune types, taking account of the leading driving force and the intended development of tourism products.
Source: Kosmaczewska (2013).

driver of tourism development,[5] and should be oriented at developing tourism products based on values which are specific to that very type of commune (Figure 4.1).

In turn, the commune's potential for tourism development—and especially the commune's capacity to develop through the implementation of its tourism function—should be assessed in three steps:

- identifying the commune type;
- identifying the leading drivers of tourism development and their key characteristics;
- verifying the importance of tourism in rural development at commune level using the Data Envelopment Analysis (Kosmaczewska, 2013).

Furthermore, the review of Polish literature from the third post-transformation decade suggests that consumers of rural tourism services increasingly value an authentic experience, silence, peacefulness, and genius loci (a unique, outstanding climate, nature, and spirit of a destination). At the same time, the above is also what constitutes the slow tourism philosophy (Majewski, 2010; Duda, 2019). In that period, the literature also started to indicate that tourists may contribute to the creation of agritourism products (Kuźniar & Surmacz, 2015). As part of aligning the concept of customer contribution to product value with the particularities of rural tourism, Kuźniar and Surmacz (2015) indicated the theoretic existence of different prosumer groups driven by economic, social, and environmental benefits. The authors believe that in the Polish realities, prosumers driven by economic benefits are people who choose rural destinations as a cheaper form of tourism. This is a relatively large group of low-income tourists, usually represented by families with small children who cannot afford more expensive forms of leisure. To pay less for their stay, visitors can help with the farm work, assist in the preparation of meals or consume fruits and vegetables they harvested themselves at the farm free of charge. Prosumers motivated by social and/or environmental benefits are usually educated tourists aware of the unique features of a rural landscape. They use the offering to compose their own mix of services, with emphasis placed on environmentally friendly farm solutions and activities such as questing or geocaching (Kuźniar & Surmacz, 2015). In that period, broad research was carried out on consumer expectations for quality and scope of the offering (cf. Table 4.4), and some papers were published that addressed the profiling of service providers (Golian, 2016; Kurtyka, 2011) and the assessment of whether individuals are capable of entering the broadly defined rural tourism sector on a self-employment basis (Kosmaczewska, Barczak, & Rudnicki, 2016). A survey carried out with young people (aged 18–29) who are not in education, employment or training (NEETs, Arnold & Baker, 2012) from the Kujawsko-Pomorskie voivodeship ($n = 380$) found that females aged 25–29 with a tertiary education were the most likely to set up an economic activity related to tourist services (Kosmaczewska et al., 2016a). Also, a broader approach was adopted in addressing a previously neglected research problem, namely the attitudes of local communities who do not reap direct benefits from the development of the tourism function in rural areas (Sikorska-Wolak & Zawadka, 2011; Kuźniar, 2015; Maćkowiak, 2010b). When it comes to stakeholders, the literature called for continuous efforts to increase tourism awareness among the local population, for increased social participation, and for the development of a local tourism partnership (Kuźniar, 2015). Making use of the Internet was already discussed in earlier publications. However, the literature from the third post-transformation decade clearly shows that a more in-depth approach was adopted in exploring this topic. The authors focus on aspects such as the efficiency, quality, and optimization of websites (Król, 2016, 2017, 2018; Stepaniuk, 2010a, 2010b; Sammel, 2010; Kosmaczewska, 2010a; Pisarek et al., 2013; Dziechciarz, 2013; Gralak, 2016). Papers were published that indicated the reasonability and efficiency of using social media as a modern marketing tool for agritourism farms but also for agritourism associations or local government units (Krzyżanowska, 2014; Kuźniar, 2013; Matuszewska et al., 2016). There were changes in the approach to financial support from the EU for rural tourism and agritourism development, too. Most researchers shifted their interest from papers focused on the availability itself of funds (in the second post-transformation decade) to assessing how efficiently the funds are used by agritourism farms, micro-enterprises, local government units, associations or local action groups in developing the tourism function in rural areas (Zawadka, 2015; Siemiński & Poczta, 2014; Korczak & Tomaszewski, 2015; Sammel & Prochorowicz, 2013; Przeorek, 2014; Patoła, 2016; Kozłowska-Burdziak, 2012; Marks-Bielska & Zielińska-Szczepkowska, 2011).

The decade 2010–2019 saw a considerable improvement in the availability of English-language literatures dealing with the development of the tourism function in rural areas. Actually, most publications featured at least an English abstract which allowed exploration of their main assumptions regarding research problems mentioned earlier (cf. Table 4.4).

The review of literature pertaining to research on the development of the tourism function in Polish rural areas, as carried out during 1989–2019, leads to a conclusion that researchers representing different fields of science act independently, and an interdisciplinary approach is still far from common practice. A quantitative analysis of available literatures reveals a considerable deficiency of English-language papers, especially in the first post-transformation decade. Also, the Web of Science gives grounds for concluding that the papers published in the second post-transformation decade (2000–2009) were mainly focused on research areas such as the agricultural economics policy and environmental studies, whereas in the next decade (2010–2019), a large number of scientists tackled research areas such as environmental sciences, ecology, business economics, and social sciences. The above also shows the evolution of how the rural tourism function is perceived. The differences in assets between communes should be the basis for selecting the functions to be implemented and for setting their priorities of relevance in the current and future development of the territorial unit concerned. Meanwhile, the development itself—especially in the context of the tourism function—should primarily be based on economic diversification; conversely, less emphasis should be placed on the economies of scale and on concentration efforts. Another clearly noticeable aspect is the evolution in how the different fields of science approach these topics and view their scope. Also, the two last post-transformation decades (2000–2009; 2010–2019) saw a more in-depth investigation into these issues. What additionally needs to be emphasized is the considerable difficulty in comparing research findings both between periods and between territories, primarily because the research samples are often small and non-representative. Furthermore, due to lack of continuous, homogeneous statistical data, the papers rarely address the whole national territory.

Having in mind the general trends followed by the tourism industry around the world, it can be assumed that when it comes to the tourism function in Polish rural areas, issues investigated by the scientists in the next decade will include overtourism, dissatisfaction, conflicts, and suspicion, on the one hand, and responsible tourism and local community satisfaction, on the other. The likely purpose of these papers will be to determine how much and how fast the tourism function should be developed in Polish rural areas to enable preserving their unique nature.

4.2 Attitudes of the rural population towards rural tourism development: research overview

While tourism development could be one of the rural economic functions, its role and importance in multipurpose rural development depend on the availability of endogenous factors and on how much it is linked to other economic fields. Quite often, tourism aggressively competes with other possible uses of land. Its increased importance usually provides a stimulus for reducing other economic functions, including agriculture, in the rural economy mix (Ziółkowski, 2006; Adamowicz & Zwolińska-Ligaj, 2009; Wilkin, 2011). In extreme cases, this could mean shifting from an agricultural monoculture to a tourism monoculture. It could result in reinforcing hostile attitudes of the local population towards the development of the tourism function in rural areas. These adverse phenomena could particularly affect rural communes which experience a sharp, uncontrolled, and often only seasonal development of tourism. In Poland, research on the attitudes of residents of tourist destinations started only in the second post-transformation decade. Initially, this was a marginal topic, mainly because the rural tourism function was at an early development stage and researchers focused their attention on the classification of territories according to their usefulness in the development of rural tourism and agritourism. Note also that the tourism function develops in a rural community environment where people have specific mindsets and follow different lifestyles and customs. In Poland, the characteristics of the rural living model include traditionalist behaviors, conservative viewpoints, and a religiously oriented system of attitudes. Although this can considerably restrict the development of tourism, it plays a much lesser role today than in the two first post-transformation decades. Moreover, social capital components such as the populations' entrepreneurship, self-organization and collaboration capabilities, the system of social ties, and mutual trust can result in the establishment of new forms of tourism activity, and can trigger a process where early adopters are followed by others which provide momentum for rural tourism development (Kosmaczewska, 2007).

The literature explains residents' attitudes towards tourism in a number of ways. Most studies explored residents' attitudes using the social exchange theory (SET) (Sharpley, 2014). Accordingly, those who reap direct benefits from the activity tend to have more positive attitudes towards tourism development than those who do not (Chuang, 2010; Rasoolimanesh et al., 2017). On the other hand, the Tourism Area Life Cycle model (TALC), which interprets tourism development as a series of stages through which a destination evolves proved that residents' attitudes are positive during the initial stages of tourism development but become negative in the later stages (Ambroz, 2008; Reid et al., 2011; Latkova & Vogt, 2012). From the perspective of the Theory of Reasoned Action (TORA), behavior is influenced by the behavioral intent which, in turn, is influenced by attitudes, and those attitudes are influenced by beliefs (Hadinejad et al., 2019). Moreover, a systematic review of the literature reveals that age (Faccioli, 2011), gender (Dadvar-Khani, 2012; Huh & Vogt, 2008) education level (Korça, 1998), and knowledge about tourism (Andereck et al., 2005) appear to influence the residents' perception of tourism development. For a systematic assessment, see Harrill (2004), Sharpley (2014), and Hadinejad et al. (2019).

As shown by a literature review of research carried out in Poland, rural residents' attitudes towards tourism development in the territory where they live should be considered in the context of several aspects, i.e.:

- residents' attitudes towards the tourism enterprise;
- the way the residents perceive the measures taken by local government units to develop the tourism function;
- residents' attitudes towards tourists (creation of what can be referred to as a hospitable climate) and towards the existing or foreseeable consequences of tourism development in their commune.

Potential financial benefits are the most commonly cited reason for delivering tourist services in rural areas, including agritourism services (Sikora & Karczewska, 2004; Kosmaczewska, 2007; Kurtyka, 2010; Zawadka, 2010a; Gąsiorowska-Mącznik, 2011; Siedlecka & Wielogórska, 2018). As shown in research carried out by Kosmaczewska (2007) with rural residents who did not deliver any tourist services ($n = 765$), the presence of agritourism farms in the commune boosts the entrepreneurial mindset of other residents, especially if well educated. In most cases, the respondents who considered setting up an agritourism activity were people with a secondary (30.3%) or tertiary (19.4%) education. Those with a primary education were the least inclined to do so (15%) (Kosmaczewska, 2007). Research carried out by Kosmaczewska a few years later found that only 35% of residents of Polish rural communes ($n = 1000$) see a positive influence of tourism on their incomes (Kosmaczewska et al., 2016b). Moreover, the type of commune was found to have a statistically significant impact on how the residents perceive the economic effects of tourism. Residents of "aspiring communes" showed less positive attitudes towards tourists. This was explained by the fact that in these communes, tourist traffic is relatively light, investments in tourism infrastructure are not so noticeable, and actions taken to increase demand for tourist services failed to produce the desired effects (Kosmaczewska et al., 2016b). In their research, other authors diagnosed the rural population's intent to set up a tourism activity. Usually, this meant starting an agritourism business, offering private accommodation, providing catering services or rental of leisure equipment (Sikorska-Wolak & Zawadka, 2011). Also, a survey carried out with rural NEETs[6] found that females aged 25–29 with a tertiary education were the most likely to set up an economic activity related to tourist services (Kosmaczewska et al., 2016a). Another conclusion from this research was that a moderate positive correlation ($r = 0.53$) exists between having a close friend or family member who earns money from tourist services and considering the possibility to set up such a business (Kosmaczewska et al., 2016b). Moreover, an in-depth analysis of data collected clearly showed that the respondents perceive their own commune as being, or not being, attractive to tourists depending on how they view their own business opportunities in that context (Kosmaczewska et al., 2016b). The district where the population surveyed had the highest percentage of replies affirming the attractiveness of the commune they live in (the Grudziądz district) was also where the highest percentage of respondents (50%) consider the possibility to set up an economic activity related to tourism services. Conversely, according to the respondents' replies, in the district at the lowest level of attractiveness to tourists (the Mogilno district), none of the NEETs surveyed declared that they were interested in starting an economic activity related to tourism (Kosmaczewska et al., 2016a). The study also found that the higher the education level of rural residents, the more frequently they perceive their commune as being attractive to tourists (Kosmaczewska et al., 2016a).

As regards the way the residents perceive the activity of local government units in developing the tourism function, Sikorska-Wolak and Zawadka (2011) demonstrated that if the residents themselves believe their commune to be attractive to tourists, they are more likely to approve the purposefulness of tourism development measures taken by the LGU[7]. The opinions differed depending on the respondents' education level, i.e. the more they are educated, the more positively they view the purposefulness of tourism development (Sikorska-Wolak & Zawadka, 2011). The study on the population's attitude towards the development of the tourism function as the dominant one was carried out with residents of thematic villages[8] located in the Zachodniopomorskie voivodeship (Maćkowiak, 2010b). Findings revealed that only 26% of respondents believe their expectations for improvements in their financial standing were met after the establishment of the thematic village. However, it should be emphasized that according to 60% of respondents, the thematic village construction project had a positive effect on relationships within the rural community, including through the establishment of new contacts (86.7%), more frequent social events (73.3%) and the sense of community (71.7%) (Maćkowiak, 2010b). Moreover, Idziak et al. (2015) investigated the local community's attitudes towards and engagement in the creation of thematic villages after five years of their establishment. They indicated the importance of leaders and external experts in an efficient development of the tourism function in previously collectivized rural areas.

The rural population's feedback on their attitude towards tourists provides a basis for analyzing the climate that surrounds this kind of enterprise. When using tourism as a local development tool, it is essential that the residents see the attractiveness to tourists and be willing to create a climate of hospitability that fosters initiatives taken to build a local tourism/agritourism product. The hospitability of the destination, as the surroundings of an agritourism farm, can be considered in two dimensions: a hospitable climate and a hospitable community. Kaczmarek et al. (2008) define a hospitable climate as managing the facilities in a way to provide the visitors with a safe and comfortable stay, and as using the hospitality formula to link land organization with the attitudes of the destination's residents. The greater the percentage of the destination's residents who are, or are believed to be, hospitable hosts to visitors, the more hospitable is the community. The above indicates that the hospitability of a destination is the combined impact of facilities management and of the local community's attitudes towards tourists. Also, reputation based on the hospitability of the service provider and of the destination can considerably contribute to the intangible value of an agritourism or rural tourism operator, and may be decisive for its competitive edge (Kosmaczewska, 2012). Research by Kosmaczewska (2007) found that a positive attitude to tourists was mostly exhibited by rural residents with a tertiary or secondary education ($n = 765$). Conversely, a negative attitude to tourists visiting the commune was usually declared by rural residents with a primary education. Furthermore, this study revealed a moderate correlation ($r = 0.329$) between a commune's residents' attitude towards tourists and the perceived positive impact of agritourism on incomes earned by the respondent and his/her family. People who declared that they saw a positive impact of agritourism development on their incomes had a positive (82%) or neutral (8%) attitude towards tourists. Another conclusion was that the perceived impact of agritourism on selected aspects of life is directly proportional to the education level of residents of rural communes and inversely proportional to their age (Kosmaczewska, 2007). Based on a survey carried out with rural residents not engaged in tourism activities in the Podlachia region ($n = 900$), Kuźniar (2015) identified three categories of residents, using their attitudes towards tourists as the criterion: "hostile observers" (7%), "neutral neighbors" (54%), and "hospitable residents" (39%). The largest category ("neutral neighbors") defines people who show little interest in local community issues and believe rural tourism development to be only one of many development paths for their commune. According to Kuźniar (2015), the percentage share of particular population categories defined by their attitude towards tourists will change over time depending on the maturity of the tourism function in the commune.

As shown by a review of research on the attitudes of the Polish rural population towards tourism development in the territory where they live, the authors of the studies referred to above used some determinants of the local population's attitudes which were already known in the international literature, and redefined them in the Polish realities. Tourism can be reasonably viewed as a driver of local development only if the local community has the necessary minimum of capital so that the exogenous stimulus (i.e. tourists in a tourist destination) may become an enabler of changes in the physical and institutional structure of the local economy. The conclusion from the literature review presented above is that the development of the tourism function is only recommended in rural areas where most of the local community accept it and the local government units know how to manage their resources in such a manner that while providing every resident of the commune with improved access to different kinds of capital, they also contribute to positive attitudes towards the tourism business and towards tourists themselves.

4.3 Transformation of tourism in rural areas: a demand and supply approach[9]

Over the last 30 years, changes in the country's socioeconomic situation have affected the supply of tourism goods and services and the level and structure of their consumption. Tourism demand generated by the Polish population in the real socialism era was a component of a quasi-market for tourism, and was mostly addressed by accommodation facilities operated under employee holiday schemes. Due to the collective and non-rivalrous nature of tourism consumption, the tourism price system did not result from the production costs of different services. As a consequence, tourism development was focused on quantity rather than quality. The quasi-demand for tourism was subsidized with occupational social funds; this made the society strongly believe that access to tourism should be guaranteed and financed by the government. Viewed as a state's social policy tool, tourism enabled the consumption of goods in the times of scarcity. The principles for institutional and legal regulations introduced in the tourism market included the following: abolishing the requirement for foreign visitors to exchange currency; making the zloty a domestically convertible currency; changing the zloty/U.S. dollar exchange rate; equalizing the prices of tourism services provided to foreigners and Polish nationals; eliminating the state monopoly in the organization of international tourism; new principles for the establishment of the tax system; and reducing the contribution of state-owned enterprises in the social fund (Szubert-Zarzeczny, 1996). The implementation of market economy principles resulted in an increase in the macroeconomic importance of tourism. Undoubtedly, this was also enabled by political decentralization which provided grounds for future strategic planning of tourism development at local government level. With the transition from a centrally planned to a market economy, it was possible for private operators to compete in how they meet consumer needs, including tourism needs which, in the Polish People's Republic, were addressed on a collective basis. Initially, the transformation triggered growth in economic freedom, resulting in an entrepreneurial boom. Combined with limited social protection from the government, this caused a reduction in the number of holiday trips organized by employers, with individual tourism gaining in importance. Also, the economic liberalization resulted in income disparities and, thus, in changing the volume and structure of consumption. On a per-capita basis, average rural incomes were, and continue to be, lower than those earned by urban households for reasons which include the greater average number of rural household members (Grzywińska-Rąpca, 2014). In addition to having an effect on the volume of structure of tourism service consumption, this makes people seek new sources of income, including in the tourism sector. Another important aspect in the context of developing the tourism function of rural areas is the fact that the rural employment structure has progressively changed over the last 30 years. This is manifested in the declining share of households which rely on their own farm as the main revenue stream (see Section 2.3.1 for details). Certainly, the transformation occurred at the expense of reducing the tourism activity of social groups at lower income levels and in the polarization of consumption, including in the tourism sector (Podolec, 2000; Kieżel, 2004). The changes that accompanied the economic transformation (for a broader description, see Chapters 1–2), and factors such as the transformation of the ownership structure of accommodation and catering facilities, made the tourism market shift from a producer market to a consumer market. Also, the structure of the households' consumption expenditure has undergone a major transformation within the last 30 years. Since 1990, the share of food in total expenditure has been on a consistent decline in all household types, including farms (Figure 4.2). The strong relationship between tourist trips, the household's financial standing and professional status of family members has exacerbated the disparities in tourism consumption. This resulted in farmers and pensioners having a marginal contribution to tourism demand. The expenditure on leisure declined by ca. 4% in the first post-transformation period (1990–1993) in all household groups as a consequence of the economic crisis. However, in late 1990s, households of self-employed and employed persons went back to a level similar to what had been recorded before the reforms (Kosmaczewska, 2010b). Only the third post-transformation decade witnessed significant, consistent growth in leisure expenditure (recreation and culture; restaurants and hotels), including in farmers' households (Utzig, 2018) (Figure 4.2). Hence, the territorial disparities in tourism consumption narrowed down, especially when it comes to short-term travels.

Note also that while the protective role of the state in the area of tourism consumption was considerably reduced by the implemented reforms, even in

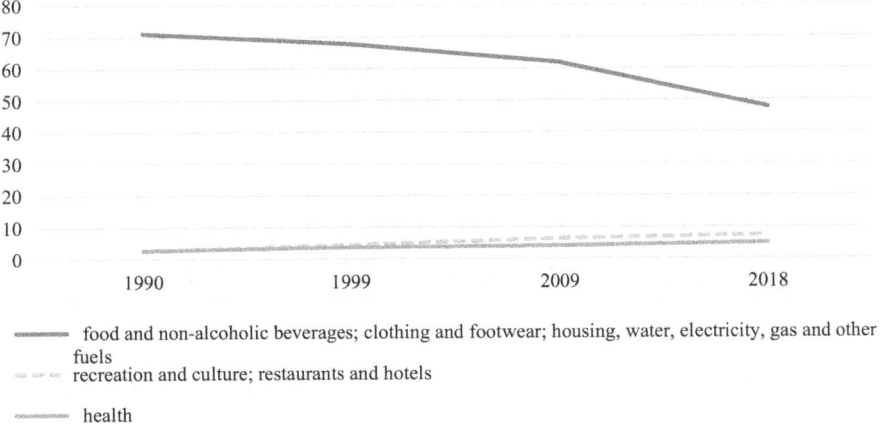

Figure 4.2 Tourism development in different commune types, taking account of the leading driving force and the intended development of tourism products.
Source: Own study based on household budgets, Central Statistical Office.

the late 1990s, nearly 40% of tourist trips were subsidized by employers and institutions (Alejziak, 2009). The initial years of the economic transformation were characteristically marked by a decline in the relatively high levels of tourism activity which, however, was not directly proportional to the reduction in households' real incomes (Kosmaczewska, 2010b). This is because in the Polish People's Republic, participation in tourism activities was considered a public good; after the transformation, this was manifested by a reduction in tourism expenditure rather than by not going on any trips at all. In the first post-transformation decade, trips made by the Polish population were less frequent but relatively longer (two-week holidays continued to be a highly popular option). In the second post-transformation decade, the trips were more frequent but shorter; this was largely due to the fact that in a developing economy, the society has limited amounts of free time. In the third post-transformation period, that trend became even stronger. This, in turn, was caused by factors such as good economic conditions, stabilization in the labor market, increasing income levels, and the households (especially large families) being supported with social benefits (Figure 4.3). Also, the popularity of what is referred to as bridging days contributed to long-term trips becoming shorter (Kryczka, 2014a; Balińska, 2014a, 2014b; Berbeka, 2016).

The reduction in the Polish population's participation in tourist trips, witnessed during 1998–2002, was mainly due to economic growth becoming slower and to a strong increase in unemployment (Kryczka, 2014b). Moreover, in the second post-transformation period, car ownership became common; combined with relatively high accommodation prices and family

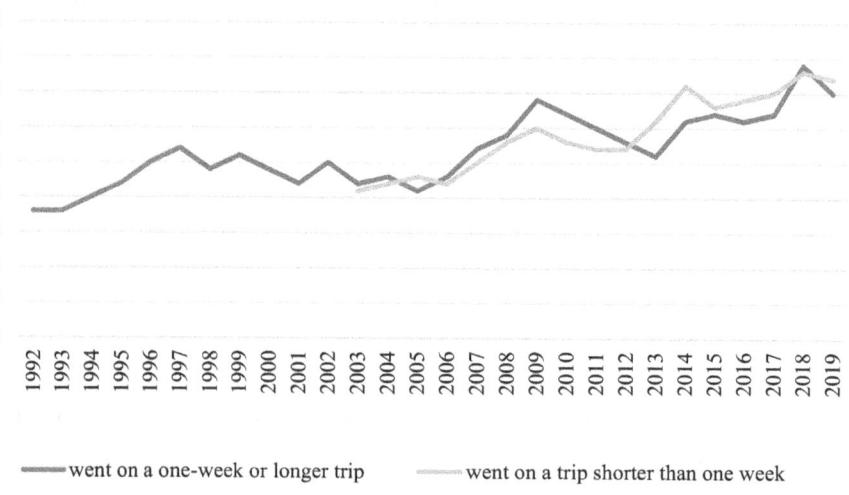

Figure 4.3 Participation of the Polish population in tourist trips, 1992–2019 (%).
Note: Participation in short-term trips has been surveyed from 2003 onwards.
Source: Public Opinion Research Center, Leisure trips of the Polish population in 2010, 2018, 2019, Warsaw.

ties between urban and rural areas becoming progressively looser, this resulted in a substitution of tourist trips with one-day breaks (Łaciak, 2011). The economic recovery recorded in 2003 (although unemployment continued to be high) as a consequence of the general improvement in economic outlooks around the world and of the upcoming accession of Poland to the European Union, triggered a further increase in the Polish population's participation levels in tourism activities. In turn, the declining rates of participation in (mainly long-term) tourist trips recorded since 2009 were the consequence of the impact the global economic crisis had on the level and structure of household consumption. It was also observed that the lower the rank of the town of residence, the higher the higher the share of households not engaged in any tourism activity. The highest percentage of households who do not travel on holidays was recorded in rural areas (Kryczka, 2014a). Over the years, in addition to lack of funds, farm chores have remained the most frequent reason preventing individual farmers and their family members from going on tourist trips (Łaciak, 2001; Łaciak, 2011; Łaciak, 2013). Note that the percentage of farmers who do not go on tourist trips keeps fluctuating around 70–80%, with the overall non-participation rate being ca. 50–60%.[10] Over the years, the Polish population has also changed their preference for tourism destinations. According to a 2015 study[11] (Sala, 2016), there is smaller interest in rural destinations (a drop by 19 percentage points compared to 2009). However, it must be assumed that in the context of the pandemic unleashed by COVID-19, rural areas will see their importance considerably grow, whereas city breaks will become less popular (Figure 4.4).

The development level of the tourism function in Polish rural areas was measured with metrics of intensity and density of tourism traffic (e.g.

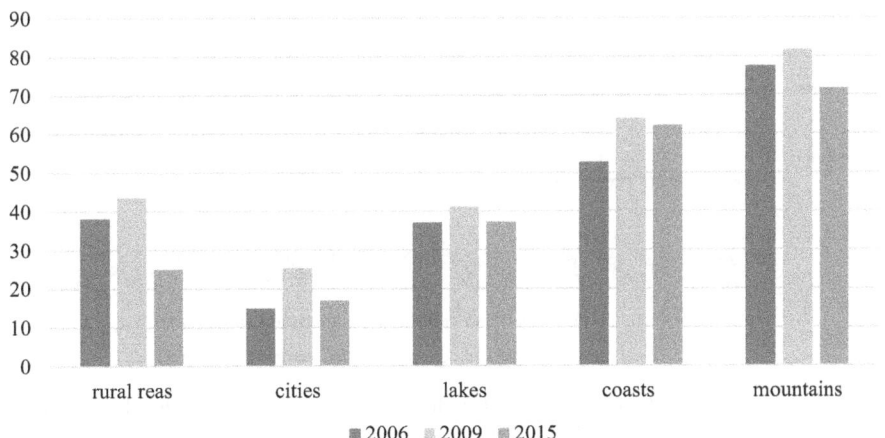

Figure 4.4 Domestic tourism destinations of the Polish population in 2006, 2009, and 2015. Source: Sala, J. (2016).

Schneider's index, Charvat's index, tourism density index)[12] and tourism management indicators (e.g. Baretje–Defert index, tourist accommodation density index)[13] in order to represent both the demand and supply sides.

The change in demand for tourism services offered in Polish rural areas was described with indicators of the number of users of collective living quarters (Schneider's index) and the number of overnights sold (Charvat's index) in relation to the population of a territory. Another indicator used in this context is the tourism density index, expressed as the number of accommodation users in relation to the size of a territorial unit. The analysis of 1995–2014 data suggests that the number of users of collective living quarters in Polish rural areas grew at a sluggish rate (Figure 4.5).

The average growth rate (1995–2014) of the number of accommodation users in relation to the population was more than twice as high in cities than in urban areas (Figure 4.6). However, it needs to be emphasized that the growth rate for Schneider's index disaggregated by the urban/rural dimension was affected by the reducing urban population and the growing population of rural areas, especially those surrounding large urban agglomerations (cf. Figure 4.19). Moreover, there were changes in the tourists' preferences for the type of accommodation facilities. Note also that the number of visitors to agritourism farms grew considerably (39%) in the third post-transformation decade, and the share of foreign tourists increased from 3% to 5.5% (cf. Figure 4.7).

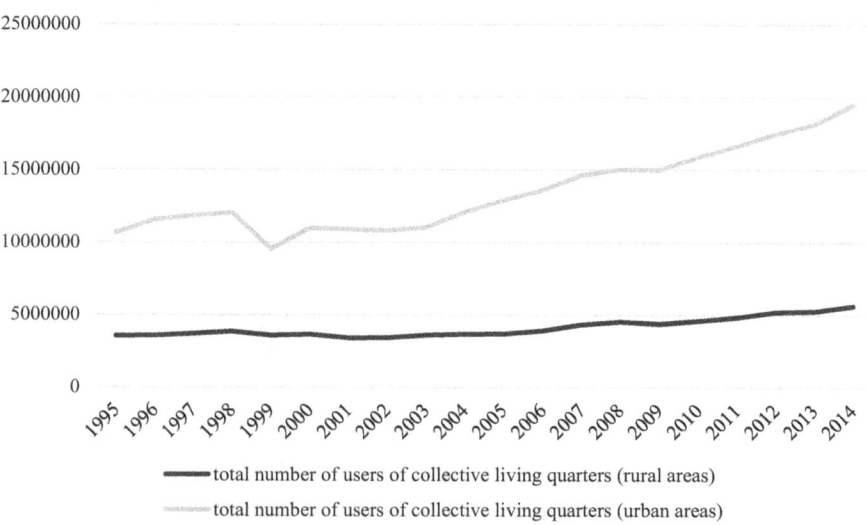

Figure 4.5 Number of users of collective living quarters, disaggregated by the urban/rural dimension, 1995–2014.

Note: No statistical data is available for subsequent years that would enable a disaggregation by the urban/rural dimension.

Source: Own study based on the Local Data Bank of the Central Statistical Office.

Figure 4.6 Schneider's index in rural and urban areas, 1995–2014.
Note: Index = (number of accommodation users / number of permanent residents) × 100.
Source: Own study based on the Local Data Bank of the Central Statistical Office.

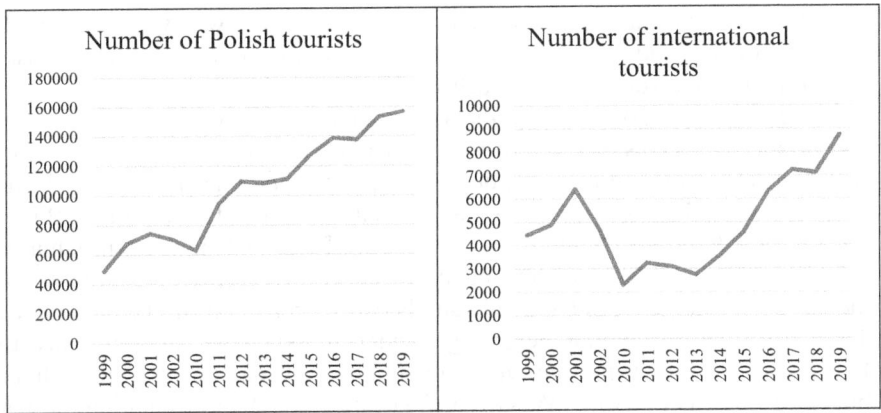

Figure 4.7 Total number of tourists who stayed in agritourism farms.
Source: Local Data Bank, Central Statistical Office.

The analysis of changes in the tourism function of Polish rural areas by region (voivodeship)[14] was based on 1999 and 2014[15] data. In both periods covered by this analysis, the highest levels of Schneider's index were recorded in the Zachodniopomorskie voivodeship because it is a coastal region located close to the German border. Compared to the overall pace of changes recorded in Polish rural areas, the lowest regional pace of changes (reflected by the number of accommodation users in relation to the population of each voivodeship) was recorded in the northern part of the country (Warmińsko–Mazurskie and

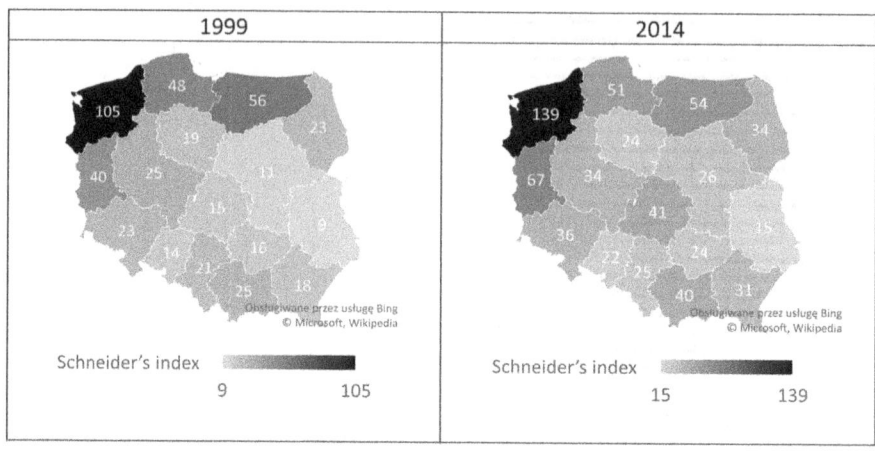

Figure 4.8 Comparison of Schneider's index for rural areas between voivodeships, 1999 and 2014.

Note: Index = (number of accommodation users / number of permanent residents) × 100.
Source: Own study based on the Local Data Bank of the Central Statistical Office.

Pomorskie voivodeships). Conversely, the highest pace was witnessed in the central (Łódzkie and Mazowieckie voivodeships) and south-eastern (Lubelskie and Podkarpackie voivodeships) parts of the country.

In the early 1990s, the total number of overnights sold was very similar in rural and urban areas. This was due to the government still playing a protective role with respect to tourism consumption at that time (National Paid-Leave Fund) and to the specific mix of accommodation facilities with a dominant role of large resorts, including those built by state-owned establishments in extra-urban areas attractive to tourists. In a longer time perspective, the analysis of overnights sold indicates the emergence of a growing gap between urban and rural areas (Figures 4.9, 4.10). It was becoming increasingly wider in 2004, when Poland joined the EU. This reveals that international tourists, the vast majority of whom choose big cities (such as Krakow, Warsaw or Gdansk) as a tourist destination, have an impact on domestic tourism demand. The average growth rate (1995–2014) of the number of overnights sold in relation to the population was more than six times higher in cities than in urban areas; the growing rural population was among the contributing factors.

The greatest decline in the number of overnights sold in relation to the population was recorded in the Kujawsko-Pomorskie (by 40%) and Pomorskie (by 37%) voivodeships. Growth was experienced in only four voivodeships located in the southern (Podkarpackie voivodeship, +89%) and central (Łódzkie voivodeship, +57%; Mazowieckie voivodeship, +38%; Świętokrzyskie voivodeship, +48%) parts of the country (Figure 4.11).

The demand for rural tourism services is reflected by an indicator referred to as the tourism density index, expressed as the number of

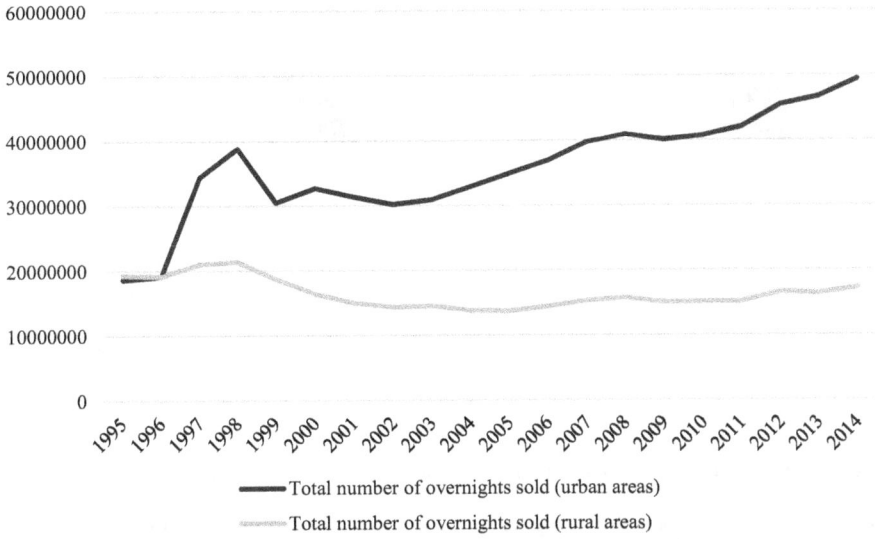

Figure 4.9 Number of overnights sold, disaggregated by the urban/rural dimension, 1995–2014.
Source: Own study based on the Local Data Bank of the Central Statistical Office.

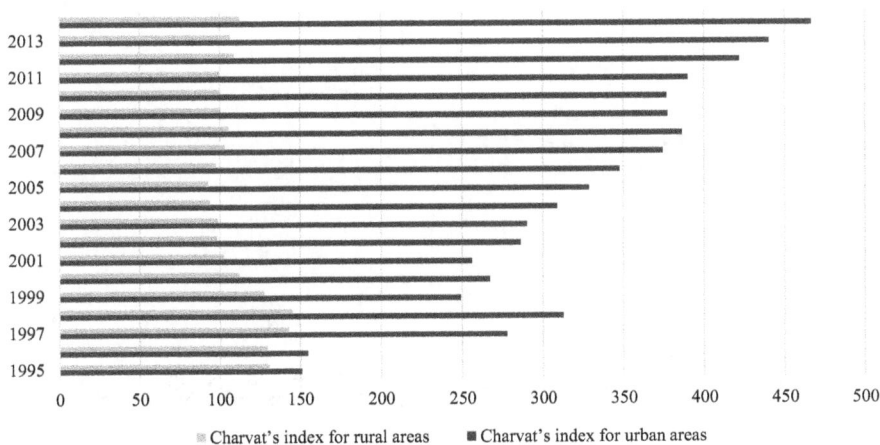

Figure 4.10 Charvat's index for rural and urban areas, 1995–2014.
Note: Index = (number of overnights sold / number of permanent residents) × 100.
Source: Own study based on the Local Data Bank of the Central Statistical Office.

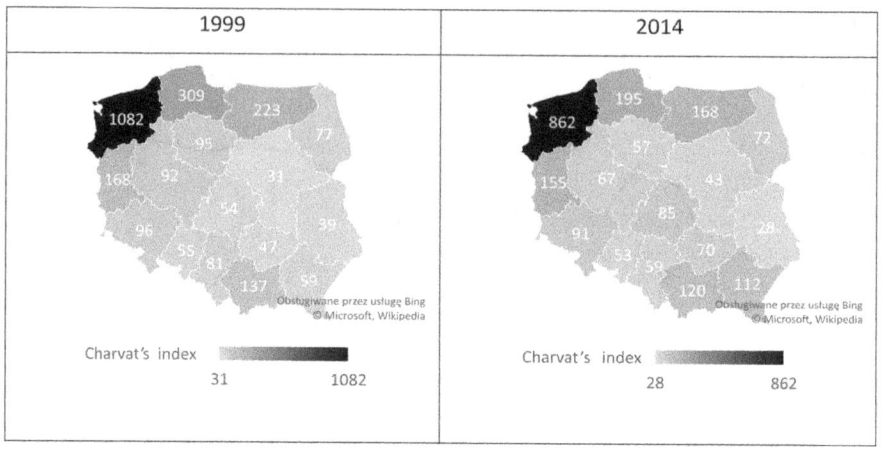

Figure 4.11 Comparison of the Charvat's index for rural areas between voivodeships, 1999 and 2014.
Note: Index = (number of overnights sold / number of permanent residents) × 100.
Source: Own study based on the Local Data Bank of the Central Statistical Office.

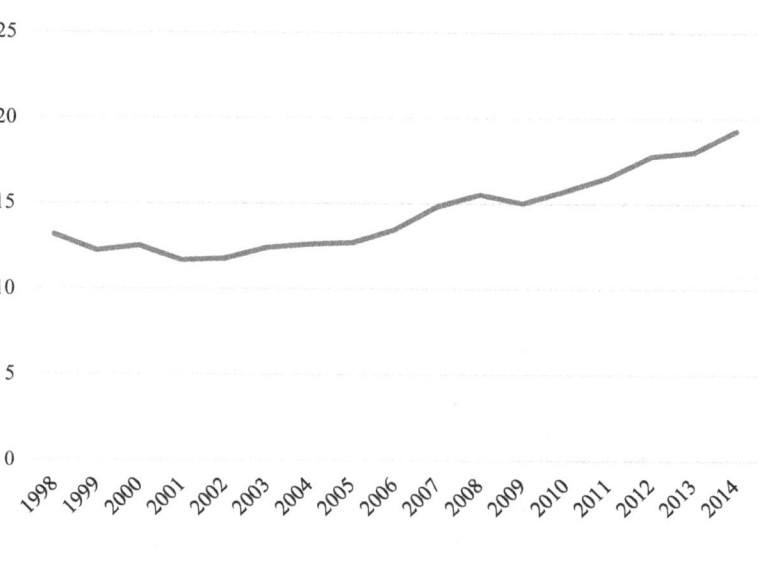

Figure 4.12 Tourism density index in Polish rural areas, 1998–2014.
Note: Index = (number of tourists / size of the territory [km²]).
Source: Own study based on the Local Data Bank of the Central Statistical Office.

accommodation users in relation to the size of a territorial unit. During 1999–2014, the growth rate for tourism traffic density in cities (+98%) was nearly twice as high as in rural areas (+57%). A detailed analysis of changes in that indicator for rural areas reveals that growth accelerated in 2009, at the time of the global financial crisis (Figure 4.12). It had an effect on the level and structure of consumption, including of tourism services in Poland. Tourism demand went down, and so did the expenditure on private and business trips, although there was an increase in the contribution of tourism to GDP (Kotra & Ruszkowski, 2012). In an effort to find cheaper forms of leisure, the Polish population used domestic tourism services, including in rural areas. In the context of the crisis engendered by COVID-19, rural service providers can again be reasonably expected to reap the greatest benefits of this situation.

Both in 1998 and 2014, the highest value of the tourism density index was recorded in the Małopolskie voivodeship. However, the highest growth rate was experienced in the central part of the country (Łódzkie voivodeship, +172%; Mazowieckie voivodeship, +153%).

In summarizing the analysis of changes to Schneider's, Charvat's, and the tourism density indexes, it should be emphasized that the fact alone of tourism traffic growing slower in rural areas than in cities is not an unfavorable process; in certain situations, it may enable a more sustainable development of tourism in the territory considered. The rapid changes in these indexes could trigger conflicts between residents and tourists, and lead

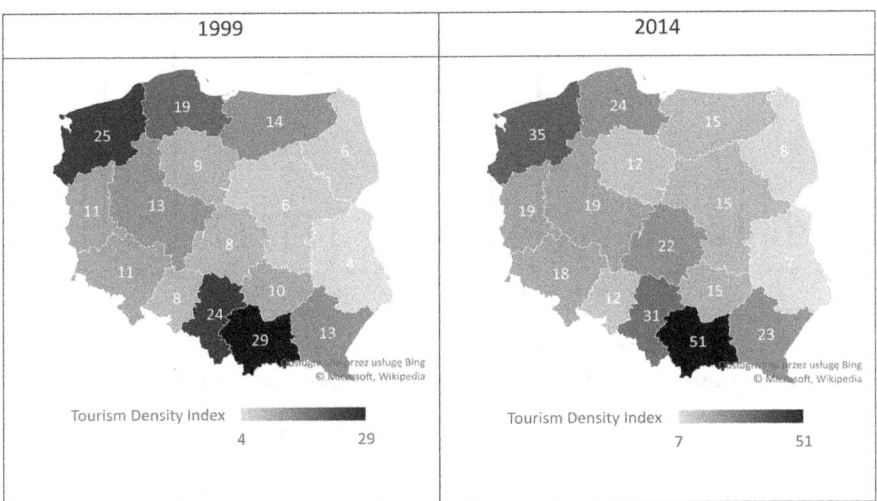

Figure 4.13 Comparison of the tourism density index in Polish rural areas, 1999 and 2014.
Note: Index = (number of tourists / size of the territory [km^2]).
Source: Own study based on the Local Data Bank of the Central Statistical Office.

to overtourism which, so far, is only characteristic of cities (Seraphin et al., 2018; Szromek et al., 2019). In the study period, the fastest favorable changes in demand for rural tourism services were recorded in the central (Łódzkie and Mazowieckie voivodeships) and southern (Podkarpackie voivodeship) parts of the country. This partly results from large conference (training and leisure) centers being located in rural areas which are easily accessible and conveniently located to big cities.

After the transformation, Poland experienced tremendous changes in the structure and size of accommodation facilities in both urban and rural areas. Until 1989, only 20% of all tourism facilities in Poland were operated on a commercial basis (Sala, 2017). The privatization and restructuring of state-owned enterprises and cooperatives resulted in a rapid decline in the number of social facilities financed mainly with social funds, such as large holiday resorts (a decline by 69%) and holiday youth centers (a decline by 62%), and an increase in the number of such facilities as hotels (by 610%), guest rooms (by 57%), and agritourism farms (by 24%) (Figure 4.14).

Currently, guest rooms and agritourism farms hold an important position in the mix of accommodation facilities in Polish rural areas. Large holiday

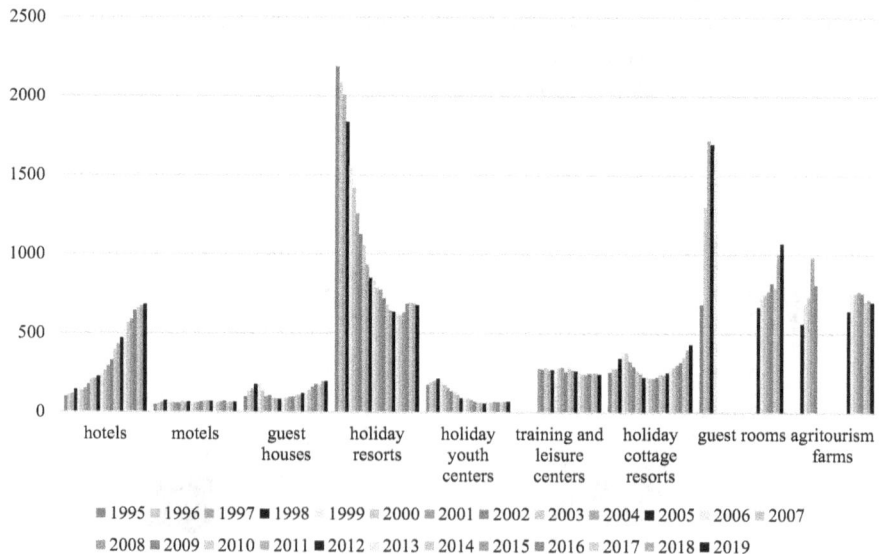

Figure 4.14 Change in the number of selected accommodation facilities in Polish rural areas, 1995–2019.
Note: In the study period, incomplete data was available for training and leisure centers (data for 1995–2000 was missing). Also, as regards guest rooms (missing data for 2000–2011) and agritourism farms (missing data for 1995–1997 and 2003–2011), there were changes in the data collection methodology. As a consequence, the data presented above is for illustrative purposes only.
Source: Local Data Bank, Central Statistical Office.

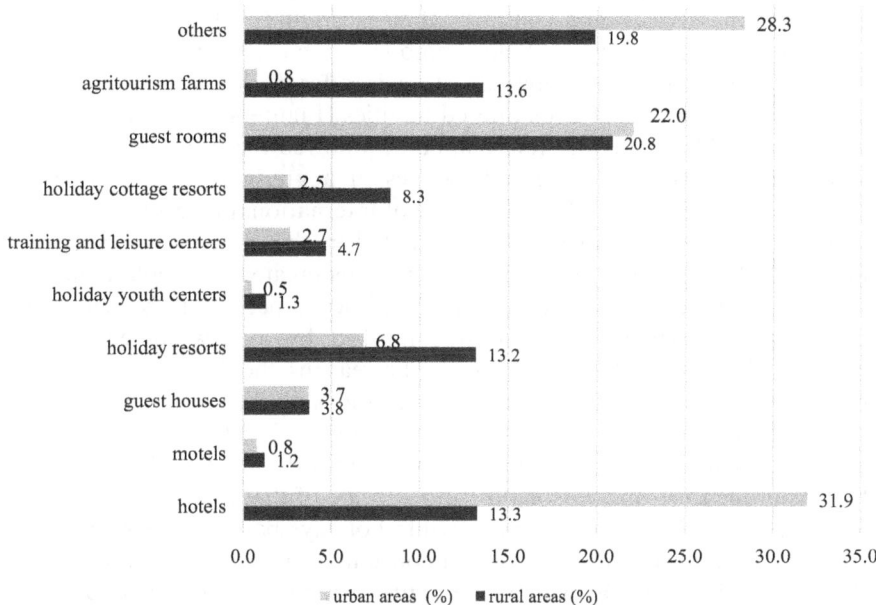

Figure 4.15 Share of selected accommodation facilities, disaggregated by the urban/rural dimension, in 2019 (%).
Source: Local Data Bank, Central Statistical Office.

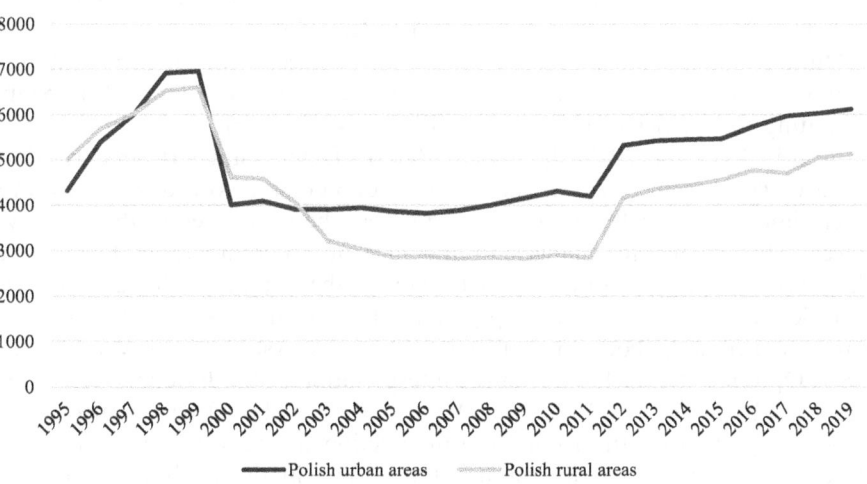

Figure 4.16 Number of tourist accommodation facilities in Poland, disaggregated by the urban/rural dimension.
Source: Local Data Bank, Central Statistical Office.

resorts keep their importance, too, although most of them are held privately and offer their services on a commercial basis (Figures 4.15, 4.16).

The changes in the mix of accommodation facilities referred to above caused a reduction in the number of beds offered in rural areas and an increase in the number of beds offered in cities (Figure 4.17). This was particularly noticeable in the first few years that followed the transformation. The second post-transformation decade witnessed a gradual integration of the Polish economy with the EU. The inflow of international investments and the interest in the new member country resulted in the growing importance of international leisure and business tourism. This created favorable conditions for the development of accommodation facilities, especially in big cities. The third post-transformation decade was marked by a slow increase in the number of beds offered in rural areas. The reasons should be sought in two opposite phenomena, namely overtourism and slow life. Rural and urban–rural communes located in the immediate vicinity of such natural attractions as sea or mountains are particularly exposed to a dysfunctional development of tourism due to strong seasonality and/or scale of that process. On the other hand, there is growing interest in slow-life holidays based on seeking a quiet and calm experience and an increased consumption of organic products (e.g. offered in agritourism farms). In the context of the present COVID-19 epidemic risks, it needs to be assumed that this trend will continue and may even strongly accelerate within the next 2–3 years.

The analysis of changes in accommodation facilities that followed the Polish economic transformation will be illustrated by the Baretje–Defert index (Baretje & Defert, 1972), otherwise known as the tourism function index and tourist accommodation density index, expressed as the number of beds in relation to the size of an administrative unit.

During 1995–2018, the Baretje–Defert's index declined in Polish rural areas (–9%) and increased in cities (+11%). In this case, it was impacted by people migrating from cities to urban areas (Figures 4.18, 4.19).

Both in 1999 and 2018, the highest Baretje–Defert's index (nearly five times the average) was recorded in the Zachodniopomorskie voivodeship. However, after almost 30 years from the transformation, the value itself of the index was lower. This is due to changes in the mix of accommodation facilities, as described above. A positive growth rate in the Baretje–Defert's index was recorded in only four voivodeships, mainly located in the southern and central part of the country (Podkarpackie, +15%; Świętokrzyskie, +15%; Mazowieckie, +8%; Opolskie, +4%). Following the interpretation of the Baretje–Defert index proposed by M. Boyer (1972) and Pearce (1995), it should be assumed that in most Polish regions, the rural tourism function is not a significant driver of development, and plays an important role only in the Zachodniopomorskie and Pomorskie voivodeships.[16] Even if reduced values of the Baretje–Defert index were used in the interpretation (due to them referring to rural rather than urban areas), it still would be clear that the development of Polish rural tourism is only a component of the rural multi-purpose development process.

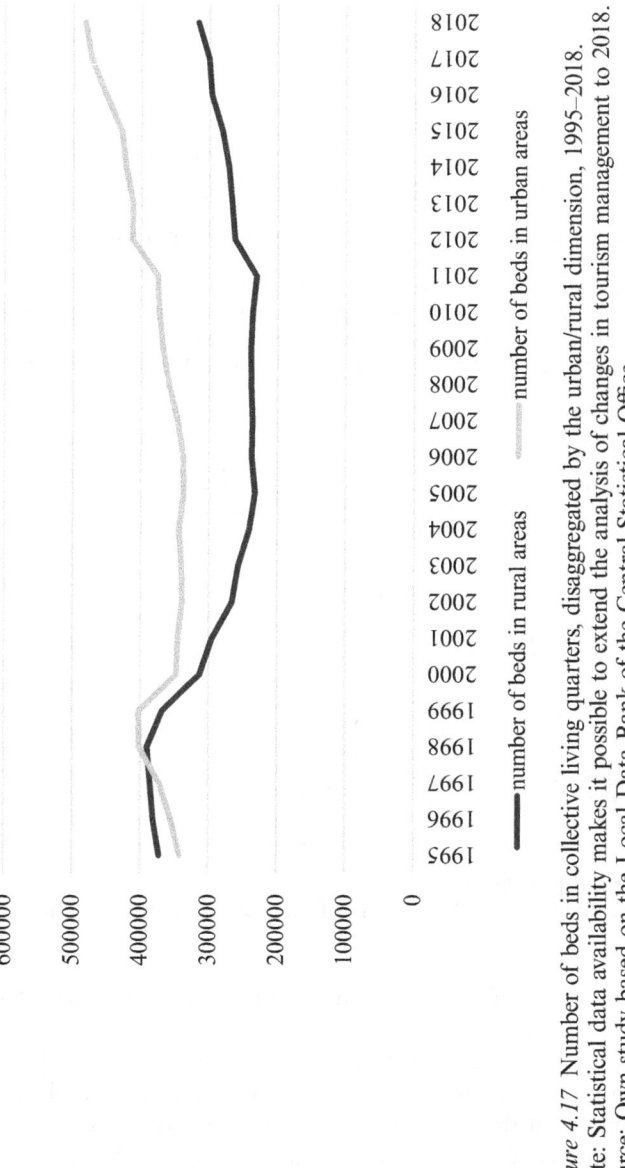

Figure 4.17 Number of beds in collective living quarters, disaggregated by the urban/rural dimension, 1995–2018.
Note: Statistical data availability makes it possible to extend the analysis of changes in tourism management to 2018.
Source: Own study based on the Local Data Bank of the Central Statistical Office.

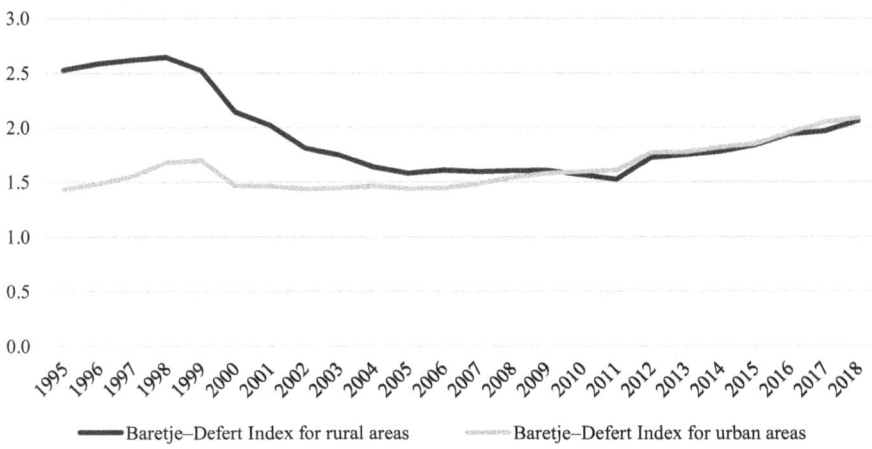

Figure 4.18 Baretje–Defert index in rural and urban areas, 1995–2018.
Note: Index = $(N/R) \times 100$, where N is the number of tourist accommodation establishments and R is the number of permanent residents of the area.
Source: Own study based on the Local Data Bank of the Central Statistical Office.

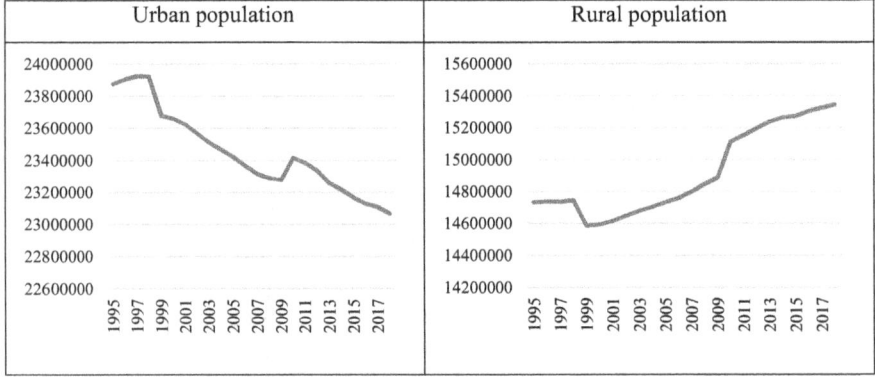

Figure 4.19 Urban and rural population, 1995–2017.
Source: Own study based on the Local Data Bank of the Central Statistical Office.

Usually, above-average indexes of the rural tourism function are due to extraordinary local values or proximity to big cities.

The tourist accommodation density index (number of rural accommodation facilities in a voivodeship per km^2) was also used in illustrating the supply-side changes in the development of the tourism function in Polish rural areas.

The average growth rate (1999–2018) for accommodation density across all Polish rural areas was negative (–7%). This was the consequence of changes in

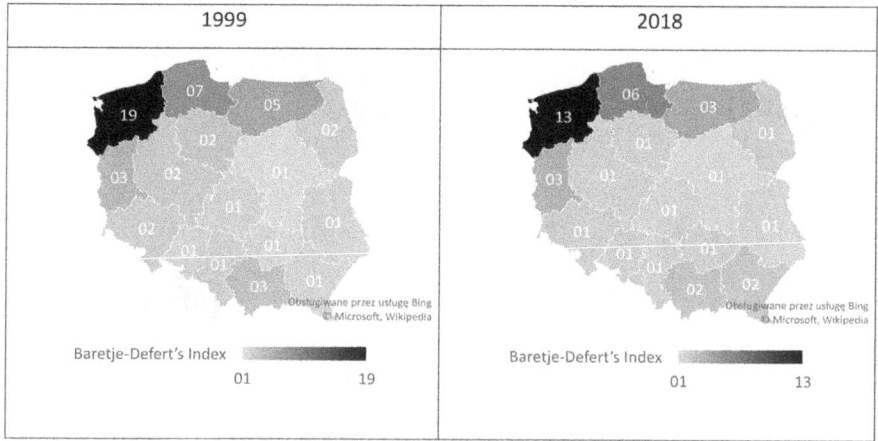

1999	2018

Figure 4.20 Comparison of the Baretje–Defert index for rural areas between voivodeships, 1999 and 2018.
Note: Index = $(N/R) \times 100$, where N is the number of tourist accommodation establishments and R is the number of permanent residents of the area.
Source: Own study based on the Local Data Bank of the Central Statistical Office.

the mix and number of accommodation facilities in Poland, as mentioned earlier.
The highest average growth rate was recorded in the southern (Podkarpackie
voivodeship, +17%) and central part of the country (Świętokrzyskie
voivodeship, +14%; Mazowieckie voivodeship, +11%) (Figure 4.21, 4.22).

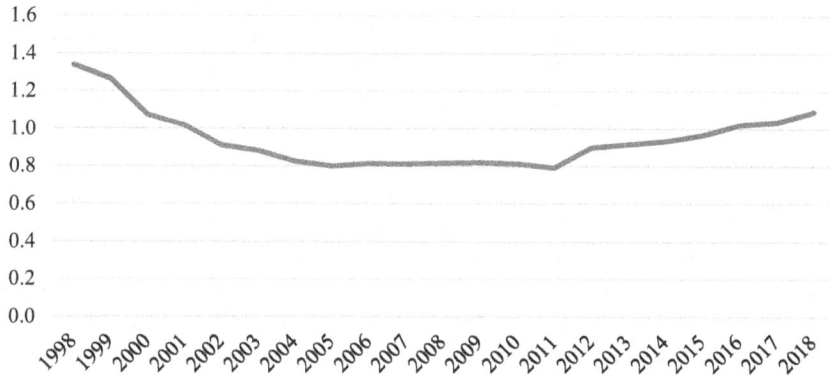

Figure 4.21 Tourist accommodation density index for rural areas, 1998–2018.
Note: Index = (number of tourist accommodation establishments / size of the territory
[km^2]). No territory size figures are available at commune level for 1995–1997 which
would enable data to be aggregated by the urban/rural dimension.
Source: Own study based on the Local Data Bank of the Central Statistical Office.

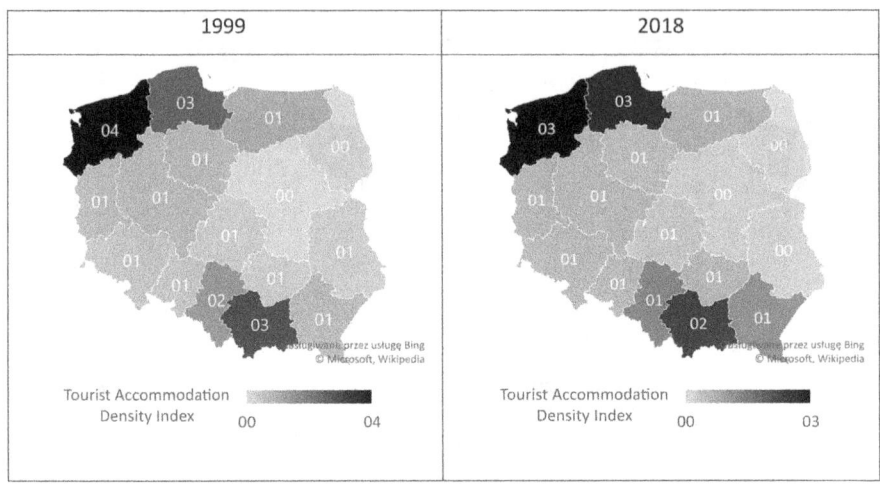

Figure 4.22 Tourist accommodation density index for rural areas, 1999 and 2018, disaggregated by region.
Note: Index = (number of tourist accommodation establishments / size of the territory [km^2]).
Source: Own study based on the Local Data Bank of the Central Statistical Office.

According to Wojciechowska (2006, 2018), changes in accommodation resources and in facility management practices resulted in the emergence of two different styles of tourism in Polish rural areas. The first one, currently on a decline, is a continuation of what was seen in previous historical periods regarding both the structure and the organization method of accommodation facilities. The second, currently in its development phase, is based on new tourist facilities intended to accommodate residential school trips during spring and winter,[17] and uses the existing individual accommodation facilities, such as agritourism farms (Wojciechowska, 2006). Over the last 10 years, the number of agritourism farms[18] in Poland has nearly doubled.[19] The greatest increase in agritourism farm numbers was recorded in the first and third post-transformation decade.

Furthermore, Poland witnesses the succession of the leisure function; indeed, the location of agritourism farms overlaps with that of holiday villages active in mid-1970s (Wojciechowska, 2018) (Figure 4.24).

However, despite continuous growth in the number of rural accommodation providers, agritourism activities are largely unstable; many providers discontinue their agritourism businesses and are replaced by newcomers (Wojciechowska, 2002; Wiatrak, 2003; Przezbórska, 2005a; Wojciechowska, 2018). Note also that due to lack of legislative solutions for the legal protection of the notion itself of "agritourism farm," the offering of some agritourism farms does not include anything related to farm resources, which is sometimes viewed as an important downside. Therefore, the "Hospitable

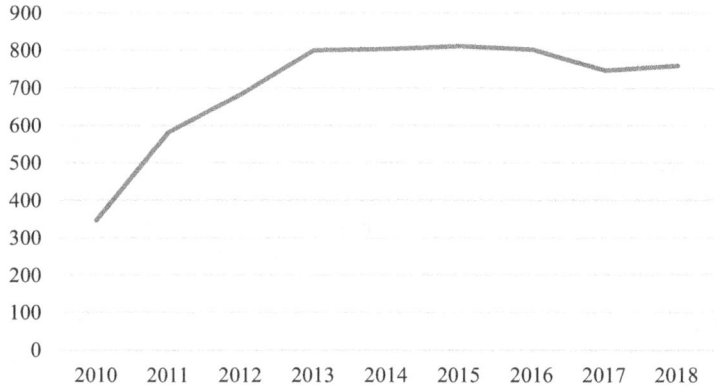

Figure 4.23 Number of agritourism farms in Poland, 2010–2018 (see notes 18 and 19).
Source: Local Data Bank, Central Statistical Office.

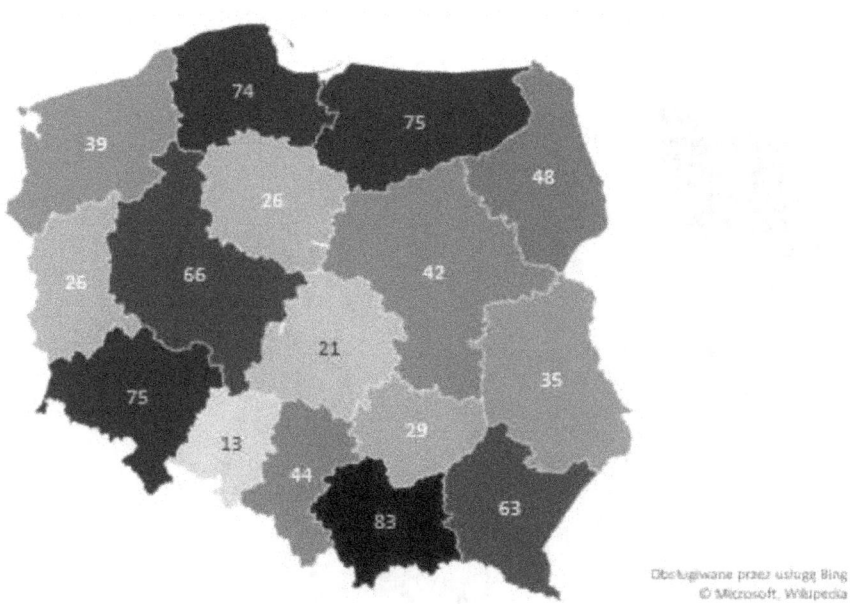

Figure 4.24 Number of agritourism farms in 2018, disaggregated by region.
Note: This should be viewed as estimate data because information delivered by Agricultural Consultancy Centers is slightly different. However, as this chapter refers to the total number of accommodation facilities, data from the Central Statistical Office's Local Data Bank is presented (although these values may be assumed to be underestimated).
Source: Local Data Bank, Central Statistical Office.

Farms" Polish Federation of Rural Tourism, as part of a voluntary classification scheme, assigns the farms to one of two categories: farm holidays or rural holidays.

In the first two post-transformation decades, the main areas of improvement were the equipment of agritourism farms and the scope of services offered. In the third post-transformation decade, efforts made by agritourism farms to pave their way to success largely consist in specialization measures and in offering recreation and caretaking services to urban dwellers. The analysis of average growth in the number of agritourism farms during 2010–2018 suggests that the regions considerably differ from one another in that respect (Figure 4.25).

In the Zachodniopomorskie and Lubelskie voivodeships, the number of agritourism farms grew more than twice as fast as the average national level. It should be assumed that in the case of the Zachodniopomorskie voivodeship, this is the consequence of a convenient location with respect to seaside and the German border. When it comes to the Lubelskie voivodeship, a factor that contributes to agritourism development is the traditional, fragmented farming system; because of their poor cost-efficiency, small farmers are somehow forced to seek non-agricultural revenue streams.

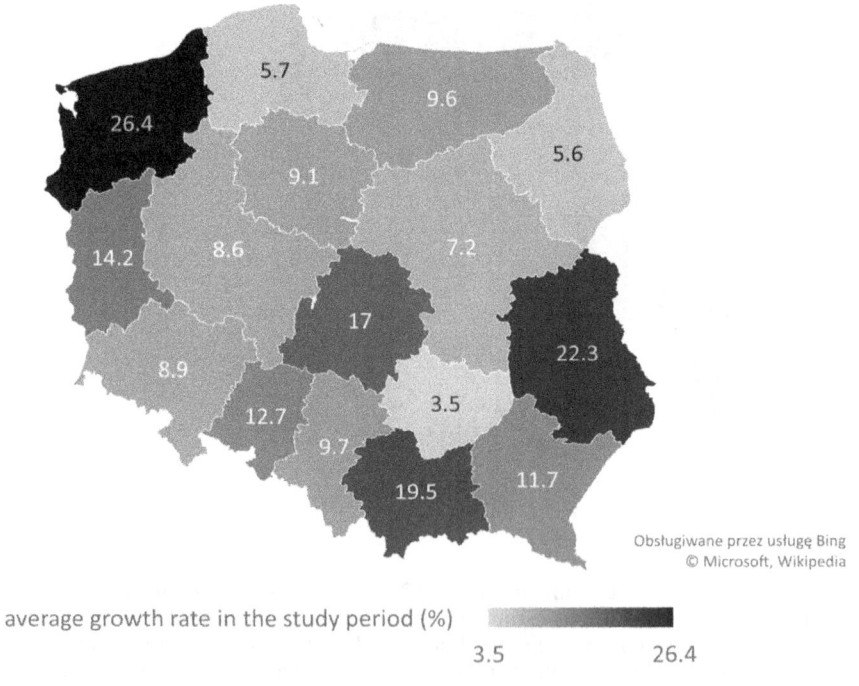

average growth rate in the study period (%)

3.5   26.4

Figure 4.25 Average growth in the number of agritourism farms, 2010–2018 (%).
Source: Local Data Bank, Central Statistical Office.

In summarizing growth of the number of beds offered in all types of collective living quarters in rural areas, it should be emphasized that the fastest advantageous developments in this respect were witnessed in the central-south part of the country (Podkarpackie and Świętokrzyskie voivodeships).

According to Central Statistical Office data, the average annual occupancy rate for agritourism farms is 15.3%. Although it has grown consistently within the last 10 years, it remains by far lower than what is recorded in facilities such as guest rooms (26.6%) or holiday resorts (40.9%). Moreover, the highest occupancy rate is observed in agritourism farms located in the north-western and south-eastern part of the country. This can be explained by reasons of transport accessibility and natural assets, as mentioned above (Figure 4.26).

For a more complete picture of territorial differences in the tourism function of Polish rural areas and of changes they underwent during 1999–2018,[20] this study also used Hellwig's development pattern, a widely used method (Hellwig, 1968; Stec, 2015; Pomianek, 2019). As a synthetic metric of the combined effect of variables covered by the analysis, it allows the units to be ordred by distance to a defined artificial ideal development scenario. The distance between each unit and P_0, the point which corresponds to the ideal

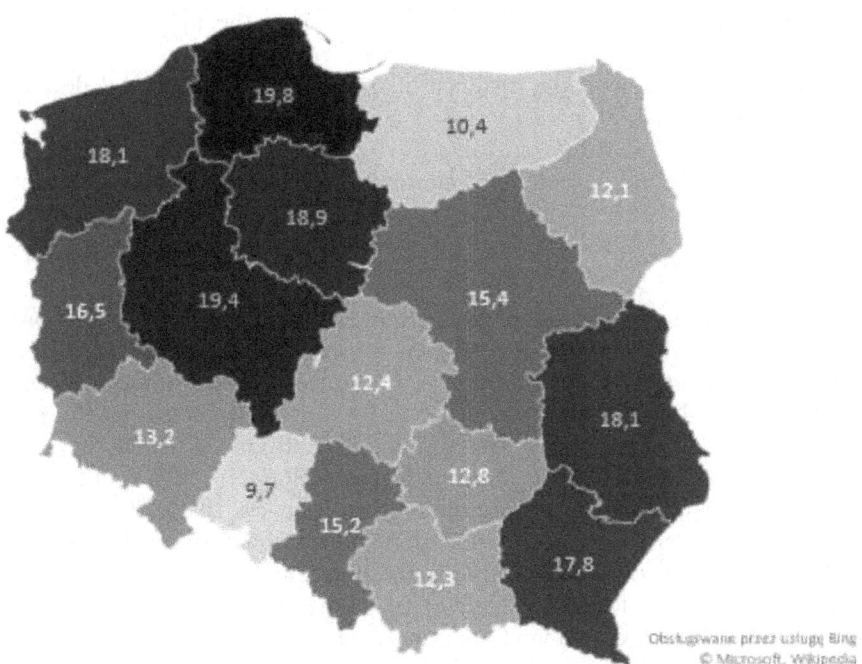

Figure 4.26 Occupancy rate (%) for agritourism farms in 2018, disaggregated by voivodeship.
Source: Local Data Bank, Central Statistical Office.

development scenario, is designated C_{i0} and is calculated as follows:

$$C_{i0} = \left[\sum_{j=1}^{n} \left(z_{ij} - z_{0j}\right)^2\right]^{\frac{1}{2}} \quad (i = 1, ..., \omega).$$

$$C_0 = \bar{c}_0 + 2S_0,$$

$$\bar{c}_0 = \frac{1}{\omega}\sum_{i=1}^{\omega} c_{i0},$$

$$S_0 = \left[\frac{1}{\omega}\sum_{i=1}^{\omega}(c_{i0} - \bar{c}_0)^2\right]^{\frac{1}{2}}$$

where:

z_{ij}—standardized value of the variable j in the unit i;
z_{0j}—standardized value of the variable j in the best development scenario;
c_{i0} —distance of the unit i from the ideal development scenario;
$\bar{c}_0$ —arithmetic mean of the distance of unit i from the ideal development scenario;
S_0—standard deviation of the distance of the unit i from the ideal development scenario.

Defined as such, the development metric was used in the following form:

$$d_i = 1 - \frac{c_{i0}}{\bar{c}_0}$$

The selected research procedure gives grounds for concluding that the more a unit is developed (in this case: the more its tourism function is developed), the closer the synthetic development metric d_i is to one. Of all the indexes used earlier, the Baretje–Defert index was retained for further analyses (as the most commonly used). An indicator was additionally introduced which shows the number of economic operators entered to the REGON[21] register per 1,000 population. Usually, when assessing the tourism function, that indicator is narrowed down to providers of accommodation and catering services. This leads to the discarding of operators engaged in organizing tourism, cultural, leisure, and recreational activities, for instance. Over the study period, the types of activity were rearranged between different sections of the Polish Classification of Activity. Hence, the calculations used an indicator expressed as the number of all rural economic operators entered to the register per 1,000 population, assuming that it is not supposed to reflect the development of the tourism function in a strict sense (as the Baretje–Defert's index was used for that purpose) but that tourism is only a component of the multipurpose development process.

The next step consisted in standardizing the variables and ensuring that no correlation exists between them. Both variables covered by this analysis had a stimulating effect. Following this, based on the value of the synthetic indicator \overline d_i, five development levels (classes) of the rural tourism function were established for 1999 and 2018. When setting the boundaries of class intervals, account was taken of the arithmetic mean $(\overline{d_i})$ and standard deviation (SD), as per the routine proposed by Pomianek (2019):

1st class (very high development level of the tourism function): $d_i > \overline{d_i} + SD$;
2nd class (high development level of the tourism function) $\overline{d_i} + SD \geq d_i > \overline{d_i} + \frac{1}{2}SD$;
3rd class (medium development level of the tourism function): $\overline{d_i} + \frac{1}{2}SD \geq d_i > \overline{d_i} - \frac{1}{2}SD$;
4th class (low development level of the tourism function): $\overline{d_i} - \frac{1}{2}SD \geq d_i > \overline{d_i} - SD$;
5th class (very low development level of the tourism function): $d_i \leq \overline{d_i} - SD$.

Table 4.5 presents the voivodeships ranked by development level of the tourism function in Polish rural areas.

The analysis based on the ideal development scenario suggests that the tourism function in Polish rural areas develops at a sluggish pace. The vast majority of voivodeships (11 of them, i.e. 68.7%) were assigned to the 3rd class both in 1999 and 2018, which means their tourism function was at a low development level. Moreover, the top five voivodeships that reached the highest development levels of the tourism function in 1999 continue to be top-ranked after nearly 20 years. The gap in development level of the tourism function between the Zachodniopomorskie and Pomorskie voivodeships reduced by 11%. The latter moved up to the 1st class, which is composed of voivodeships at a very high development level of rural tourism. The greatest improvement in the ranking of the development of the rural tourism function was observed in the Podkarpackie (five ranks up) and Łódzkie (four ranks up) voivodeships (Figure 4.27).

The use of Hellwig's ideal development indicator allowed us to demonstrate that the territorial distribution of rural areas with a dominant tourism function did not substantially change between 1999 and 2018, and that the changes were a sluggish process evenly distributed across the territory. The share of voivodeships where the rural tourism function is at a medium development level remained high throughout the study period. This proves that tourism is not a fundamental driver of development for most Polish rural areas. Despite the political, economic, and structural changes experienced in Poland, the basic economic functions continue to be centered around agriculture and forestry in the largest group of rural areas (Bański & Stola, 2002; Bański, 2004a, 2006, 2012; Stanny et al., 2018). The rural

Table 4.5 Voivodeships ranked by development level of the tourism function in 1999 and 2018 based on Hellwig's indicator[a]

Voivodship	Value of d_i 1999	Class	Place	Value of d_i 2018	Class	Place	Change in place (2018 vs. 1999)
Zachodniopo-morskie	0.850	1	1	0.930	1	1	0
Pomorskie	0.438	2	2	0.584	1	2	0
Warmińsko-mazurskie	0.372	3	3	0.409	3	3	0
Lubuskie	0.314	3	4	0.380	3	4	0
Małopolskie	0.284	3	5	0.311	3	5	0
Podkarpackie	0.236	3	11	0.301	3	6	+5
Dolnośląskie	0.247	3	9	0.281	3	7	+2
Łódźkie	0.235	3	12	0.275	3	8	+4
Podlaskie	0.249	3	8	0.273	3	9	−1
Wielkopolskie	0.254	3	7	0.272	3	10	−3
Kujawsko-Pomorskie	0.264	3	6	0.271	3	11	−5
Śląskie	0.239	3	10	0.262	3	12	−2
Świętokrzyskie	0.220	4	15	0.260	3	13	+2
Opolskie	0.228	3	14	0.250	4	14	0
Lubelskie	0.232	4	13	0.241	4	15	−2
Mazowieckie	0.213	4	16	0.234	4	16	0

Source: Own compilation.

a Data in the table was sorted in descending order by the indicator of the development of the tourism function in 2018.

tourism function develops and becomes stronger mostly in traditional tourism regions which offer valuable natural assets (Zachodniopomorskie, Pomorskie, and Podkarpackie voivodeships), as also confirmed in studies by other authors (Wiatrak, 2003; Wojciechowska, 2009; Durydiwka, 2006, 2008, 2009, 2012, 2015; Wojciechowska, 2018). The greatest limitations to the development of the rural tourism function are found in territories which used to be home to state-owned agricultural holdings. With the initiation of the economic transformation process and the accession to the EU, the last three decades have played a particular role in the economic development of Poland. The implementation of free market principles and openness to international capital brought an opportunity for ownership transformation on the supply side of tourism services. This, in turn, triggered a series of changes on the demand side. The changing realities also resulted in a slow evolution of development levels of the rural tourism function, and of how

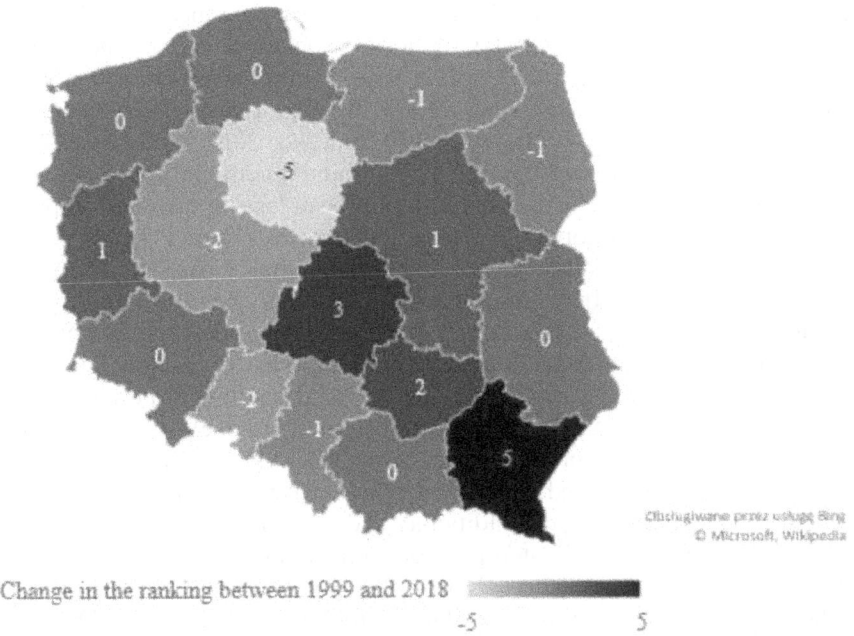

Change in the ranking between 1999 and 2018

-5   5

Figure 4.27 Changes in the ranking of the development level of the tourism function based on Hellwig's indicator (2018 vs. 1999).
Source: Own calculations based on the Local Data Bank of the Central Statistical Office.

the importance of multipurpose development processes is viewed. Rural areas started to enhance their functions; in addition to the traditional agricultural and tourism function, they now offer educational, recreational and caretaking functions which allow them to expand their customer base with same-day visitors (in addition to tourists).

4.4 Using the potential behind regional and traditional food in the development of tourism products

As the European consumers become wealthier and increasingly interested in environmental protection and food safety, the consumption culture is slowly undergoing a process referred to as the "quality turn" (Goodman, 2003; Biénabe et al., 2011). This means that regional and traditional food can be viewed as one of the determinants of regional competitiveness. Due to the particularities of food, competitiveness could mean rivalry both between producers of regional/traditional foods (including agritourism farms) and between regional culinary cultures. Hence, a competitive edge in the market for rural tourism services can be built upon attractive natural resources and natural heritage values (including culinary culture), and by highlighting their role as a component of tourism products. Also, the relevant literature indicates many functions of traditional regional products, such as: being an additional non-agricultural revenue stream for the local population; creating new jobs in the production, processing and distribution of traditional products; supporting rural development; and preserving cultural heritage (Kuźniar, 2010; Sieczko, 2014; Jęczmyk et al., 2014; Orłowski & Woźniczko, 2015; Grębowiec, 2017). The capacity to use the potential of regional and traditional foods was additionally enhanced under the national and European legal systems which protect them. This allows a monopoly in local markets to be ensured and more earnings in return for efforts made to preserve product quality based on recipes and technological processes that have long remained unchanged. The first regulations for the protection of geographical indications and designations of origin for agricultural products came in the early 1990s.[22] This was mainly initiated by France, which had already established the National Institute of Designations of Origin (INAO, Institut National des Appellations d'Origine) in 1935 to control the quality of agri-food products (Lipińska, 2008). As a consequence, three designations were introduced to the European market to ensure protection and emphasize the uniqueness of traditional and regional products (Table 4.6). Product registration is a two-step process, with the first step being executed at national level.

The analysis of the numbers of Polish products registered as PDO, PGI, and GTS (Figure 4.28) shows that their structure changes over time. A total of 15 products (4 PDOs; 5 PGIs; 6 TSGs) were registered in the second post-transformation decade, and were nearly equally distributed between the designations. During 2010–2019, 27 more products were registered, with a noticeable majority labeled as PGI. Currently, this is the most commonly used food product labeling around the world (of those listed above).[23] Note also that the first Polish product registered in the European Union was *Bryndza podhalańska* (2007), a PDO cheese.

The analysis of the total number of registered CEE products, taking the two sub-periods (2000–2009 and 2010–2019) into account, shows that Poland was among the few countries that successively increased the number of legally labeled foodstuffs. The fastest growth in the number of products registered during 2010–2019 was recorded in Croatia and Slovenia (Figure 4.29).

However, the growing interest in the labeling system among Polish producers does not have an impact on consumer behavior. Surveys carried out to investigate the recognition of food product designations among Polish buyers revealed a low level of awareness. Oleksiuk and Werenowska (2019) proved that TSG, PDP, and POG labels were recognized by only 38%, 16%, and 16% of respondents, respectively, even though nearly all of them declared that they purchase traditional (92%) and regional (89%) products. Similar results were reported by other authors, too (Grębowiec, 2010; Tomaszewska et al., 2014).

Furthermore, pursuant to the Act of December 17, 2004[24] on the registration and protection of names and designations of agricultural products and foodstuffs and on traditional products, Polish producers can enter their

Table 4.6 Specification of product name protections in the EU

Labels	EU quality schemes
Protected Designation of Origin (PDO)	Name of the region (specific place/country) designating an agricultural product or a foodstuff which originates from that region (specific place/country): • product quality or properties are determined by the geographical environment, including natural and human factors; • harvesting, processing and preparation (all phases, until obtaining a final product) take place within the determined geographical area
Protected Geographical Indication (PGI)	Name of a region, a specific place or, in exceptional cases, the name of a country, used as a description of an agricultural product or a foodstuff which originates from that region (place or country): – the product or a foodstuff has a specific quality, goodwill or other characteristic property, attributable to its geographical origin; – at least one of the stages of harvesting, processing or preparation takes place in the designated geographic area; – raw materials used in the production process can originate from other areas
Traditional Specialities Guaranteed (TSG)	Label carried by products with a traditional name which refers to its specific nature or is traditionally used to designate that product: • the product has been manufactured with the use of traditional raw materials, in accordance with a traditional recipe handed down from generation to generation or with traditional methods for no less than 30 years; • the production process has some characteristic properties which make the product clearly stand apart from similar products of the same category

Source: https://ec.europa.eu/info/food-farming-fisheries/food-safety-and-quality/certification/quality-labels/quality-schemes-explained_en

products to the Traditional Products List kept by the Minister of Agriculture and Rural Development in cooperation with voivodeship marshals. The list includes products (Figure 4.30) whose quality or unique features and properties result from the use of traditional production methods (for no less than 25 years), which are part of their region's cultural heritage and an element of the identity of the local community. The purpose of entering a product to the Traditional Products List is not to ensure legal protection but only to identify and promote traditionally made products and to prepare the producers, as and if

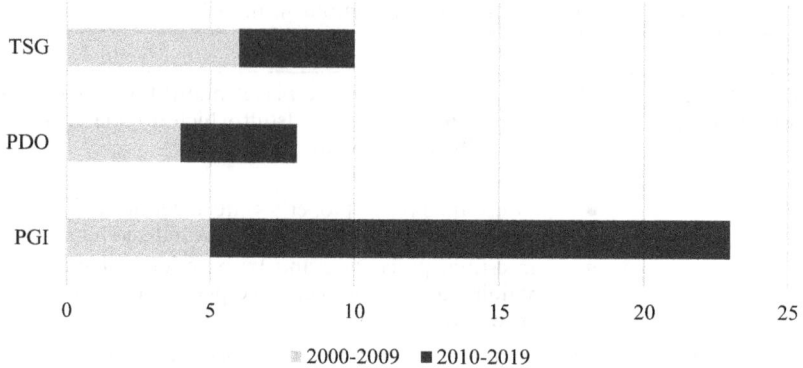

Figure 4.28 Number of products registered as PDOs, PGIs, and GTS in Poland.
Source: DOOR database, as of December 20, 2019.

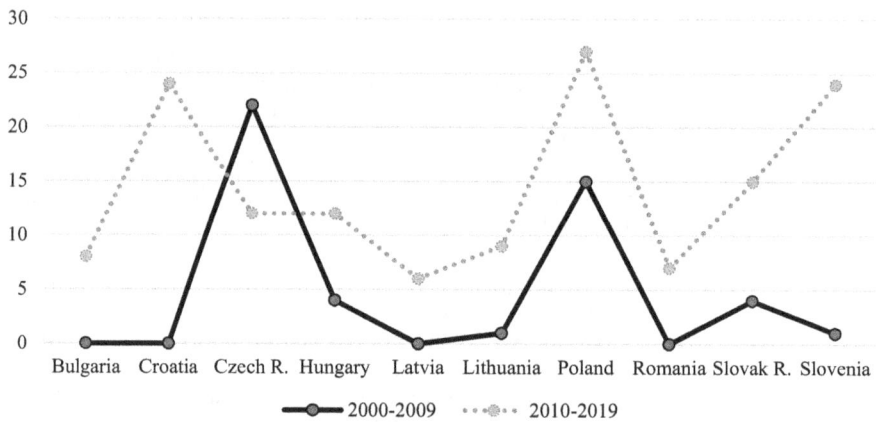

Figure 4.29 Total number of products registered as PDOs, PGIs, and GTS in CEE
          countries.
Note: The analysis covered only the countries which reported products for registration.
The countries were those selected in Krstić et al. (2017).
Source: DOOR database, as of December 20, 2019.

needed, to meet the formal and organizational requirements for registering a
product under the EU legal protection schemes.

As shown by the map (Figure 4.31), the distribution of the numbers of products entered to the Traditional Products List differs strongly between voivodeships. There is a noticeable increase in the number of traditional products in the
south-east of the country (Podkarpackie voivodeship: 241; Małopolskie voivodeship: 220; Lubelskie voivodeship: 216) which has an extensive agriculture and a
strongly preserved regional identity of local communities. The Traditional Products List includes 179 products from the Pomorskie voivodeship whose today's

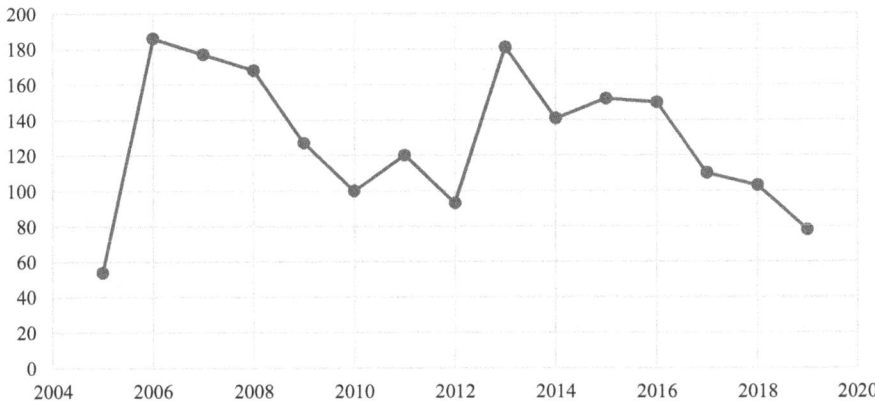

Figure 4.30 Number of products entered annually to the Traditional Products List.
Note: currently, the list includes 1967 products.
Source: Own compilation based on data from the Ministry of Agriculture and Rural Development.

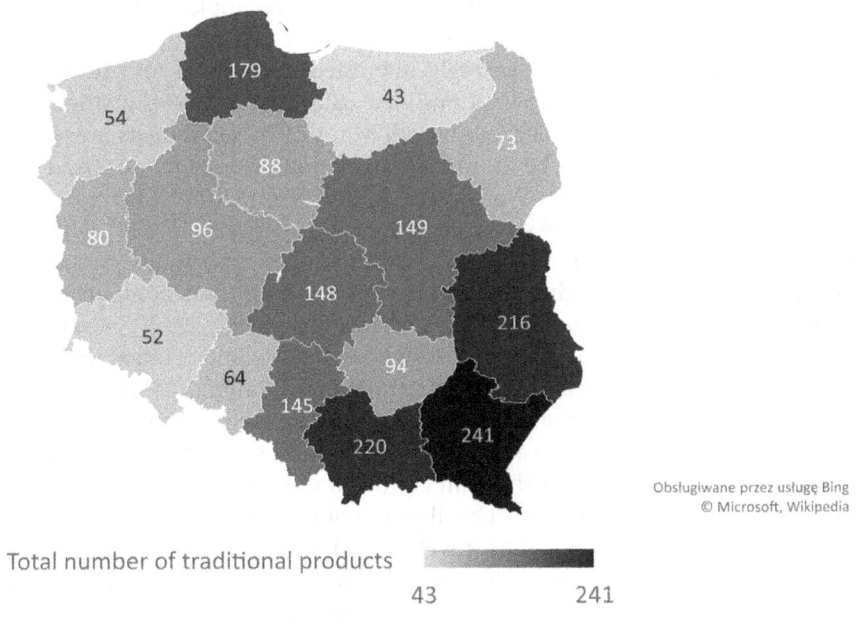

Total number of traditional products

43   241

Figure 4.31 Territorial distribution of the number of products entered to the Traditional Products List.
Source: Own compilation based on data from the Ministry of Agriculture and Rural Development, prepared in Microsoft Excel 365.

local cuisine was strongly influenced by German colonization and the era of Prussian partition. What also can be seen is the small concentration of traditional products in the western and north-eastern part of Poland. This results from the relocation of the population after World War 2 and the interrupted cultural continuity of these regions. The above can be illustrated by the example of the Dolnośląskie voivodeship which, of all the Polish voivodeships, has the smallest number of traditional products (52) entered to the list. The analysis of these products reflects the different origins of the post-war population of Lower Silesia. After World War 2, nearly all the residents were relocated and the territory was repopulated by people coming from different parts of Poland and Europe, including the Borderlands, the Kielce region, the Rzeszów region, Mazovia, and Polesia. Returnees from Bosnia, Čadca mountaineers, and Lemkos (Szczepańska & Szczepański, 2019) also formed a large group.

Also, certain regions (e.g. the Podlaskie voivodeship) are multicultural as they combine the native traditional Polish culture with ethnic minorities (e.g. Lemkos, Tatars, Crimean Karaites, Romani people). For the regional cuisine, this is an enriching mix which can provide rural tourism products with a competitive advantage in market. Traditional foods can become part of regional tourism brands and be considered in regional strategies for tourism development (Królikowska & Pijet-Migoń, 2019). Moreover, in Podkarpackie and Dolnośląskie voivodeships, traditional foods (as an endogenous regional resource) were identified as a priority for the agricultural sector in the implementation of smart regional specializations (Wiatrak, 2017).

As the tourists become increasingly interested in regional and traditionally made food products, many destinations view catering services as a strategic element when defining their image and brand. Many of these products are made on farms, and can therefore be a tourist attraction, especially in rural areas (Mokras-Grabowska, 2009; Woźniczko & Orłowski, 2011; Marzejon-Frycz, 2012; Pisarek et al., 2017). As revealed by surveys carried out with Polish consumers, the availability of regional cuisine and the opportunity to taste and purchase products made directly on farms is an important component of an agritourism offering (Zawadka, 2012; Uglis & Krysińska, 2012; Świstak et al., 2013). Moreover, 83% of foreign tourists who visited Poland claimed that it has a potential for the development of culinary tourism (Dominik, 2019).

"Culinary tourism" is the sole term consistently used in this chapter (without discussing an alternative terminology), and is defined as an activity focused on farmers' markets that can contribute to rural sustainability (Silkes, 2012). The Polish literature on the subject views culinary tourism as a part of cultural tourism and defines it as traveling with the intent to explore local and regional food products and dishes while also getting to know the culinary traditions, considered to be the main tourism attraction (Durydiwka, 2013; Orłowski & Woźniczko, 2015; Stasiak, 2015; Woźniak & Batyk, 2017). The forms of culinary tourism include culinary trails, festivals, thematic villages, and local and open-air markets. As shown in previous discussions, each Polish voivodeship has a relatively high potential for the creation of culinary tourism products. However, that potential differs between regions. Established in 2015 on the initiative of the Polish Tourism Organization, the "Polish Culinary Trails" consortium is a voluntary informal association.[25] The consortium's activity meets the assumptions of the Poland's 2012–2020 Marketing Strategy for the tourism sector, and places the main focus on creating branded tourism products underpinned by the culinary heritage, and on creating better conditions for selling them domestically and internationally.

The culinary trails operated by the consortium differ in nature and territorial scope, e.g.:

- they promote a city (Białystok, Gdynia, Gdansk, Kalisz, Poznań);
- they promote a region (Culinary Stopovers of the Suwałki Region and Masuria, the Tastes of Podhale);
- they promote regional traditions of manufacturing specific products (the Wine-Making Trails of Sandomierz, the Taste of Prunes, the Oscypek Cheese Trails, the Goose Meat Trail of Kuyavia and Pomerania).

These trails were created and are financed by different operators: regional and local tourism organizations, local government units, foundations, and associations (Woźniak & Batyk, 2017). Other entities engaged in promoting and protecting regional products in Poland include the National Rural Network, the Polish Chamber of Regional and Local Products and—in broader terms—the European Network of Regional Culinary Heritage and Slow Food International.

In addition to the trails listed above, Poland also offers a number of cultural trails which, to a different degree, relate to local traditions of manufacturing and consuming foods, including as part of other, different cultures. These include the Tatar Route in Podlasia and the Mennonite Cultural Trail of Lower Vistula which promotes the tradition of Mennonites who settled in these territories in the 16th and 17th century (Makała, 2014; Woźniak & Batyk, 2017). Selected examples of culinary trails in Polish rural areas are presented in Table 4.7.

In Poland, the potential behind regional and traditional products is widely used as a basis for organizing occasional or periodic culinary events. They largely take place in rural areas as a regional tourism attraction. The Farmer's Wives' Associations[26] play an important role in organizing such events. They take integration measures for the whole local community, including efforts made to sustain the region's culinary heritage (Szczepańska & Szczepański, 2019). Culinary events are also initiated by open-air museums which, due to their resources, are in a position to organize presentations of historical methods of harvesting, processing and preserving food products. In addition to visiting the open-air museum and watching the traditional rural cuisine exposition, the tourists can use historical cooking utensils and equipment (or their replicas) (Orłowski & Woźniczko, 2007). Selected examples of culinary events in Polish rural areas are presented in Table 4.8.

The region's culinary heritage is also viewed by Polish local communities as an endogenous resource which underpins their concept of thematic villages. In

Table 4.7 Selected examples of culinary trails in Polish rural areas

Year established	Name and characteristics of culinary trails
2013	Subcarpathian Tastes: the concept of these trails was prepared by the "Pro Carpathia" association together with the Marshall's Office of the Podkarpackie Voivodeship and the Subcarpathian Regional Tourism Organization. The Subcarpathian Tastes Culinary Trails include three culinary routes, namely the Bieszczady Route, the Beskidy-Pogórze Route, and the Northern Route, and a total of 50 facilities (guest houses, taverns, inns, roadhouses, farms, recreation lefts, agritourism establishments, bars/cafés, cafés, restaurants, villas, and mansions) specializing in regional, ethnic, courtly, and hunting cuisine
2005	The ca. 200 km long "Sandomierz Apple Trail" is located in the Świętokrzyskie voivodeship. Its main path runs through 30 orchard farms with many years of horticultural tradition as producers of apples, pears, stone fruit, berries, and hazelnuts. The trail is an opportunity to acquire horticultural know-how and learn about crops and fruit-based production processes. The visitors may also see how to cultivate, store and process fruit, or attend trainings on fruit tree pruning and caretaking. The tourists may taste and buy fresh fruit (including both old, forgotten, and modern varieties) and home-made processed products. The trails include 16 agritourism farms with six pre-arranged options of stay
2011	Located in the Śląskie voivodeship, the Bielsko Land Carp Route was created on the initiative of 12 local fishery groups based in southern Poland as part of measure 4.2 "Support for interregional and international cooperation" of the 2007–2013 Operational Program for the Sustainable Development of the Fisheries Sector and Coastal Fishing Areas. The trails project has 75 members, including fish farms which produce carp and other freshwater fish, angling facilities, restaurants, and one agritourism farm
2009	The Lesser Poland Herbal Village Trail is the joint initiative of the Nowy Sącz Tourism Organization and the Małopolskie voivodeship, and consists of 19 agritourism farms with herbal gardens. The farmers know much about growing herbs and of their types, uses, and medicinal properties. They also sell herbal souvenirs (such as herbal bags, herbal cushions, and herb seedlings) made by themselves or by local folk artists. All farms were classified in accordance with the requirements defined by the Polish Rural Tourism Federation. Most of them are holders of organic certificates

Source: Own compilation based on websites of each trail and of the Polish Tourism Organization.

Poland, thematic villages have been in place for more than two decades and have developed quite rapidly (Kłoczko-Gajewska, 2015; Jankun et al., 2016; Sala, J., 2016). In the early years, thematic villages were established in areas which struggled with development challenges and social problems such as unemployment and depopulation (Bielski, 2011; Czapiewska, 2012). They first emerged in voivodeships where local authorities or local government

Table 4.8 Selected examples of culinary events in Polish rural areas

Exhibitors/ visitors[a]	Name and characteristics of culinary events
150/ 40,000	The Festival of Taste in Gruczno (Kujawsko-Pomorskie voivodeship) is an event held to present the region's culinary heritage. It includes the Taste of the Year competition, the Liqueur Tournament and the Bee-keepers' Convention (the Honey Feast). During the festival, the visitors can watch a presentation of local food production, meet the producers and taste local products
160/ 50,000	Held in the Pomorskie voivodeship, the Strawberry Harvest Celebration at the Golden Hill is the biggest open-air event in Kashubia. It focuses on locally grown strawberries. During the event, the visitors can buy fresh strawberries and many processed products. The strawberry cake contest and the biggest strawberry contest are also organized
45/ 7,000	"Powidlaki," the Plum Jam Frying and Tasting Celebration, has been held in Krzeszów (Podkarpackie voivodeship) since early 1990s. Using a traditional method of making jam with locally grown plums in copper vats is the main attraction. The event also hosts the fair of Sub-carpathian food products entered to the Traditional Products List, and folk band concerts. The best plum product contest is also organized
30/ 900	"Cooking at a glade": an event organized by the Rural Mazovian Open-Air Museum in Sierpc (Mazowieckie voivodeship). The purpose of this event is to popularize traditional Mazovian dishes. It also includes presentations of butter making, weaving, mending, spinning, mangling, shoemaking, ropework, making hay toys, embroidery and lace. The tastiest Mazovian dish contest is also organized

Source: Own study based on data provided by the organizers.

a Estimated values. Admission to all events (except for "Cooking at a glade") is free.

organizations showed interest in organizing dedicated thematic training sessions (Kłoczko-Gajewska & Markiewicz, 2018). According to estimations, ca. 140 attempts were made to establish a thematic village in Poland, of which 100 continue to be active (Kłoczko-Gajewska et al., 2015). Thanks to successful pioneering projects and financial support from the European Union, at the current development stage of this concept, the establishment of thematic villages in Poland has become a model solution for rural tourism and social activation of rural residents (Czapiewska, 2012; Głuszak, 2012; Kłoczko-Gajewska, 2013, 2014; Idziak & Idziak, 2015). Good practices from Poland were used in establishing thematic villages in Lithuania, Latvia, and Ukraine (Idziak & Idziak, 2015). However, according to a study by Kłoczko-Gajewska et al. (2015), thematic villages should not be equated solely with the need for economic activation. Instead, social and cultural aspects of this process should be sought because in most cases covered by analyses, the creation of a thematic village did not cause any major change in the development of the tourism function at commune level. Nevertheless, it is quite common in Poland to use the elements of culinary heritage as the key idea behind the establishment of thematic villages. According to estimations by Kłoczko-Gajewska et al. (2015), ca. 15–20% thematic villages operating in Poland build their offering based on food products, and are therefore consistent with the rural culinary tourism pattern (Table 4.9). Note also that those who largely rely on the potential of regional and traditional products in building their offering and competitive edge are mostly individual farms (e.g. the Bread Hut, an agritourism farm in the Śląskie voivodeship; the Honey Apiary with a generations-long history in the Dolnośląskie voivodeship) which could be members of the Countrywide Network of Educational Farms run by the Agricultural Consultancy Center (Jęczmyk et al., 2011). The purpose of the Network is to promote the concept of farm education and to protect the rural cultural heritage, including the preservation of traditional methods of harvesting, processing and preparing regional and traditional products.

As shown above, Polish regional and traditional products provide great opportunities for developing the tourism offering in rural areas. However, the potential for creating such products differs across regions. This is mostly because of socioeconomic processes which take place in different parts of the country and carry a heavy burden of historical events. In a context of general globalization and unification, folk culture resources and traditional farming, which is quite well preserved (especially in southern and eastern Poland), could be a highly appreciated and desired advantage of rural tourism products. Therefore, regional and traditional products could be the main theme of a trail, a culinary event, or a thematic village, or just an aspect that makes the agritourism product stand apart.

4.5 Training for rural tourism development professionals in Poland, 1989–2019

In Poland, the education levels of the rural population, including farmers, gained in importance in the 1960s. At that time, completing a formal agricultural vocational course became a prerequisite for the succession of a farm (Wawrzyniak & Wojtasik, 2004). The 1990s were marked by the systemic transformation which required different sectors (including agriculture) to accommodate the new requirements brought by the free market economy. The decentralization of education management in the first post-transformation decade and the growing educational aspirations of rural dwellers were the key factors that affected the role of education in rural development. Local government units were vested with the responsibility for school management, and were allowed to use their own funds for that purpose (in addition to the education subsidy from the state budget). This provided momentum for a profound and territorially heterogeneous transformation (Rosner & Stanny, 2017). In Polish rural areas, the education system became a social empowerment tool in the first post-transformation decade (Domalewski, 2010). However, the teaching quality of rural schools was by far lower than in an urban environment, and the place of residence played a role in whether an individual could be educated enough to attain a certain position in society

Table 4.9 Selected examples of thematic villages related to food products

Product/ key idea	Village name	Offering and its implementation	Location / year established
Mush-rooms	The Mush-room Village	At the Mushroom Feast held each year in the Mushroom Village, the visitors can taste different mushroom dishes. The Village offers the following attractions: • mushroom drying rooms; • mushroom lectures	Krzywogoniec (Kujawsko-Pomorskie voivodeship)/ 2005
Eggs	Village at the Begin-ning of Life	The village is focused on eggs. The guests can explore the secrets of eggs from both a natural/biological and a culinary/cul-tural perspective. As a symbol of being reborn, eggs appear in the myths of var-ious cultures. In the village, the visitors can assist in a joint preparation of egg meals and taste the local products. The offering also includes handicraft work-shops, including Easter egg making	Jajkowo (Kujawsko-Pomorskie voivodeship)/ 2015
Honey	The Honey Village	Apiaries are the village's biggest attrac-tion. The guests can enhance their knowledge of beekeeping and of bees and their medicinal uses, and can taste and buy honey-based products. The village offers wax candle making workshops and "Exploring the Honey Village," an out-door game	Wielki Mędro-mierz (Kujawsko-Pomorskie voivodeship)/ 2008
Bread	The Bread Village	In the village, the tourists can watch a presentation of traditional threshing methods and can attend bread-making workshops: preparing the bread dough in accordance with an original old recipe and assisting in baking the bread in a genuine bread oven. The village also organizes a role-playing game with a sce-nario centered around the bread village legend	Jania Góra (Kujawsko-Pomorskie voivodeship)/ 2011
Milk	The Milky Village	The village offers four guided tours with plenty of attractions, such as the pre-sentation of manual milking, butter and cheese making, and visiting a calf farm. Also, there is the Milky Village Museum, established on the initiative of the Asso-ciation of Local Initiatives	Leszczynka (Lubelskie voivodeship)/ 2009

Source: Own compilation.

(Domalewski, 2010). Changes triggered by the transformation process resulted in a profound socio-occupational restructuring, including the emergence of unemployment, an issue previously unheard of. During that period, people who lost their jobs due to restructuring of establishments based in cities could be observed to migrate back to rural areas. The influx of population had a positive impact on rural education levels because those who migrated back were better educated and not afraid of engaging in non-farming activities (Wawrzyniak & Wojtasik, 2004; Wrzochalska, 2005). The transformation had an effect on education levels of Polish rural dwellers: there was a considerable decline in the number of people with a primary education and an increase in the population with a tertiary education. This is related to the process of farm succession and to the growing mobility of young people. However, despite some positive changes in rural access to education (e.g. the establishment of remote branches of universities and the development of private tertiary schools), the educational gap between the urban and rural population continues to be noticeable (Figure 4.32).

The pace of changes in the proportion of rural population aged 15+ with a secondary or tertiary education varies between regions, too. The slowest improvement in rural education levels was witnessed in the Opolskie voivodeship which is largely explained by young people migrating to cities.

The analysis of education levels of the farming population also revealed that the share of people with a tertiary education is higher in those regions where agriculture is not the sole occupation, i.e. in south-east and south-west part of Poland (Zwoliński, 2007). A greater proportion of people with a tertiary education is concentrated in communes surrounding large cities. Compared to residents of remote communes, they face fewer barriers in accessing education (Klonowska-Matynia, 2017). Moreover, the Monitoring of Rural Areas

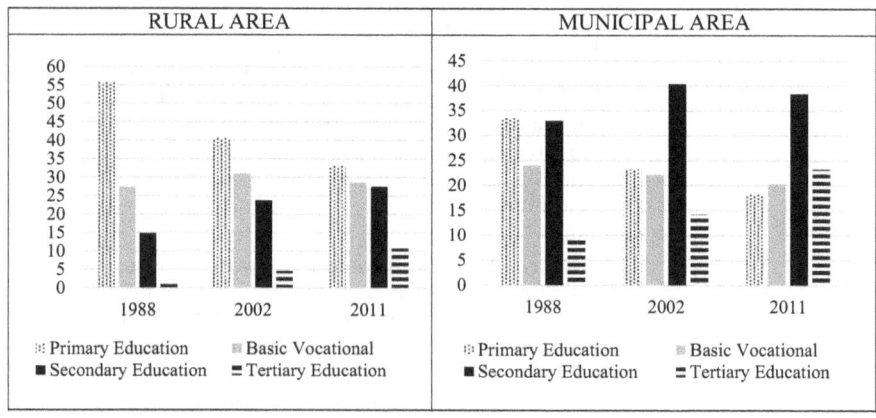

Figure 4.32 Changes in the education structure of the urban and rural population (percent share).
Source: Own compilation based on the 1988, 2002, and 2011 National Census.

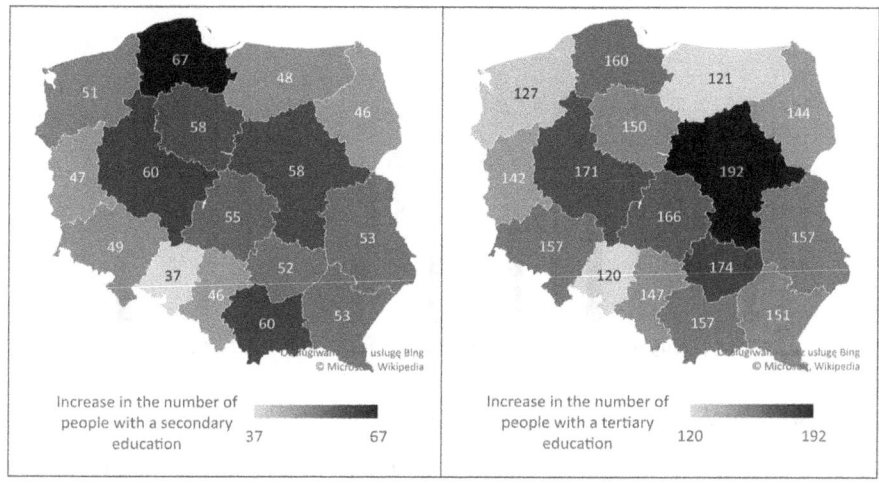

Figure 4.33 Increase in the number of people with a secondary or tertiary education in rural areas between 1988 and 2011 (%), grouped by voivodeship.
Source: Own compilation based on data from the 1988, 2002 and 2011 National Census.

(Stanny & Rosner, 2014; Rosner & Stanny, 2016; Stanny, Rosner &Komorowski, 2018), a project which includes comparing the educational status between 2014 and 2018, demonstrated that the largest compact region with a below-average pace of changes is found in Central Pomerania, i.e. in a region previously dominated by state-owned agricultural holdings.

The political and socioeconomic change which followed the transformation resulted in vocational education being better suited to address the expectations of the emerging free market. Papers and reports developed in the first transformation period devoted particular attention to: the need for building new relationships between vocational schools and employers; the changing role of the artisanal sector; the development of professional training and training institutions; and the need for developing the career counseling system (Adamski et al., 1993; Chłoń-Domińczak et al., 2011). The second and the third post-transformation decades were marked by the aging of the population and, thus, by a declining share of people in education, especially including secondary vocational education (Figure 4.34). Furthermore, despite the decrease in the number of secondary vocational schools, the number of pupils willing to follow that education path remained quite high. This was the consequence of numerous reforms of the Polish vocational education system, implemented to improve its overall flexibility, link it to the local labor market, and offer vocational courses. The classification of occupations developed by the Ministry of National Education (which includes a total of 213 items) specifies the professions taught in Poland, the corresponding types of schools, the skills defined for each profession, the period of education, and the opportunities for adults

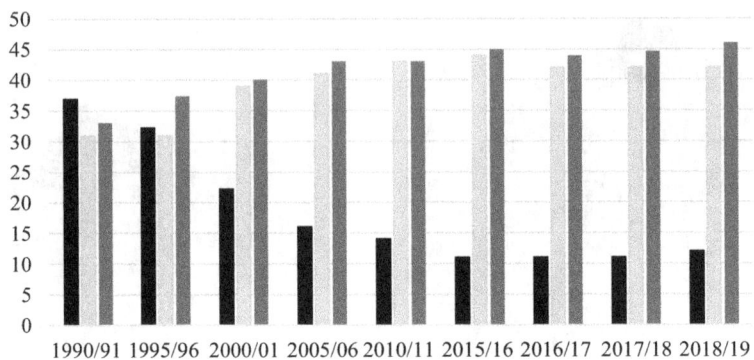

1990/91 1995/96 2000/01 2005/06 2010/11 2015/16 2016/17 2017/18 2018/19

■ basic vocational school   ▨ technical secondary school   ■ general secondary school

Figure 4.34 Number of pupils grouped by most popular school types, 1990–2019 (%).
Note: After the last reform, the Polish vocational education system includes 1st and
2nd degree industry-oriented schools, technical schools and post-secondary schools.
Source: Own study based on Central Statistical Office data (Education yearbooks).

to acquire the relevant skills on an extramural basis (Chłoń-Domińczak et
al., 2011).

Note, however, that the educational framework defined by the government
should be modified as per regional needs and resources. Although the development of tourism provides new employment opportunities, attracting and
retaining skilled employees becomes a major problem facing today's tourism
industry, including in Poland. Well-educated and experienced human resources
are of extreme importance, especially in those industries where the perceived
service quality particularly depends on the employee (Lundberg et al., 2009).
Human capital is crucial to the development of the tourism function. The
Polish system of tourism education is very extensive[27] and includes:

1 formal education, including vocational and tertiary schooling;
2 non-formal education: regulated occupations and activities, and professional training and development;
3 lifelong learning as continuous professional development.

Formal secondary education includes a separate group of tourism and
catering-related occupations such as: hospitality technician, tourism service
technician, rural tourism technician and hotel service auxiliary.[28] Rural tourism technicians graduate from 5-year technical schools, post-secondary
schools or vocational courses. Skills acquired by the students include the calculation, sale, conduct and settlement of tourism events and services; organization of farm work; and making farms suitable for the delivery of
agritourism services. Education on rural tourism techniques has been offered

in Poland since 2008. Today, two specializations are available, i.e. the pursuit of rural tourism activities and operation of an agritourism farm. In the school year 2018/2019, education on rural tourism techniques was offered by 139 schools in Poland, including nine technical schools (all of them public) and 130 post-secondary schools (including only 25 public establishments). Data presented in Figure 4.35 suggests that the territorial distribution of the number of rural tourism technician students between Polish regions does not coincide with the distribution of the number of agritourism farms. The largest number of rural tourism technician students was recorded in the Lubelskie voivodeship which, according to the Central Statistical Office, is home to only 35 agritourism farms. This suggests that educational choices are not necessarily backed up by a strong rationale, and that the youth do not always have enough reliable information on job opportunities matching their future professional profile and sometimes are unable to tell whether they are the right people for that job. This means that career counseling services offered to the youth need to be rethought.

The analysis of the numbers of students taking the qualifying examination shows that hundreds of people trained at secondary level to develop the rural tourism function enter the market each year. Also, since 2015, students have become more interested in being specialized in operating an agritourism farm than in pursuing a rural tourism activity, and the pass rate of the qualifying examination has become higher (Figures 4.36, 4.37).

The fields, coverage, and organizational forms of vocational education largely depend on the local labor market and on educational aspirations of the society (Chłoń-Domińczak et al., 2011). The changes in vocational education that took place in the 30-year post-transformation period were largely driven by the

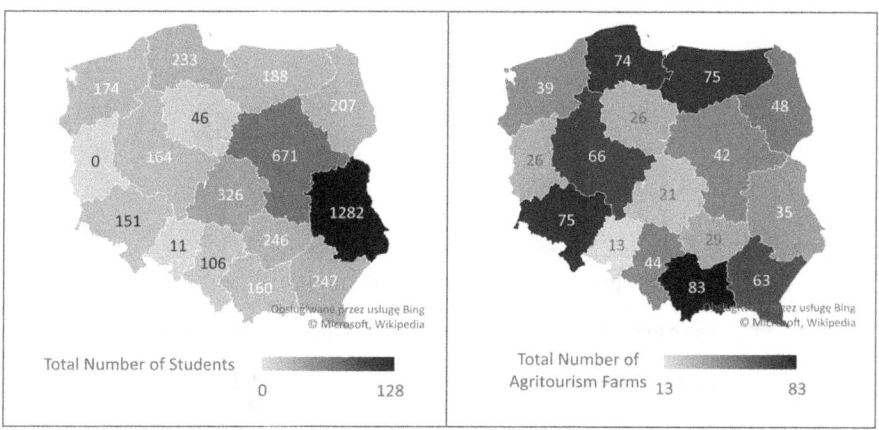

Figure 4.35 Number rural tourism technician students vs. the location of agritourism farms.
Source: Own compilation based on Vocational schools: number of pupils in the school year 2018/2019 grouped by occupation and school, Ministry of National Education.

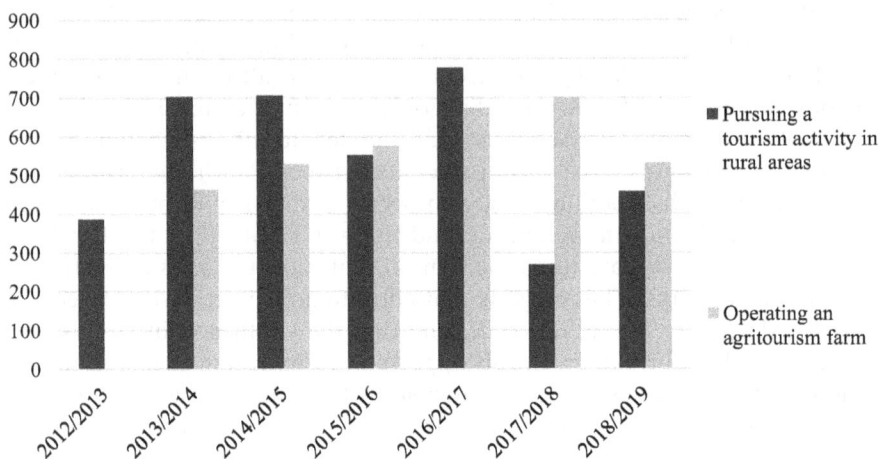

Figure 4.36 Number of students taking the qualifying examination, 2012–2019.
Note: The first qualification examination on the operation of agritourism farms took place in the school year 2013/2014.
Source: Own compilation based on data from the Central Examination Committee.

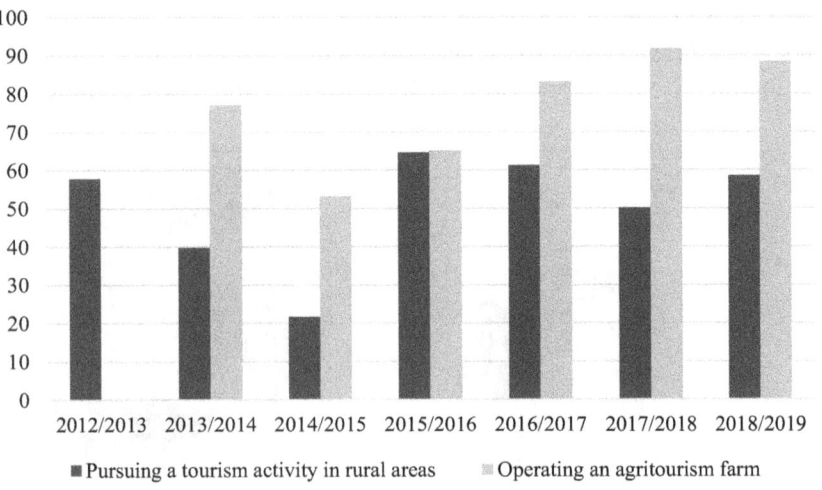

Figure 4.37 Pass rate of the qualifying examination, 2012–2019 (%).
Note: The first qualification examination on the operation of agritourism farms took place in the school year 2013/2014.
Source: Own compilation based on data from the Central Examination Committee.

decline in the numbers of upper secondary school students and by the need to adjust the educational offering to the changing needs of the labor market.

In Poland, tertiary tourism education started in the 1960s at Physical Education Universities, under a field of study referred to as "tourism and recreation" (Kruczek, 1998; Zamelska & Kaczor, 2015; Owsianowska & Winiarski, 2017). The tourism education system was developed as part of physical culture because physical culture, mass sports, and tourism were viewed as a whole. In 1992, the Krakow University of Physical Culture and the Warsaw Institute of Tourism organized an international conference to review the existing forms of tourism staff education in Eastern European countries, to compare them with the views and experiences of representatives of Western European countries, and to seek new methods for staff training in a changing socioeconomic context (Dziegięć & Liszewski, 1993). The attendees concluded that the tourism education system needs to be rapidly reformed, and that education itself should be highly diversified in terms of organizational forms and programs (Dziegięć & Liszewski, 1993). Currently, the Polish higher education system is strongly determined by the rapidly changing global external factors which are primarily related to technological advancements, the internationalization of the teaching process, and the growing aging of the population in European Union countries. Today, tourism-related higher education streams are offered as bachelor's and master's degree programs at physical education universities, universities of economics, universities of agricultural sciences, and non-state universities, which strongly differ in the scope of what they teach (Uglis & Kozera-Kowalska, 2016). The analysis of long-term graduation data (2006–2014) reveals a declining interest in social sciences, economics and law (–7.1%), and education (–2.2%) and an increased share of graduates from fields of study such as technology, industry, construction (+5.2%), services (+3.7%), and health and social welfare (+3.3%) (Figure 4.38). This reflects the way the higher education system adapts to a free market economy mostly based on services, including the developing tourism services. Note that in Poland, tertiary agritourism programs are primarily offered by universities of life sciences because agritourism activities are recommended to be pursued by active farms.

In Poland, tertiary rural tourism and agritourism programs are usually offered by agricultural universities; in the vast majority of cases, this is a specialization stream within fields of study such as agriculture, animal science or horticulture (Uglis & Kozera-Kowalska, 2016). The average growth rate of the share of students enrolled in technical and life sciences programs[29] during 2007–2018, grouped by region, suggests that interest in these fields of study exists mostly in eastern and central Poland (Figure 4.39). In this context, it should be explained that the high growth rate of students enrolled in technical and life sciences programs in the Podkarpackie voivodeship is largely related to the existence of the Rzeszów University of Technology, the sole Polish university which offers master's degree programs for aspiring aviation industry professionals.

Note that in accordance with the findings from a survey on the future of graduates from tourism schools and universities,[30] 36.1% of graduates from

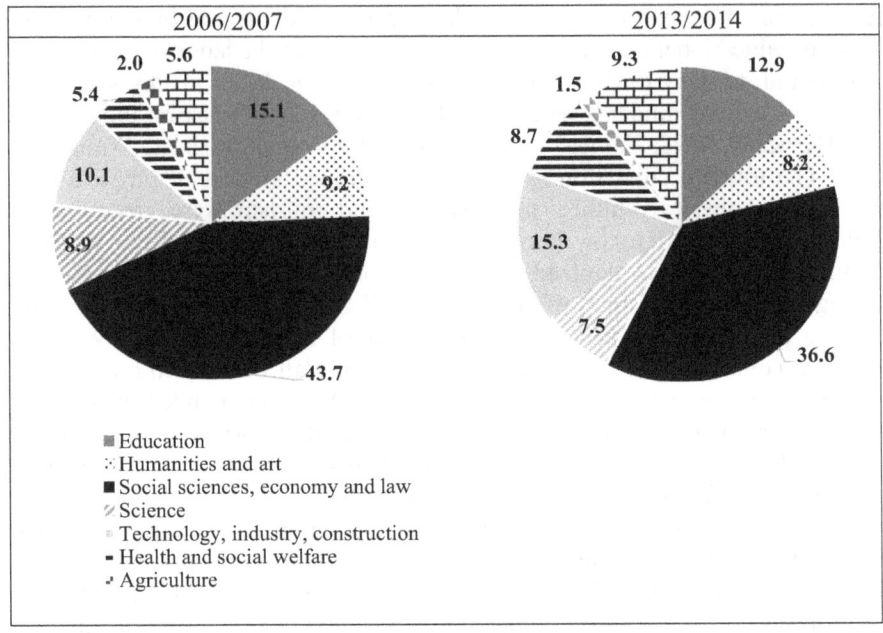

Figure 4.38 Graduates of higher education institutions by broad field of education in Poland (%).
Note: The fields of education are consistent with the International Classification of Fields of Education (ISCED'97).
Source: Own elaboration based on the Local Data Bank, Poland.

secondary schools and 45.3% of graduates from tertiary establishments found a job matching their professional profile.

Considering the constantly changing economic realities, particular focus should be placed on lifelong learning which can also include non-formal education or autodidacticism. Participating in education outside the formal system includes all organized educational activities which are not supported by formal educational establishments. Also, non-formal education activities do not result in upgrading one's education level. Usually, education outside the formal system makes an individual develop, enhance and acquire skills related to different walks of professional, social and cultural life. Unlike formal and non-formal education, autodidacticism generally means learning alone (without the involvement of a teacher). Based on data collected in the Adult Education survey carried out in 2009, 2011, and 2016, it may be observed that the adults' participation in different forms of formal, non-formal and autodidactic education depends on their age, education and place of residence (Adult Education, 2009, 2011, 2016). Young urban dwellers with a tertiary education usually engage in one of the forms of training listed above. Note however that the average growth rate in the share of active learners in rural areas (calculated for the entire study period) is more than double

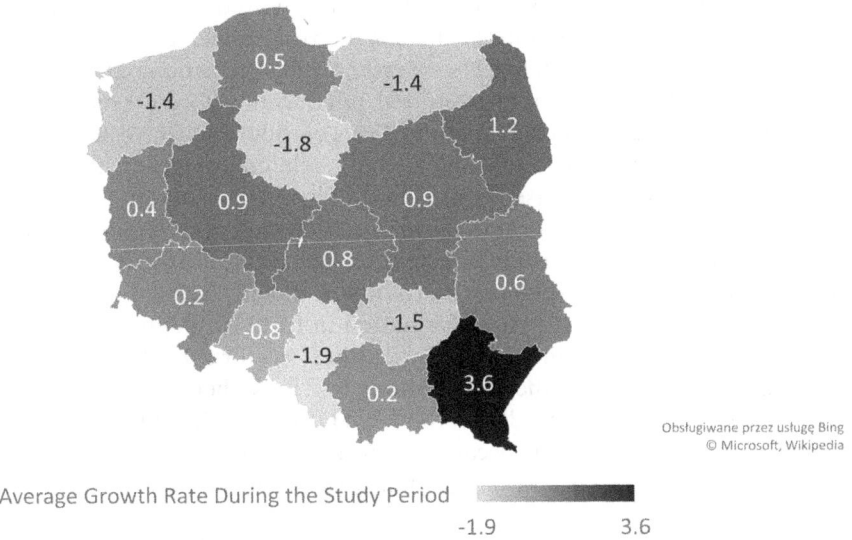

Figure 4.39 Average growth rate of the share of students enrolled in technical and life sciences programs, 2007–2018 (%).

Source: Own compilation based on Higher Education, Central Statistical Office.

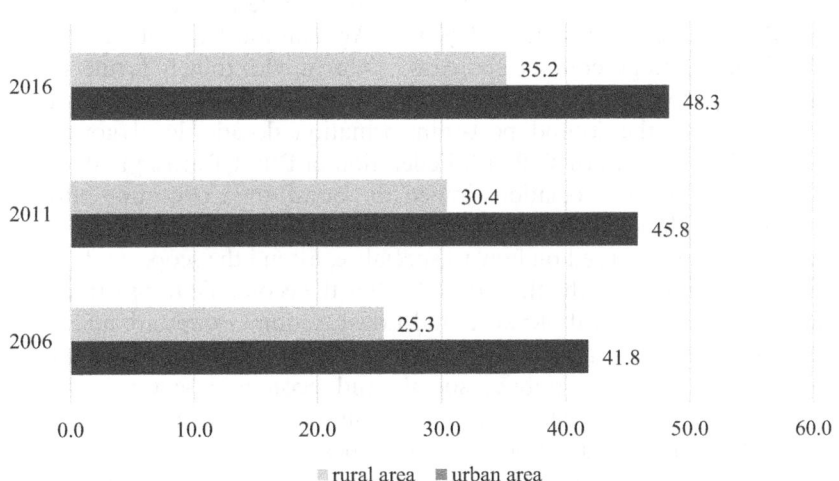

Figure 4.40 Fractions of the population aged 25–64 grouped by participation in any form of education and by place of residence (%).

Note: Participation in formal, non-formal or autodidactic education.

Source: Own elaboration based on data from the 2006, 2011, and 2016 Adult Education survey, Statistical Analyses, Polish Statistics.

that recorded in cities; this suggests that the urban-rural gap is consistently closing (Figure 4.40). Moreover, urban residents showed the greatest educative interest in arts and humanities, whereas the rural population were mostly oriented towards broadly defined services (Adult Education, 2009, 2011, 2016). Rural dwellers were more interested in education forms which enable obtaining a certificate of specific qualifications empowering them to practice a profession or pursue an economic activity related to technical or construction services or craft-based production. Therefore, activation trainings for the rural population were mostly focused on technical, industrial, and construction-related topics (Adult Education, 2009, 2011, 2016). The low participation rates of rural residents in trainings directly related to farming (7.0%) were a reflection of the transformation in rural employment structure and of the workforce drain to non-agricultural sectors. Furthermore, the rural population financed their education activities themselves more frequently (18.4%) than urban residents (16.9%). This reflects the differences in the structure of employment status because a considerable part of the rural population work on their own account, including as farmers (Adult Education, 2009, 2011, 2016).

Auxiliary institutions play a significant role in non-formal education of tourism development professionals. This includes the Agricultural Consultancy Center and Provincial Agricultural Consultancy Centers, the National Rural Network, the "Hospitable Farms" Polish Federation of Rural Tourism, regional and local agritourism associations, and Local Action Groups. The importance and role of different institutions in non-formal education for rural tourism development changed from one post-transformation decade to another. In the early 1990s, the greatest role was played by Agricultural Consultancy Centers, which offered trainings, courses, and case-by-case advice to help farmers in setting up an agritourism activity (Wojciechowska, 2007a; Mazurek-Kusiak & Golian, 2011). In the second post-transformation decade, local government units, the "Hospitable Farms" Polish Federation of Rural Tourism and regional and local agritourism associations joined the non-formal education efforts to promote the development of the tourism function in rural areas. The trainings offered were largely focused on how to specialize, extend the scope, and improve the quality of services. In the third decade after the economic transformation in Poland, the National Rural Network and Local Action Groups are major players in non-formal education. They embrace bottom-up initiatives based on a partnership between the public, social, and economic sectors to enable rural activation and exchange of best practices. These efforts were also encouraged by increased amounts of European Union funds allocated to subsequent generations of the Leader program (Borowska, 2009; Kiryluk-Dryjska & Hadyński, 2016). In the third post-transformation decade, non-formal education is largely focused on enhancing interpersonal skills which enable cooperation in such forms as networking, on looking for state-of-the-art tools to reach potential customers, and on defining new fields of specialization, e.g. social economics. Also, the socioeconomic transformation which started in Poland in the 1990s has been accompanied by a continuous increase in the number of Universities of the Third Age which, as their primary goal, seek to promote education, integration, and activation of the elderly (Figure 4.41). Indeed, the number of Universities of the Third Age increased six times between the second and the third post-transformation decade. This can be explained by aging of the society, dedicated financial support and the emergence of "next-generation retirees' who spent most of their professional life in a free market economy. Note however that only 11% of all Universities of the Third Age operate in rural areas (Pawlikowska-Łagód et al., 2017).

Moreover, in addition to the role and scope of institutions established to support non-formal education, the form itself of acquiring knowledge and skills has changed too over the last 30 years in Poland. The non-formal education process has evolved from lectures and simulations to imitation-based learning (study tours) and learning through action. The systemic transformation which took place in Poland after 1989, the post-modernization processes witnessed in the second and third decade, and the gradual economic integration with the EU were the factors that made human capital an increasingly important driver of economic development processes. People aged 50+ were the dominant group of participants to different forms of lifelong learning until mid-1990s. The generational shift occurred in the second post-transformation decade. At that time, trainings were attended by young people, with a growing share of domestic migrants (Romanowski, 2005; Wojciechowska, 2009; Zawadka, 2010a). As a consequence, the development of the rural tourism function (especially including the development of agritourism) involved the emergence of two behavioral models of accommodation providers, largely determined by their assets, including the type of farm and education. In the first model, the business strategy is to survive. A typical owner is an old or middle-aged person with a secondary education who derives his/her financial security from another

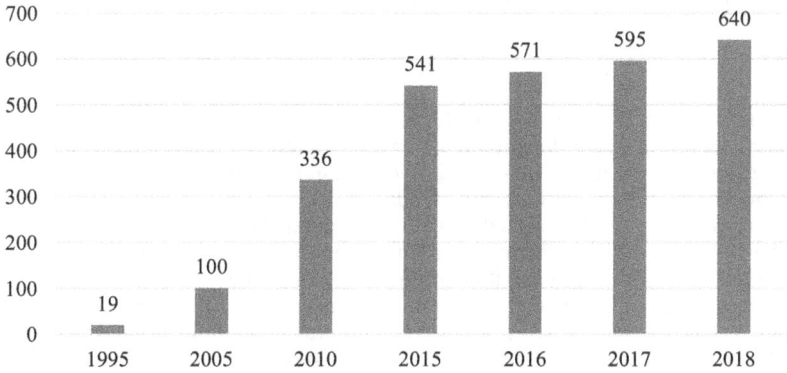

Figure 4.41 Total number of Universities of the Third Age in Poland, 1995–2018.
Source: Universities of the Third Age in Poland, Polish Statistics.

source than agritourism. The main reasons for engaging in this activity are non-economic aspects. In the second behavioral model, geared towards development goals, the entrepreneurs are young people with a tertiary education who intentionally chose tourism as their field of study. Note, however, that Polish agritourism was pioneered by those currently aged 50+ (Wojciechowska, 2018). Today, older, less-educated owners are believed to be the prevailing profile in eastern Poland. Conversely, younger, more educated agritourism farmers are more likely to be found in western Poland (Wojciechowska, 2018).

The intensity of socioeconomic changes that have taken place in Poland within the last 30 years, including the development of the rural tourism function, was considerably determined by the capacity to tap into unique endogenous factors derived from social, economic, and land resources. The changes experienced in Poland resulted in attempts being made to adapt the structure of vocational education and the development of higher education to the growing educational aspirations of the society and to the changing labor market. The pace of change differed between regions of the country; this is due to reasons which include differences in historical events and environmental conditions between regions, their location, and cultural background. Even though the pace of change in access to education differed between regions, the educational gap between the urban and rural population consistently declined throughout the 30-year post-transformation period. Thus, in the context of the urban–rural divide, the place of residence has significantly lost its importance as a determinant of education levels. In the third post-transformation decade, factors such as social and economic status and cultural capital of families play a stronger role in the education process. From the perspective of education for professionals in charge of developing the tourism function in rural areas, it is important that individuals be able to find a job matching their acquired profession. This, in turn, is largely determined by the location of rural areas in relation to the core–periphery structure and by how much are rural areas open to regional transmission (diffusion) processes.

Notes

1 Journal of Laws of 1993, No. 90, Item 416.
2 Today, there are 222 Educational Farms in Poland, with the largest number (27) being located in the Dolnośląskie voivodeship, and the smallest (5) in the Zachodniopomorskie voivodeship (Duda, 2019).
3 The respondents were selected with a Computer-Aided Telephone Interviewing (CATI) system which generated random phone numbers ($n = 1000$). The sample included 234 residents of satellite communes, 336 residents of star communes, 203 residents of aspiring star communes and 227 residents of peripheral communes.
4 WZF $= (p/L) - ((s + d)/L)$, where p: the commune's share in taxes which are state budget revenue; s: general subsidies; d: total subsidies; L: population.
5 The leading driver is defined as the one with the highest β value of the standardized regression coefficient within the type of communes considered.

6 Research carried out under the project was financed by the National Rural Network. The purpose of this research was to analyze whether young people (aged 18–29) not in employment, education or training (referred to as NEETs) are capable of working on a self-employment basis and what their attitudes are towards entrepreneurship, including in the context of tourism development ($n = 384$).

7 LGU: local government unit.

8 $n = 100$; research was carried out in five thematic villages in Sierakowo Sławieńskie (the Hobbit Village), in Dąbrowa (the Healthy Living Village), in Podgórki (the Fairytale and Biking Village), in Iwięcino (the Village of the End of the World) and in Paproty (the Labyrinth and Spring Village).

9 In some cases, it was necessary to restrict the time scope of this analysis due to lack of information. This was caused by changes in data presentation levels, changes in the list of territorial units, or modifications to the list of characteristics during a reporting period. Statistical data used in describing rural areas is combined data for rural communes and rural areas located in urban-rural communes. This type of data aggregation is not available for all characteristics covered by this analysis throughout the 1990–2019 period.

10 Public Opinion Research Center, Holiday trips in 1992–2007, Holiday trips of the Polish population in 2008, 2012, 2013, 2014, 2018, 2019.

11 Although this was a countrywide project, the study was mostly carried out in the southern part of the country as a direct interview survey.

12 These indexes are widely used in the literature, e.g. Schneider's index (Gogonea, et al., 2017; Hacia, 2014); Charvat's index (Przybyla & Kulczyk-Dynowska, 2017; Stefko, et al., 2018); Tourism Density index (Vojnovic, 2018). Schneider's index was calculated based on the following formula: $Ts = (N/P) \times 100$, where Ts is Schneider's index, N is the number of tourists, and P is the local population. Charvat's index was calculated based on the following formula: $Tch = (N/P) \times 100$, where Tch is Charvat's index, N is the number of overnights, and P is the local population. The tourism density index was calculated as the ratio of the number of tourists to the size of the territory (km^2). The Baretje–Defert index was calculated based on the following formula: $Tbd = (N/R) \times 100$, where Tbd is the Baretje–Defert index, N is the number of tourist accommodation establishments, and R is the number of permanent residents of the area. The tourist accommodation density index was calculated as a ratio of the number of tourist accommodation establishments to the size of the territory (km^2) (Marković et al., 2017; Korzeniewski & Kozłowski, 2019).

13 These indexes are widely used in the literature, e.g. the tourist accommodation density index (Marković et al., 2017; Simancas Cruz & Peñarrubia Zaragoza, 2019), the Baretje–Defert index (Podhorodecka & Dudek, 2019; Borzyszkowski, et al., 2016; Korzeniewski & Kozłowski, 2019).

14 The 1999 administrative reform introduced a three-level (commune, district, voivodeship) territorial division structure and reduced the number of voivodeships from 49 to 16. The purpose of the reform was to establish and empower the local government, and to make the voivodeships larger and stronger units (for a broader description, see Chapter 1).

15 This study chose the first and the last year for which data could be disaggregated by the urban/rural dimension. In 1999, the territorial division was changed to that in force today; and some data collected has not been disaggregated by the urban/rural dimension after 2014.

16 Interpretation of the Baretje–Defert index on the M. Boyer's scale: <4: tourism activity is virtually non-existent; 4–10: the tourism function is only one of many functions; 10–40: the tourism function is important but not dominant; 40–100: the tourism function dominates; 100–500: a known tourist center; >500: a renowned tourist center (Boyer, 1972; Pearce, 1995).

17 In Poland, residential school trips during spring and winter mean organized trips for pupils where they can combine learning with leisure. This also includes one-day stays.

18 In this case, agritourism is defined as an activity which consists in offering rooms (no more than five rooms at a time) for rent to holidaymakers in residential buildings located in rural areas, owned by farms which derive no less than 50% of their incomes from agricultural activities (Marciniak, 2011, p. 34).

19 That period was narrowed down due to changes in statistical reports on the number of agritourism farms and due to missing data from 2003–2009. According to the Local Data Bank of the Central Statistical Office, there were 759 agritourism farms and 2291 guest rooms in Poland in 2018. Data published by the Central Statistical Office considerably differs from what is held by Agricultural Consultancy Centers. Therefore, it has to be concluded that the actual number of Polish agritourism farms is unknown. The number of beds offered by agritourism farms from the 1990s to date accounts for 4–6% of total beds offered in Poland (Wojciechowska, 2018).

20 The transformation cannot be traced back to 1989 because of changes in the country's administrative division which took place in 1999.

21 REGON is the Official Register of Operators of the National Economy kept by the President of the Central Statistical Office.

22 Regulation (EEC) No. 2081/92 on the protection of geographical indications and designations of origin for agricultural products and foodstuffs; and Regulation (EEC) No. 2082/92 on certificates of specific character for agricultural products and foodstuffs. The UE system for protecting regional and traditional products is based on two regulations that govern the principles for their registration and protection: Council Regulation No. 510/2006 on the protection of geographical indications and designations of origin for agricultural products and foodstuffs; and Council Regulation No. 509/2006 on agricultural products and foodstuffs as traditional specialities guaranteed. The UE system for protecting regional and traditional products was transposed into Polish law through the act on the registration and protection of names and designations of agricultural products and foodstuffs and on traditional products (Journal of Laws of 2005, No. 10, Item 68). The act also governs the national level of application evaluation.

23 Source: eAmbrosia electronic register (PGI = 766; PDO = 644; TSG = 63), as of April 16, 2020.

24 Journal of Laws of 2005, No. 10, Item 68; and of 2008, No. 171, Item 1056; and of 2009, No. 216, Item 1368.

25 The consortium has the following members: the Tastes of Podhale; the Carp Route in the Carp Valley; the Oscypek Cheese Trails; the Taste Carnival, Krakow, Lesser Poland; Culinary Prestige of Pomerania, Gdansk; Silesian Tastes; Bielsko Land Carp Route; At the Prune Trails; the Noble Bowl of Mazovia—the Mazovian Folklore and Taste Trails; the Lesser Poland Herbal Village; the Lesser Poland Gourmet Route; the Goose Meat Trail of Kuyavia and Pomerania; the Gdynia Uptown Culinary Trails; the Taste Trails of the Land of Loess Canyons; Culinary Stopovers of the Suwałki Region and Masuria; the Wine-Making Trails of Sandomierz; the Subcarpathian Tastes Culinary Trails; the Tradition and Taste Trails; the Culinary Experience of Poznań; the Kitchen Dresser of Opole; the Subcarpathian Food and Wine Trails; the Culinary Trails of Białystok.

26 The Farmer's Wives' Associations (FWA) are the most popular form of female self-organization in Polish rural areas. According to a study by "Stocznia," the Foundation for Social Research and Innovation, at least one FWA exists in 75% of Polish communes (Mencwel, et al., 2014).

27 In 2014, measures were taken to establish the Sectoral Qualifications Framework for Tourism (SQFT) applicable to hospitality, catering, and tourism professionals who acquire their qualifications in the non-formal education system. The SQFT is

linked to the European and National Qualifications Frameworks previously developed and put in place in EU member countries (Kruczek, 2015).
28 Regulation of the Minister of Education on the classification of professions covered by the vocational education system (Journal of Laws, Items 622 and 2356; and Journal of Laws of 2019, Item 1536).
29 No countrywide statistical data is available that makes a distinction between rural tourism and agritourism education because it is part of many different fields of study. Data which illustrates the education in technical and life sciences (available as a combination only) can only be an indirect indicator of interest in agritourism education.
30 Ministry of Sports and Tourism (2014). The survey (of 2012/2013 graduates) covered all voivodeships and was carried out with respondents from all over Poland. A stratified proportional sampling procedure was used in each of the respondent groups covered by the survey.

References

Adamski, W. (1993). *Edukacja w okresie transformacji: analiza porównawcza i propozycje modernizacji kształcenia zawodowego w Polsce* [*Education in the transformation period: a comparative analysis and proposed modernization measures for the Polish vocational education system*], Publishing House of the Institute of Philosophy and Sociology of the Polish Academy of Sciences.

Adult Education (2006, 2011, 2016). *Statistical Analyses*, Statistics Poland.

Alejziak, W. (2009). *Determinanty i zróżnicowanie społeczne aktywności turystycznej* [*Determinants of and social differences in tourism activity*], Krakow University of Physical Culture, Krakow.

Ambroz, M. (2008). Attitudes of local residents towards the development of tourism in Slovenia: The case of the Primorska, Dolenjska, Gorenjska, and Ljubljana regions, *Anthropological Notebooks*, 14, 63–79.

Andereck, K. L., Valentine, K. M., Knopf, R. C., & Vogt, C. A. (2005). Residents' perceptions of community tourism impacts, *Annals of Tourism Research*, 32 (4), 1056–1076.

Abrham, J. (2014). Clusters in tourism, agriculture and food processing within the Visegrad Group. *Agricultural Economics*, 60 (5), 208–218.

Adamowicz, J. (2010). Towards synergy between tourism and nature conservation. The challenge for the rural regions: The case of Drawskie Lake District, Poland. *European Countryside*, 2 (3), 118–131.

Adamowicz, M., & Zwolińska-Ligaj, M. (2009). Koncepcja wielofunkcyjności jako element zrównoważonego rozwoju obszarów wiejskich [The concept of multifunctionality as an element of sustainable development of rural areas], *Polityki Europejskie, Finanse i Marketing* [*The Scientific Journal European Policies, Finance and Marketing*], 2 (51), 11–38.

Arnold, Ch., & Baker, T. (2012). Transitions from school to work: Applying psychology to "NEET". *Educational and Child Psychology*, 29 (3), 67–80.ref id="ref470" Augustyn, M. (1998). National strategies for rural tourism development and sustainability: The Polish experience. *Journal of Sustainable Tourism*, 6 (3), 191–209.

Bajgier-Kowalska, M., & Tracz, M. (2019). Bariery rozwoju przedsiębiorczości na przykładzie agroturystyki w polskich Karpatach [Barriers to entrepreneurship development based on the example of agritourism in the Polish Carpathian Mountains]. *Przedsiębiorczość-Edukacja* [*Entrepreneurship -Education*], 15 (1), 158–172.

Bajgier-Kowalska, M., Tracz, M., & Uliszak, R. (2016). Uwarunkowania rozwoju przedsiębiorczości na obszarach wiejskich na przykładzie gospodarstw agroturystycznych województwa małopolskiego [Conditions for development of entrepreneurship in rural areas on the example of agritourism farms in the Małopolska Voivodeship]. *Przedsiębiorczość-Edukacja* [*Entrepreneurship-Education*], 12, 256–273.

Balińska, A. (2001). Agroturystyka jako forma przedsiębiorczości na obszarach wiejskich na przykładzie rejonu nadbużańskiego. *Zeszyty Naukowe Akademii Rolniczej im. H. Kołłątaja w Krakowie*, 78 (2), 479–485.

Balińska, A. (2010). Konkurencyność produktu turystyki wiejskiej w opinii turystów [Agrotourism as a form of business activity in rural areas on an example of Bug Valley Region. Competitiveness of rural tourism product in respondents' opinion], *Acta Scientiarum Polonorum Oeconomia*, 9 (4), 5–14.

Balińska, A. (2014a). Determinanty popytu mieszkańców Warszawy na usługi agroturystyczne [Determinants of demand inhabitants of Warsaw for agritourism]. *Ekonomiczne Problemy Turystyki*, 1 (25), 251–264.

Balińska, A. (2014b). Aktywność turystyczna mieszkańców wsi w kontekście przemian społeczno-gospodarczych–rzeczywisty i potencjalny popyt turystyczny mieszkańców wsi [Tourist activity in the context of the rural population socio-economic transformation – actual and potential tourists demand of rural residents]. *Roczniki Naukowe Ekonomii Rolnictwa i Rozwoju Obszarów Wiejskich* [*Annals of Agricultural Economics and Rural Development*], 101 (2), 112–122.

Balińska, A. (2017). Sieciowe produkty turystyczne jako przykład przedsiębiorczości na obszarach wiejskich [Networked tourism products as an example of entrepreneurship in rural areas]. *Turystyka i Rozwój Regionalny* [*Journal of Tourism and Regional Development*], 8, 5–14.

Balińska, A., & Sikorska-Wolak, I. (2009). *Turystyka wiejska szansą rozwoju wschodnich terenów przygranicznych na przykładzie wybranych gmin* [*Rural tourism as a development opportunity for the eastern borderland: a case study of selected communes*]. Publishing House of the Warsaw University of Life Sciences, Warsaw.

Bański, J. (2004a). The development of non-agricultural economic activity in Poland's rural areas. *Rural Areas and Development*, 2 (740-2016-50996), 31–43.

Bański, J. (2004b) Możliwości rozwoju alternatywnych źródeł dochodu na obszarach wiejskich [Possibility of development of alternative income sources in rural areas], Pałka, E. (red.), *Pozarolnicza działalność gospodarcza na obszarach wiejskich* [*Non-agricultural economic activity in rural areas*], Studia Obszarów Wiejskich [Rural Studies], 5, 9–22.

Bański, J. (2006). *Geografia polskiej wsi* [*Polish rural geography*], PWE Publishing House, Warsaw.

Bański, J. (2012). *Turystyka wiejska, w tym agroturystyka, jako element zrównoważonego i wielofunkcyjnego rozwoju obszarów wiejskich* [*Rural tourism, including agritourism, as an element of sustainable multipurpose rural development*], research manager, implementing scientist: Agrotec Sp. z o.o., Warsaw.

Bański, J., & Stola, W. (2002). Przemiany struktury przestrzennej i funkcjonalnej obszarów wiejskich w Polsce [Transformation of the spatial and functional structure of rural areas in Poland], *Studia Obszarów Wiejskich* [*Rural Studies*], 2, 7–113.

Bański, J., Rudolf, A., Przybył, C., Bednarek-Szczepańska, M., Czapiewski, K., Mazur, M., & Pieniążek, W. (2012). *Turystyka wiejska, w tym agroturystyka, jako element zrównoważonego i wielofunkcyjnego rozwoju obszarów wiejskich* [*Rural tourism, including agritourism, as an element of sustainable multipurpose rural development*]. Agrotec Polska Sp. z o.o. and Institute of Geography and Spatial Organization, Warsaw.

Baretje, R., & Defert, P. (1972). *Aspects economiques du tourisme.* Paris: Berger-Levrault.

Basińska, A., & Poczta, J. (2009). Rola współpracy międzygminnej w zrównoważonym rozwoju turystyki na obszarach wiejskich–studium przypadku [Role of inter-district cooperation in the development of balanced tourism in rural areas – case study]. *Folia Pomeranae Universitatis Technologiae Stetinensis. Oeconomica,* 54, 5–16.

Baum, S., & Gramzow, A. (2009). Rural tourism: an opportunity for the development of rural areas in Poland?. *Rural Areas and Development,* 6, 61–80.

Bednarek-Szczepańska, M. (2010). Rola podmiotów lokalnych w rozwoju turystyki wiejskiej na wybranych obszarach Lubelszczyzny. Komisja Obszarów Wiejskich, Polskie Towarzystwo Geograficzne [Presentation].

Berbeka, J. (2016). Teoretyczne podstawy zachowań turystycznych [Theoretical foundations of tourism behaviors], *Zmiany zachowań turystycznych Polaków i ich uwarunkowań w latach 2006–2015 [Changes in and conditions of the Polish population's tourism behavior in 2006–2015],* Berbeka, J. (ed.), Foundation of the University of Economics, 34–46.

Bielski, P. (2011). Karwno. W stronę wsi uczącej się [Karwno — towards a rural learning community], *Zrównoważony Rozwój — Zastosowania [Sustainable Development Applications],* 2, 37–49.

Biénabe, E., Vermeulen, H., & Bramley, C. (2011). The food quality turn in South Africa: An initial exploration of its implications for small-scale farmers' market access. *Agrekon,* 50 (1), 36–52.

Boczar-Rozewicz, K. (1997). Rola doradztwa a rozwoj agroturystyki w wojewodztwie krosnienskim [Role of consulting vs. agritourism development in the Krośnieńskie voivodeship]. *Krośnieński Magazyn Rolniczy [Krosno Agricultural Magazine],* (12), 16–17.

Bodak, A., Jagoda, A., & Pietroń-Pyszczek, A. (2006). Stowarzyszenie jako przykład instytucjonalizacji w kreowaniu wizerunku przedsiębiorstwa agroturystycznego [Association as an example of institutionalization in creating the image of a farm tourism company]. *Agrobiznes,* 1118 (1), 98–102.

Bogucka, A. (2010). Twórczość ludowa produktem turystycznym obszarów wiejskich [Folk products as a tourism product of rural areas]. *Zrównoważony rozwój obszarów wiejskich – wybrane aspekty społeczne [Sustainable rural development: selected social aspects],* Kryk, B. (ed.), Szczecin University, Szczecin, 37–49.

Bogucki, J., Woźniak, A., & Zątek, W. (1995). Proekologiczne tendencje w turystyce. Agroturystyka i ekoturystyka [Environmentally-friendly trends in tourism. Agritourism and eco-tourism], *Turystyka i rekreacja jako czynnik integracji europejskiej [Tourism and recreation as a factor of European integration],* Bosiacki, S. (ed.), Poznań University of Physical Culture.

Bogusz, M., & Wojcieszak, M. (2018). Zagrody edukacyjne jako przykład markowego produktu turystyki wiejskiej [Educational farms as an example of a branded product of rural tourism]. *Intercathedra,* 37 (4), 329–334.

Borowska, A. (2009). Lokalne grupy działania czynnikiem stymulującym rozwój obszarów wiejskich w Polsce [Local action groups a factor stimulating the development of rural reas in Poland], *Acta Scientiarum Polonorum, Oeconomia,* 8 (4), 13–22.

Borkowska-Niszczota, M. (2015). Tourism clusters in Eastern Poland-analysis of selected aspects of the operation. *Procedia-Social and Behavioral Sciences,* 213, 957–964. Borzyszkowski, J., Marczak, M., & Zarębski, P. (2016). Spatial diversity of tourist function development: The municipalities of Poland's West Pomerania Province, *Acta Geographica Slovenica*, 56 (2), 267–276.

Bott-Alama, A. (2004). *The economic and social benefits of rural tourism development in Poland*. Rural tourism in Europe: experiences, development and perspectives, Belgrade, Serbia and Montenegro24–25 June 2002, Kielce, Poland, 6–7 June 2003, Yaremcha, Ukraine, 25–26 September 2003, 101–109.

Boyer, M. (1972). *Le Tourisme*. Paris: Le Seuils.

Brelik, A. (2005). Rozwój produktu markowego turystyki na terenach wiejskich [Development of branded product of tourism in rural areas], *Prace Naukowe Akademii Ekonomicznej we Wrocławiu* 1070 (1), 70–77.

Brelik, A., & Bogusz, M. (2013). Działania Centrum Doradztwa Rolniczego na rzecz turystyki wiejskiej [Activities of agricultural advisory centre for rural tourism]. *Centrum Doradztwa Rolniczego w Brwinowie Oddział w Poznaniu*, 1, 90–97.

Brodziński, Z. (2006). Uwarunkowania zintegrowanego rozwoju obszarów wiejskich w opiniach przedstawicieli lokalnych grup działania [Determinants of integrated development of rural areas in opinions of the representatives of local activity groups]. *Zeszyty Naukowe Akademii Rolniczej we Wrocławiu*, 540, 105–111.

Chiriţescu, V. (2011). European funds available for agrotourism development in Romania and Poland. *Agricultural Economics and Rural Development*, 241–254. Chłoń-Domińczak, A., Dębowski, H., Drogosz-Zabłocka, E., Dybaś, M., Holzer-Żelażewska, D., Maliszewska, A., & Tomasik, M. (2011). *Edukacja zawodowa w Polsce. Raport o stanie edukacji—kontynuacja przemian [Vocational education in Poland. A report on the condition of the education system: the transformation goes on]*, Wojciuk, A., & Fedorowicz, M. (eds.), Warsaw, Institute of Educational Research.

Chmielewska, B. (2018a). Gospodarstwa opiekuńcze odpowiedzią na potrzebę społeczną [Nursing homes respond to social needs]. *Nierówności społeczne a wzrost gospodarczy*, (54), 242–251.

Chmielewska, B. (2018b). Gospodarstwa opiekuńcze jako forma pozyskiwania dochodów przez małe gospodarstwa rolne w Polsce. *Problemy Drobnych Gospodarstw Rolnych*, (3), 21–32.

Chuang, S. T. (2010). Rural tourism: perspectives from social exchange theory. *Social Behavior and Personality: An International Journal*, 38, 1313–1322.

Chudy-Hyski, D. (2009). Uwarunkowania turystycznego kierunku rozwoju górskich obszarów wiejskich [Conditions of mountain areas in Poland development through tourism]. *Infrastruktura i ekologia terenów wiejskich [Infrastructure and ecology of rural areas]*, 1, 1–309. Cichowska, J. (2014). Oferta produktów tematycznych w turystyce wiejskiej [The offer of thematic products in rural tourism]. *Ekologia i Technika*, 22 (6), 340–345.

Czachara, J., & Krupa, J. (2011). Turystyka wiejska formą rekreacji i terapii dla osób niepełnosprawnych [Rural tourism as a form of recreation and therapy for the disabled], Krupa, J., & Soliński, T. (eds.), *Turystyka wiejska, ochrona środowiska i dziedzictwo kulturowe Pogórza Dynowskiego [Rural tourism, environmental protection and cultural heritage of Dynów Foothills]*, Dynów: Association of Dynów Foothills Tourism Communes, 53–63.

Czapiewska, G. (2009). Markowe produkty turystyki wiejskiej subregionu Słupskiego [Branded rural tourism products of the Słupsk sub-region], *Marka wiejskiego produktu turystycznego [Brand of a rural tourism product]*, Palich, P. (ed.), Publishing House of the Gdynia Maritime University, 76–84.

Czapiewska, G. (2012). Wioski tematyczne sposobem na aktywizację gospodarczą i społeczną regionu [Thematic villages as a way for an economic and social mobilization of a region]. *Studia i Materiały [Studies and materials]. Miscellanea Oeconomicae*, 16 (1), 109–123.

Czaplicka-Kozłowska, I. (2009). Agroturystyka w strategiach rozwoju gmin Warmii i Mazur [Agro-Tourism in Strategies of Development of Communes from Warmia and Mazury Region]. *Roczniki Naukowe Stowarzyszenia Ekonomistów Rolnictwa i Agrobiznesu*, 11 (4), 56–61.

Dąbrowska, M. (1996). Przemiany społeczno-ekonomiczne obszarów wiejskich pod wpływem turystyki [Socioeconomic transformation of rural areas under the impact of tourism]. *Turyzm [Tourism]*, 6 (1), 49–62.

Dadvar-Khani, F. (2012). Participation of rural community and tourism development in Iran, *Community Development*, 43, 259–277.

Davies, E. T., & Gilbert, D. C. (1992). A case study of the development of farm tourism in Wales. *Tourism Management*, 13 (1), 56–63.

Dębniewska, M., & Suchta, J. (1995). Agrotourism as a factor of social and economic activation of rural areas (exemplified by the province of Olsztyn). *Acta Academiae Agriculturae ac Technicae Olstenensis, Geodaesia et Ruris Regulatio*, (26), 89–100.

Dębniewska, M., & Suchta, J. (1996). Agroturystyka-dodatkowe źródło dochodów oraz czynnik aktywizujący rozwój rolnictwa i wsi [Agritourism: an additional revenue stream and a stimulus of agricultural and rural development]. *Zagadnienia Doradztwa Rolniczego [Agricultural Consultancy Issues]*, 1 (96), 76–84.

Domagalska-Gredys, M. (2008). Perspektywy rozwoju produktów tradycyjnych Podkarpacia w kontekście opinii mieszkańców [Perspectives of the development of traditional products in the Podkarpacie region in the context of the inhabitants' opinions]. *Problemy Rolnictwa Światowego*, 4 (19), 135–144.

Domalewski, J. (2010). Edukacja a procesy rozwoju obszarów wiejskich [Education vs. rural development processes], *Przestrzenne, społeczno-ekonomiczne zróżnicowanie obszarów wiejskich w Polsce. Problemy i perspektywy rozwoju [Geographic and socioeconomic differences between Polish rural areas. Problems and development prospects]*, Stanny, M., & Drygas, M. (eds.), Warsaw, Institute of Rural and Agriculture Development of the Polish Academy of Sciences, 181–199.

Dominik, P. (2019). Analiza poziomu zapotrzebowania na turystykę kulinarną w Polsce wśród turystów zagranicznych [The analysis of the level of demand for culinary tourism in Poland among foreign tourists]. *Przedsiębiorczość i Zarządzanie*, 20 (2.2), 137–152.

Dorocki, S., Rachwał, T., Szymańska, A. I., & Zdon-Korzeniowska, M. (2013). Spatial conditions for agritourism development on the example of Poland and France. *Current Issues of Tourism Research*, 2 (2), 20–29.

Dorocki, S., Szymańska, A. I., & Zdon-Korzeniowska, M. (2012). Polskie gospodarstwa agroturystyczne jako przedsiębiorstwa rodzinne [Family Entrepreneurship in Agritourism]. *Przedsiębiorczość i Zarządzanie*, 13 (8), 45–60.

Dorocki, S., Szymanska, A. I., & Zdon-Korzeniowska, M. (2015). Family agritourist enterprises in Poland: Preliminary survey results. *Entrepreneurial Business and Economics Review*, 3 (1), 151. Dorocki, S., Szymańska, A. I., & Zdon-Korzeniowska, M. (2016). Agricultural tourism farms in Poland: How the farmers improve their businesses–case study. *Understanding innovation in emerging economic spaces: Global and local actors, networks and embeddedness*, 15, 247–262.

Drag, K., & Czerniec, W. (2004). *Rural tourism development in Poland and promotional initiatives (I) and (II)*. Rural tourism in Europe: experiences, development and perspectives, Belgrade, Serbia and Montenegro24–25 June 2002, Kielce, Poland, 6–7 June 2003, Yaremcha, Ukraine, 25–26 September 2003, 127–130.

Drzewiecki, M. (1992). *Wiejska przestrzeń rekreacyjna [Rural recreational space]*, Tourism Institute, Warsaw.

Drzewiecki, M. (1994). *Delimitacja obszarów rozwoju agroturystyki w Polsce preferowanych w polityce rolnej resortu i działalności ośrodków doradztwa rolniczego [Delimitation of preferred development paths of Polish agritourism, as set out in the agricultural policy of the Ministry and as put in practice by agricultural consultancy centers]*. Department for Tourism Regions and Towns Management in Bydgoszcz, Tourism Institute, Warsaw.

Drzewiecki, M. (1995). *Agroturystyka: założenia, uwarunkowania, działania [Agritourism: assumptions, conditions, actions]*. Świadectwo Publishing Institute.

Drzewiecki, M. (1998). Pojęcie turystyki wiejskiej [The concept of rural tourism]. *Turyzm [Tourism]*, 8 (1), 21–27.

Drzewiecki, M. (2005). Agroturystyka w Polsce – stan obecny i tendencje rozwojowe [Polish agritourism: current condition and development trends]. *Uwarunkowania rozwoju turystyki wiejskiej związanej z obszarami wiejskimi [Conditions for rural tourism development]*, Sawicki, B., & Bergier, J. (eds.). Biała Podlaska, Publishing House of the Pope John II State School of Higher Education in Biała Podlaska, 46–51.

Dubel, K. (1997). Waloryzacja przyrodniczo-krajobrazowa gmin dla potrzeb planowania i organizacji turystyki na wsi [Valuation of natural and landscape assets of communes for the purposes of rural tourism planning and organization]. *Problemy Ekologii Krajobrazu [Problems of landscape ecology]*, 1, 60–66.

Duda, T. (2019). Idea turystyki powolnej (slowtourism) i edukacji kreatywnej szansą zrównoważonego rozwoju wsi i regionu–studium przypadku wsi Bełczna w powiecie łobeskim (Pomorze Zachodnie) [The concept of slow tourism and creative education – a chance for sustainable development of the rural areas and the whole region. A case study of the village of Bełczna in the County of Łobez (West Pomerania)]. *Przegląd Zachodniopomorski*, 34 (01), 111–128.

Durydiwka, M. (2003). Tourism as a factor of the activation the rural areas in Poland. *Acta Universitatis Carolinae: Geographica*, 38 (1), 53–58. Durydiwka, M. (2006). Differences in the development of the tourist function in Poland's rural areas, *Acta Geographica Universitatis Comenianae*, 48, 179–188.

Durydiwka, M. (2008). Tourism classification of rural areas in Poland (on the example of the Pomeranian region). A methodological proposal, Makowski, J. (ed.), *On the Social, Economic and Spatial Transformations of the Regions*, Faculty of Geography and Regional Studies, 207–219.

Durydiwka, M. (2009). Tourism classification of rural areas in Pomeranian region, *Global Changes: Their Regional and Local Aspects*, Wilk, W. (ed.), University of Warsaw, 196–202.

Durydiwka, M. (2012). *Czynniki rozwoju i zróżnicowanie funkcji turystycznej na obszarach wiejskich w Polsce [Development drivers of and differences in the rural tourism function across the Polish territory]*, Warsaw University. Faculty of Geography and Regional Studies.

Durydiwka, M. (2013). Turystyka kulinarna – nowy(?) trend w turystyce kulturowej, *Turystyka kulinarna, Prace i Studia Geograficzne*, Derek, M. (ed.), 52, 9–30.

Durydiwka, M. (2015). Funkcja turystyczna obszarów wiejskich w województwie pomorskim: zróżnicowanie i zmiany [The tourism function of rural areas in Pomerania Province: diversity and change], *Turyzm* [*Tourism*], 25 (1), 39–45.

Dziechciarz, T. (2013). Wykorzystanie witryn internetowych i poczty elektronicznej w marketingu agroturystyki na Podkarpaciu [Using the websites and electronic mail in the marketing of agritourism in Podkarpacie]. Krzyżanowska, K. (ed.) *Komunikowanie i doradztwo w turystyce wiejskiej*, Wydawnictwo SGGW, 60–68.

Dziegięć, E. (1990). Przemiany wsi Cisna i Wołkowyja pod wpływem funkcji turystycznej [Transformation of Cisna and Wołkowyja villages driven by the tourism function]. *Turyzm* [*Tourism*], 6 (90), 83–99.

Dziegięć, E. (1992). Rural tourist centers in the Polish Uplands Zone. *Turyzm* [*Tourism*], 2 (1), 83–97.

Dziegięć, E. (1995). Urbanizacja turystyczna terenów wiejskich w Polsce [Tourism urbanization of Polish rural areas]. *Turyzm* [*Tourism*], 5 (1), 5–56.

Dziegięć, E., & Liszewski, S. (1993). *Kształcenie dla turystyki w krajach Europy Wschodniej (konferencja)* [*Tourism education in Eastern European countries (conference)*], Krakow, September 10–11, 1992.

Faccioli, M. (2011). Youth`s perceptions of tourism impact: policy implications for Folgaria (Italy). *International Journal Tourism Policy*, 4, 1–35.

Feczko, J. (2015a). Konsumencka ocena turystyki wiejskiej na Pomorzu Środkowym w 2015 roku [Consumer research of rural tourism on the central Pomeranian in 2015]. *Roczniki* [*Annals*], 2015 (1230-2016-99994), 79–84.

Feczko, J. (2015b). Postawy konsumentów względem usług turystycznych na wsi na Pomorzu Środkowym w sezonie 2014/2015 [Consumers attitudes to services of rural tourism in Central Pomerania in 2014/2015 season]. *Roczniki Naukowe Stowarzyszenia Ekonomistów Rolnictwa i Agrobiznesu*, 17 (2), 39–44.

Firlej, K. (2002). Rola ekoturystyki i agroturystyki przy kształtowaniu przedsiębiorczości w gminie [Ekotourism at formation of enterprise in commune]. *Agrobiznes*, 941 (1), 214–219.

Furmankiewicz, M. (2006). Współpraca międzysektorowa w ramach partnerstw terytorialnych na obszarach wiejskich w Polsce [Cross-sector co-operation in area-based partnerships in rural areas of Poland]. *Studia Regionalne i Lokalne* [*Regional and Local Studies*], 7 (24), 117–136.

Furmankiewicz, M., & Foryś, M. (2006). Partnerstwa terytorialne na rzecz rozwoju obszarów wiejskich w polskiej części Sudetów–historia powstania i pierwsze efekty działań [Area-based partnerships for rural development in the Polish part of the Sudetes – origin and first achievements], *Problemy współpracy na rzecz ekorozwoju Sudetów*, Furmankiewicz, M., & Jadczyk, J. (eds.), Akademia Rolnicza we Wrocławiu, Muzeum Przyrodnicze w Jeleniej Górze, Jelenia Góra, 109–128.

Gąsiorowska, B. (1996). Preferencyjne kredyty dla gospodarstw rolnych zajmujących się agroturystyka [Soft loans for farms engaged in agritourism]. *Informacja Rolnicza-Aktualności* [*Agricultural information: news*], 12, 1–10.

Gąsiorowska-Mącznik, E. (2011). Stan i możliwości rozwoju pozarolniczej przedsiębiorczości na obszarach wiejskich województwa świętokrzyskiego w świetle badań ankietowych [Present state and opportunities for non-agricultural entrepreneurship in the rural areas of the Świętokrzyskie voivodeship], *Wieś i Rolnictwo*, 3 (152), 153–167.

Ginda, S. (1995). Agroturystyka szansą dodatkowego dochodu dla gospodarstw wiejskich [Agritourism as an opportunity to earn additional incomes for farms]. *Wieś i Doradztwo* [*Rural areas and consultancy*], 2, 22–24.

Głuszak, B. (2012). Miejscowości tematyczne na Warmii i Mazurach [Thematic towns in Warmia and Masuria], *Tam, gdzie miejsca mają duszę. Model tworzenia miejscowości tematycznych* [*Destinations with a spirit. A model for the creation of thematic towns*], Głuszak, B. (ed.), Elbląg Association for Supporting Non-Government Initiatives, Elbląg, 14–19.

Gogonea, R. M., Baltalunga, A. A., Nedelcu, A. & Dumitrescu, D. (2017). Tourism pressure at the regional level in the context of sustainable development in Romania, *Sustainability*, 9 (5), 1–24.

Gołębiewska, B. (2008). Rural tourism as a form of non-farming business activity in the Polish countryside. *Economic Science for Rural Development*, 16, 49–55.

Gołembski, G. (2002). *Metody stymulowania rozwoju turystyki w ujęciu przestrzennym* [*Methods for stimulating the development of tourism: a spatial approach*]. Publishing House of the Poznań University of Economics.

Golian, S. (2016). Profil właścicieli gospodarstw agroturystycznych na Roztoczu [Profile of tourist farms' owners in Roztocze]. *Annales Universitatis Mariae Curie-Sklodowska, sectio B–Geographia, Geologia, Mineralogia et Petrographia*, 71 (2), 153–164.

Golik, D., & Firlej, K. (2002). Wpływ agroturystyki na rozwój przedsiębiorczości na obszarach wiejskich [Effect of agrotourism on the enterprise development in rural areas]. *Zeszyty Naukowe Akademii Ekonomicznej w Krakowie*, 596, 41–49. Goodman, D. (2003). The quality "turn' and alternative food practices: reflections and agenda. *Journal of Rural Studies*, 1 (19), 1–7.

Grabowski, J. (2008). Uwarunkowania konkurencyjności turystycznej regionów [Conditions for effective competition among tourist regions], *Ruch Prawniczy, Ekonomiczny i Socjologiczny*, 3, 149–163.

Gralak, K. (2016). Witryna internetowa jako narzędzie promocji i dystrybucji oferty gospodarstw agroturystycznych [Website as a tool for the promotion and distribution of agritourism offer], *Ekonomika i Organizacja Gospodarki Żywnościowej*, 115, 171–182.

Grolleau, H., & Ramus, A. (1986). *Espace rural, espace touristique, le tourisme à la campagne et les conditions de son développement en France*. Rapport au ministre délégué auprès du ministre de l'Agriculture chargé de l'agriculture et de la forêt et au secrétaire d'État auprès du ministre du Commerce de l'Artisanat et du Tourisme. Paris.

Grębowiec, M. (2010). Rola produktów tradycyjnych i regionalnych w podejmowaniu decyzji nabywczych przez konsumentów na rynku dóbr żywnościowych w Polsce [The role of traditional and regional products in consumer decisions in retail food market in Poland]. *Problemy Rolnictwa Światowego* [*Problems of World Agriculture*], 10 (2), 22–31.

Grębowiec, M. (2017). Produkty regionalne i tradycyjne jako element budowania konkurencyjnej oferty produktów żywnościowych w Polsce i innych krajach Europy [Regional and traditional products as an element of building a competitive food product offer in Poland and other European countries], *Problemy Rolnictwa Światowego* [*Problems of World Agriculture*], 17 (2), 65–80.

Grykień, S. (1998). Tourist farms in Lower Silesia, Poland. *GeoJournal*, 46 (3), 279–281.

Grzebyk, M., & Lechwar, M. (2007). Znaczenie samorządu terytorialnego dla rozwoju lokalnego w świetle opinii mieszkańców [Meaning of territorial autonomy for local development in light of local communities opinion]. *Zeszyty Naukowe Uniwersytetu Rzeszowskiego. Seria Ekonomiczna. Ekonomia*, 1 (42), 124–134.

Grzywińska-Rąpca, M. (2014). Analiza wydatków konsumpcyjnych rolniczych gospodarstw domowych [Analysis of agricultural consumption expenditure of households], *Firma i Rynek*, 2 (47), 91–100.

Hacia, E. (2014). The development of tourist space in Polish port cities, *Procedia Social and Behavioral Sciences*, 151, 60–69.

Hadinejad, A., Moyle, B. D., Scott, N., Kralj, A., & Nunkoo, R. (2019). Residents' attitudes to tourism: a review, *Tourism Review*, 74 (2), 150–165.

Hall, D. (2000a). Sustainable tourism development and transformation in Central and Eastern Europe. *Journal of Sustainable Tourism*, 8 (6), 441–457.

Hall, D. R. (2000b). Evaluating the tourism—environment relationship: Central and East European experiences. *Environment and Planning B: Planning and Design*, 27 (3), 411–421.

Hall, D. (2004). Rural tourism development in Southeastern Europe: Transition and the search for sustainability. *International Journal of Tourism Research*, 6 (3), 165–176.

Harrill, R. (2004). Residents' attitudes toward tourism development: A literature review with implications for tourism planning, *Journal of Planning Literature*, 18, 251–266.

Hasiński, W., Głaz, M., & Kemona, S. (2004). Agroturystyka jako alternatywne źródło dochodów gospodarstw rolnych na Dolnym Śląsku [Agricultural activity as an alternative source of income for farms in Lower Silesia]. *Studia Obszarów Wiejskich*, 5, 137–150.

Hegarty, C., & Przebórska, L. (2005). Rural and agri-tourism as a tool for reorganising rural areas in old and new member states – comparison study of Ireland and Poland, *International Journal of Tourism Research*, 7 (2), 63–77.

Hełdak, M. (2002). Identyfikacja i ocena uwarunkowań rozwoju agroturystyki na Dolnym Śląsku [Identification and assessment of the conditions for agritourism development in Lower Silesia]. *Acta Scientiarum Polonarum Administratio Locorum*, 1 (2), 33–65.

Hellwig, Z. (1968). Zastosowanie metody taksonomicznej do typologicznego podziału krajów ze względu na poziom rozwoju i strukturę kwalifikowanych kadr [Application of the taxonomic method to the typological division of countries according to the level of development and the structure of qualified Staff], *Przegląd Statystyczny*, 4, 307–328. Herbst, M. (2007). Wpływ kapitału ludzkiego i społecznego na (krótkookresowy) wzrost gospodarczy w polskich podregionach, *Kapitał ludzki i kapitał społeczny a rozwój regionalny*, Herbst, M. (ed.), Wyd. Scholar, Uniwersytet Warszawski, Warszawa, 166–203.

Huh, C., & Vogt, C. A. (2008). Changes in residents' attitudes toward tourism over time: A cohort analytical approach. *Journal of Travel Research*, 46 (4), 446–455.

Idziak, M., & Idziak, W. (2015). Potencjał innowacyjny tematyzacji w turystyce wiejskiej i rozwoju wsi [The potential of innovative thematic specialization in rural tourism and rural development], *Innowacyjność w turystyce wiejskiej a nowe możliwości zatrudnienia na obszarach wiejskich*, Kamińska, W. (ed.), Warszawa: KPZK PAN, 19–33.

Idziak, W., Majewski, J., & Zmyślony, P. (2015). Community participation in sustainable rural tourism experience creation: a long-term appraisal and lessons from a thematic villages project in Poland. *Journal of Sustainable Tourism*, 23 (8–9),1341–1362. Jagusiewicz, A. (1999). Kierunki rozwoju infrastruktury turystycznej w gminach podwarszawskich [Development trends of tourism infrastructure in communes located near Warsaw]. *Problemy Turystyki [Problems of Tourism]*, 22 (2), 51–65.

Jalinik, M. (1998). Postrzeganie turystyki i agroturystyki przez mieszkańców rejonu Puszczy Białowieskiej [The Białowieża Primeval Forest Residents' Perception of Tourism and Agrotourism]. *Ekonomia i Środowisko*, 1 (12), 101–111.

Jalinik, M. (2007). Typologia gospodarstw oraz rozwój usług agroturystycznych [The typology of rural farm and agricultural services' development]. *Roczniki Naukowe Stowarzyszenia Ekonomistów Rolnictwa i Agrobiznesu,* 2 (09), 109–116.

Jalinik, M. (2009). *Uwarunkowania i czynniki rozwoju usług turystycznych na obszarach wiejskich [Conditions and drivers of development of rural tourism services].* Publishing House of the Białystok University of Technology.

Jankun, M., Janicka, E., Woźnicki, P., & Furgała-Selezniow, G. (2016). Rozwój turystyki wiejskiej w gminie Golub-Dobrzyń na przykładzie wiosek tematycznych [Rural tourism development in the Golub-Dobrzyń commune: a case study of thematic villages], *Turystyka wiejska. Zagadnienia ekonomiczne i marketingowe [Rural tourism. Economic and marketing issues],* Graja-Zwolińska, S., Spychała, A., & Kasprzak, K. (eds.), Wieś Jutra Publishing House, 122–129.

Jaszczak, A., & Žukovskis, J. (2010). Tourism business in development of European rural areas. *Management Theory and Studies for Rural Business and Infrastructure Development,* 20 (1), 35–44.

Jaźwińska, A. (1999). Doradztwo rolnicze w rozwiązywaniu problemów związanych z rozwojem agroturystyki [Problems related to the development of the agritourism]. *Zagadnienia Doradztwa Rolniczego,* 4, 46–52. Jęczmyk, A., Maćkowiak, M., & Uglis. J. (2014). Dziedzictwo kulinarne elementem przewagi rynkowej w ofercie gospodarstw agroturystycznych [Culinary heritage as an element of market advantage in agritourism business]. *Roczniki Naukowe Stowarzyszenia Ekonomistów Rolnictwa i Agrobiznesu,* 16 (2), 103–108.

Jęczmyk, A., Graja-Zwolińska, S., Maćkowiak, M., Spychała, A., Uglis, J., & Sikora, J. (2011). Stan wiedzy na temat produktów tradycyjnych wśród właścicieli gospodarstw agroturystycznychn [Knowledge of agritourism farms owners about traditional products]. *Roczniki Naukowe Stowarzyszenia Ekonomistów Rolnictwa i Agrobiznesu,* 13 (2), 155–158.

Jeleń, B. (2003). Agroturystyka jako przejaw przedsiębiorczości na przykładzie wybranych gmin [Agrotourism like a factor enterprise for example chosen communes]. *Roczniki Naukowe Stowarzyszenia Ekonomistów Rolnictwa i Agrobiznesu,* 5 (2), 62–66.

Kaczmarek, J., Stasiak, A., & Włodarczyk, B. (2008). Przestrzeń gościnna – kilka uwag o konkurencyjności regionów [A hospitable area: a few comments on regional competitiveness], *Turystyka jako czynnik wzrostu konkurencyjności regionów w dobie globalizacji [Tourism as a driver of regional competitiveness in the era of globalization],* Gołembski, G. (ed.), University of Economics, Poznań, 136–150.

Kaleta, A. (1994). Multifunctional development of rural areas in Poland. *Anthropological Journal on European Cultures,* 3 (1), 85–93.

Kamińska, W., & Mularczyk, M. (2015). Development of agritourism in Poland: a critical analysis of students' expectations of agritourism farms. *Miscellanea Geographica,* 19 (4), 44–55.

Karbowiak, K. (2013). Agroturystyka jako forma przedsiębiorczości rolników na przykładzie wybranych gospodarstw agroturystycznych w województwie warmińsko-mazurskim [Agrotourism as a form of entrepreneurial activity for farmers based on based on selected agrotourism farm in the Warmia and Mazury]. *Roczniki Naukowe Stowarzyszenia Ekonomistów Rolnictwa i Agrobiznesu,* 6 (15), 124–129.

Karbowiak, K., & Lejmel, K. (2000). Turystyka wiejska na obszarach chronionych [Country- tourism on protected areas]. *Roczniki Naukowe Stowarzyszenia Ekonomistów Rolnictwa i Agrobiznesu,* 2 (02), 61–65. Kieżel, E. (Ed.). (2004). *Racjonalność konsumpcji i zachowań konsumentów: praca zbiorowa* [*Rationality of consumption and consumer behavior (multiple authors)*]. Polskie Wydawnictwo Ekonomiczne, Warsaw.

Kiryluk-Dryjska, E., & Hadyński, J. (2016). Ocena skuteczności działań programu Leader przez mieszkańców obszarów wiejskich w Polsce [The effectiveness assessment of the leader program in opinions of rural areas residents in Poland]. *Studia, Polska Akademia Nauk Komitet Przestrzennego Zagospodarowania Kraju*, 167, 357–366. Kłoczko-Gajewska, A. (2013). General characteristics of thematic villages in Poland. *Visegrad Journal on Bioeconomy and Sustainable Development*, 2, 60–63.

Kłoczko-Gajewska, A. (2014.) Can we treat thematic villages as social innovations?, *Journal of Central European Green Innovation*, 2 (3), 49–59.

Kłoczko-Gajewska, A. (2015). Działalność wiosek tematycznych w Polsce i wstępna ocena jej efektów [Activities of thematic villages in poland and preliminary assessment of their effects]. *Roczniki Naukowe Ekonomii Rolnictwa i Rozwoju Obszarów Wiejskich*, 102 (3), 104–111.

Kłoczko-Gajewska, A., & Markiewicz, O. (2018). Mechanizm rozprzestrzeniania się wiosek tematycznych w Polsce [The spread of thematic village idea in Poland]. *Wieś i Rolnictwo*, 3 (180), 193–209.

Kłoczko-Gajewska, A., Śpiewak, R., & Zarębski, P. (2015). Aktywizacja obszarów wiejskich w oparciu o produkty żywnościowe na przykładzie wsi tematycznych [Thematic villages as an example of rural development based on food products]. *Turystyka Kulturowa*, 12, 37–50. Kłodziński, M. (2006). Polityka rozwoju obszarów wiejskich [Policy of rural development]. *Acta Agraria et Silvestria*, 46 (2), 239–245.

Klonowska-Matynia, M. (2017). Czynniki edukacyjne a przestrzenne rozmieszczenie kapitału ludzkiego na obszarach wiejskich w Polsce [Educational factors and spatial distribution of human capital in rural areas in Poland]. *Acta Universitatis Lodziensis. Folia Oeconomica*, 1 (327), 107–127.

Kmita-Dziasek, E. (2015). Zagrody edukacyjne jako przykład innowacyjnej przedsiębiorczości na obszarach wiejskich [Educational homestead as an example of innovative entrepreneurship in rural areas]. Turystyka wiejska a nowe możliwości zatrudnienia. *Studia, Polska Akademia Nauk Komitet Przestrzennego Zagospodarowania Kraju*, 155–165.

Kmita-Dziasek, E., & Bogusz, M. (2017). Networking wiejskiej przedsiębiorczości na przykładzie Ogólnopolskiej Sieci Zagród Edukacyjnych. *Sieci współpracy w turystyce wiejskiej. Stan obecny i nowe wyzwania* [*A networking approach to rural enterprise: a case study of the Countrywide Network of Educational Farms. Collaborative networks in rural tourism: current condition and new challenges*], Wojciechowska, J. (ed.), KSOW, 116–134.

Konopka, P. (2013). Potencjał rozwoju turystyki w gminie Chojna. Analiza i propozycje [Tourism development potential in the Chojna commune. An analysis and propositions]. *Rocznik Chojeński* 5, 275–300. Korça, P. (1998). Resident perceptions of tourism in a resort town. *Leisure Sciences*, 20 (3), 193–212.

Korczak, I., & Tomaszewski, M. (2015). Wsparcie turystyki wiejskiej i agroturystyki w Polsce w ramach Wspólnej Polityki Rolnej [Supporting the rural tourism and the farm tourism in Poland as part of the common agricultural policy]. *Zagadnienia Doradztwa Rolniczego*, 1, 25–39.

Korzeniewski, J., & Kozłowski, M. (2019). Development of tourism in Polish poviats in the years 2010–2017, *Quality & Quantity*, 1–22. Kosmaczewska, J. (2007). *Wpływ agroturystyki na rozwój ekonomiczno-społeczny gminy* [*Impact of agritourism on socio-economic development at commune level*]. Bogucki Scientific Publishing House, Poznań.

Kosmaczewska, J. (2008). The relationship between development of agritourism in Poland and local community potential. *Studies in Physical Culture & Tourism*, 15 (2), 141–148.

Kosmaczewska, J. (2009a). Kapitał społeczny mieszkańców wsi jako czynnik turystycznego rozwoju obszarów wiejskich [Social capital as a factor of tourism development in rural areas]. *Acta Scientiarum Polonorum*, 8 (4), 87–96.

Kosmaczewska, J. (2009b). Ginące zawody jako atrakcja turystyczna i pozarolnicze źródło zarobkowania mieszkańców obszarów wiejskich [Extinct occupations as a tourism attraction and a non-agricultural revenue stream for rural residents], *Turystyka i kultura – wspólnie zyskać! [Tourism and culture: together, we can win!]*, Stasiak, A. (ed.), WSTH Łódź Publishing House, 61–73.

Kosmaczewska, J. (2009c). Turystyczne sieci współpracy klastrowej jako przejaw przedsiębiorczości na obszarach wiejskich [Tourist network clusters as the entrepreneurial seedbed in rural areas]. *Roczniki Naukowe Stowarzyszenia Ekonomistów Rolnictwa i Agrobiznesu*, 5 (11), 160–165.

Kosmaczewska, J. (2009d). Tworzenie markowych produktów turystyki wiejskiej w oparciu o wybrane dyscypliny plastyki ludowej [Creating branded rural tourism products based on selected disciplines of visual folk arts]. *Marka wiejskiego produktu turystycznego [Brand of a rural tourism product]*, Palich, P. (ed.), Publishing House of the Gdynia Maritime University, 203–209.

Kosmaczewska, J. (2010a). Witryna internetowa jako narzędzie kreowania konkurencyjności w agroturystyce [Internet webpage as a tool for creating competitiveness in agritourism]. *Acta Scientiarum Polonorum, Oeconomia*, 9 (4), 225–232.

Kosmaczewska, J. (2010b). Changes in the consumption of tourist services among the Poles as the sign of the transformation of the political system in Poland. *The Faces and Problems of Modern Europe. Case Studies.* Jeszka, J., & Wojciechowski, S. (eds.), 153–161.

Kosmaczewska, J. (2011). Agrotourist activity as an example of family entrepreneurship. *Acta Scientiarum Polonorum. Oeconomia*, 10 (3), 73–81.

Kosmaczewska, J. (2012). Gościnność jako istotna składowa reputacji gospodarstwa agroturystycznego [Hospitality as a significant component of the agritourist farm reputation]. *Zeszyty Naukowe Szkoły Głównej Handlowej. Kolegium Gospodarki Światowej*, 35, 124–146.

Kosmaczewska, J. (2013). *Turystyka jako czynnik rozwoju obszarów wiejskich [Tourism as a rural development driver]*, Bogucki Scientific Publishing House, Poznan. Kosmaczewska, J., & Górka, J. (2009). Ekonomiczne skutki rozwoju agroturystyki w opinii mieszkańców terenów wiejskich i lokalnych przedsiębiorców [Economic results of agrotourism development opinion of inhabitant of countryside and local entrepreneurs]. *Ekonomiczne Problemy Turystyki*, 11, 175–185.

Kosmaczewska, J., Barczak, M., & Brudnicki, R. (2016a). *Przedsiębiorczość na obszarach wiejskich: postawy i predyspozycje osób z grupy NEET [Rural enterprise: attitudes and capabilities of the NEETs]*, WSG University Publishing House, Bydgoszcz.

Kosmaczewska, J., Thomas, R., & Dias, F. (2016b). Residents' perceptions of tourism and their implications for policy development: evidence from rural Poland. *Community Development*, 47 (1), 136–151.

Kotra, K., & Ruszkowski, J. M. (2012). Popyt turystyczny w Polsce w obliczu światowego kryzysu finansowego w latach 2007–2010 [Tourist demand as economy category in the face of the global financial crisis year 2007–2010]. *Ekonomiczne Problemy Usług*, 82, 81–92.

Kowalczyk, A. (1994). *Geograficzno-społeczne problemy zjawiska „drugich domów"* [*Geographic and social problems of what is referred as "second homes"*], Warsaw University, Warsaw.

Kowalczyk, A. (1996). Sustainable development and perspectives for development of tourism in rural areas in Poland. *Miscellanea Geographica*, 7 (1), 197–202.

Kozak, M. (2006). Konkurencyjność turystyczna polskich regionów [Touristic competitiveness of Polish regions]. *Studia Regionalne i Lokalne*, 7 (25), 49–65.

Kozłowska-Burdziak, M. (2012). Rola programów unijnych w rozwoju przedsiębiorczości w rolnictwie i na obszarach wiejskich w Polsce [Role of EU programmes in the development of agricultural entrepreneurship in rural areas in Poland]. *Optimum: Studia Ekonomiczne*, 2 (56), 93–112.

Kraszewska, U. (1993). Agroturystyka we Francji [Agritourism in France]. *Informator, Ośrodek Doradztwa Rolniczego Bonin*, 11, 1–13.

Kraszewska, U. (1994). Agroturystyka. Rodzaje zakwaterowania [Agrotourism. Types of accommodation]. *Informator, Ośrodek Doradztwa Rolniczego Bonin*, 07 (08), 1–10.

Król, K. (2007). Gospodarstwa agroturystyczne w konfrontacji z klientem on-line [Agritourism farms faced with online customers]. *III Ogólnopolska Młodzieżowa Konferencja Naukowa* [*3rd Countrywide Scientific Conference of the Youth*], Rzeszów University, 46–53.

Król, K. (2016). Wpływ optymalizacji witryn internetowych na promocję turystyki wiejskiej w sieci [The impact of website optimization on online promotion of agrotourism]. *Problemy Drobnych Gospodarstw Rolnych*, (3), 57–71.

Król, K. (2017). Konwersja celu w internetowej sprzedaży produktów turystyki wiejskiej [Conversion rate in sales of tourist product]. *Problemy Drobnych Gospodarstw Rolnych*, (2), 33–40.

Król, K. (2018). Jakość witryn internetowych w zarządzaniu marketingowym na przykładzie obiektów turystyki wiejskiej w Polsce [Quality of websites in marketing management based on the example of rural tourism facilities in Poland]. *Infrastruktura i Ekologia Terenów Wiejskich*, 3 (2), 5–181.

Król, K. (2019). Forgotten agritourism: Abandoned websites in the promotion of rural tourism in Poland. *Journal of Hospitality and Tourism Technology*, 10 (3), 461–472.

Król, K., & Halva, J. (2017). Measuring efficiency of websites of agrotouristic farms from Poland and Slovakia. *Studia Ekonomiczne i Regionalne* [*Economic and Regional Studies*], 10 (2), 50–59.

Królikowska, K., & Pijet-Migoń, E. (2019). Turystyka kulinarna w wojewódzkich i miejskich strategiach rozwoju turystyki w Polsce [Culinary tourism in regional and municipal strategies of tourism development], *Folia Turistica*, (51), 35–58.

Krstić, B., Radivojević, V., & Stanišić, T. (2017). Determinants of CEE countries' tourism competitiveness: A benchmarking study, *Management: Journal of Sustainable Business and Management Solutions in Emerging Economies*, 21 (80), 11–22.

Kruczek, Z. (1998). Kształcenie dla potrzeb turystyki w regionie małopolskim [Education for the tourism industry in the Małopolska Region]. *Folia Turistica*, (8), 5–16.

Kruczek, Z. (2015). Sektorowa rama kwalifikacji w turystyce i jej znaczenie dla kształcenia i certyfikowania kadr turystycznych [Sectoral qualifications framework for tourism domain and its importance for the education and certification of tourist staff]. *Prace Naukowe Uniwersytetu Ekonomicznego we Wrocławiu* [*Research Papers of Wroclaw University of Economics*], 379, 396–404.

Krupa, J. (2010). Dziedzictwo kulinarne elementem atrakcyjności turystycznej regionu [Culinary heritage as an element of tourism attractiveness of the region]. *Problemy Ekologii Krajobrazu*, 27, 151–155.

Kryczka, M. (2014a). *Konsumpcja usług turystycznych w Polsce [Tourism service consumption in Poland]*. Bronisław Czech University of Physical Culture in Krakow.

Kryczka, M. (2014b). Makroekonomiczne uwarunkowania konsumpcji turystycznej w okresie przekształceń społeczno-gospodarczych w Polsce w latach 1990–2010 [Macro-economic determinants of tourism consumption in the socio-economic transition in Poland 1990–2010]. *Folia Turistica*, 30, 81–92.

Krysa, A., & Basaj, M. (2010). Agroturystyka jako kierunek rozwoju gminy Gdów w opinii jej mieszkańców [Agritourism as a direction of development of the municipality Gdów in the opinion of its inhabitants]. *Problemy Ekologii Krajobrazu*, 27, 225–230.

Krzyżanowska, K. (2009). Doradztwo rolnicze jako czynnik stymulujący rozwój gospodarstw agroturystycznych [Agricultural extention as a factor which stimulates development of agri-tourism farms]. *Roczniki Naukowe Stowarzyszenia Ekonomistów Rolnictwa i Agrobiznesu*, 11 (5), 189–194.

Krzyżanowska, K. (2014). Tendencje zmian w działaniach informacyjno-promocyjnych w turystyce wiejskiej i ich skuteczność [Trends in the field of effective information and promotion activities in rural tourism]. *Ekonomika i Organizacja Gospodarki Żywnościowej*, 107, 57–67.

Kubal-Czerwińska, M., & Piziak, B. (2010). Wine tourism on rural areas: Polish conditions after the transformation. *Journal of Settlements and Spatial Planning*, 1 (2), 135–143. Kurek, W. (1990). *Wpływ turystyki na przemiany społeczno ekonomiczne polskich Karpat [Impact of tourism on the socioeconomic transformation of Polish Carpathians]*, H. Kołłątaj Agricultural University in Krakow, Krakow.

Kurek, W. (1995). Gospodarstwa agroturystyczne w Karpatach [Agritourism farms in the Carpathians]. *Turyzm*, 5 (2), 77–86.

Kurek, W. (1996). Agriculture versus tourism in rural areas of the Polish Carpathians. *GeoJournal*, 38 (2), 191–196.

Kurczewski, R. (2007). Edukacja zoologiczna w agroturystyce [Zoological education in agriculture]. *Turystyka wiejska a edukacja. Różne poziomy, różne wymiary [Rural tourism vs. education. Different levels, different dimensions]*, Sikora, J. (ed.), Publishing House of the Poznań University of Life Sciences, 252–259.

Kurtyka, I. (2005). Agroturystyka i jej wpływ na rozwój społeczności wiejskiej w Sudetach [Agrotouristic and its influence on rural community development in Sudety Mountains]. *Roczniki Naukowe Stowarzyszenia Ekonomistów Rolnictwa i Agrobiznesu*, 7 (7), 185–190.

Kurtyka, I. (2008). Aktywność gospodarcza właścicieli gospodarstw agroturystycznych w powiecie kłodzkim [The economic activity of agrotourist farms in Kłodzko District]. *Roczniki Naukowe Stowarzyszenia Ekonomistów Rolnictwa i Agrobiznesu*, 2 (10), 138–141.

Kurtyka, I. (2010). Agroturystyka jako forma przedsiębiorczości na terenie parku krajobrazowego Dolina Baryczy [Agrotourism as a form of entrepreneurship in the "Barycz Valley' Scenic Park]. *Acta Scientiarum Polonorum, Oeconimia*, 9 (2), 111–119.

Kurtyka, I. (2011). Właściciele gospodarstw agroturystycznych na Dolnym Śląsku-profil usługodawców [Owners of agrotouristic farms in Lower Silesia – profile of service providers]. *Roczniki Naukowe Stowarzyszenia Ekonomistów Rolnictwa i Agrobiznesu*, 13 (3), 158–162.

Kurtyka-Marcak, I., & Kutkowska, B. (2017). Innovation in rural tourism. *Journal of Agribusiness and Rural Development*, 2 (44), 383–392.

Kuźniar, W. (2010). Rola produktów tradycyjnych w rozwoju usług agroturystycznych na przykładzie województwa podkarpackiego [Role of traditional products in development of agritouristic services: on the example of Podkarpackie Province]. *Acta Scientiarum Polonorum, Oeconimia*, 9 (4), 245–254.

Kuźniar, W. (2013). Aktywność marketingowa gmin i jej oddziaływanie na rozwój turystyki wiejskiej [Marketing activity of communes and its influence on development of rural tourism]. *Prace Naukowe Wydziału Ekonomii Uniwersytetu Rzeszowskiego. Monografie i Opracowania*, 16, 1–301.

Kuźniar, W. (2015). Postawy lokalnej społeczności wobec rozwoju turystyki wiejskiej i ich konsekwencje dla obszaru recepcji [Local community's attitudes towards development of rural tourism and their implications for a reception area]. *Nierówności społeczne a wzrost gospodarczy*, 42, 296–305.

Kuźniar, W., & Surmacz, T. (2015). Zastosowanie koncepcji współtworzenia wartości produktu w turystyce wiejskiej–wybrane aspekty [The usage of product value co-creation concept in rural tourism – selected aspect]. *Ekonomiczne Problemy Turystyki*, 30, 25–36.

Kuźniar, W., & Witek, L. (2016). Traditional regional products as part of unique sales proposition on farm tourism in Poland. *Scientific Papers-Series Management Economic Engineering in Agriculture and Rural Development*, 16 (1), 249–252.

Łaciak, J. (2001). *Uczestnictwo Polaków w wyjazdach turystycznych w 2000 roku [Participation of the Polish population in tourist trips in 2000]*, Tourism Institute, Warsaw.

Łaciak, J. (2011). *Uczestnictwo Polaków w wyjazdach turystycznych w 2010 roku [Participation of the Polish population in tourist trips in 2010]*, Tourism Institute, Warsaw.

Łaciak, J. (2013). *Aktywność turystyczna mieszkańców Polski w wyjazdach turystycznych w 2012 roku [Tourism activity of the Polish population: tourism trips in 2012)*, Tourism Institute, Warsaw2013.

Lane, B. (1992). A philosophy for rural tourism, *Tourism on the farm*, Feehan, J. (ed.), Environmental Institute, University College, Dublin.

Lane, B. (1994). What is rural tourism? *Journal of Sustainable Tourism*, 2 (1–2),7–21.

Latkova, P., & Vogt, C. A. (2012). Residents' attitudes toward existing and future tourism development in rural communities, *Journal of Travel Research*, 51, 50–67.

Lewczuk, A., Chylek, E. K., & Rzepinski, W. (1995). Doradztwo i edukacja w wielofunkcyjnym rozwoju obszarow wiejskich polnocno-wschodniej Polski [Consultancy and education in multipurpose rural development in northeast Poland]. *Zeszyty Problemowe Postępów Nauk Rolniczych [Advancements in Agricultural Sciences and related Problems, Scientific Journals]*, 420, 123–134.

Lipińska, I. (2008). System chronionych nazw pochodzenia i oznaczeń geograficznych produktów rolnych [System of protected designations of origin and geographical indications for agricultural products], *Journal of Agribusiness and Rural Development*, 4 (10), 1–10.

Liszewski, S. (1995). Przestrzeń turystyczna [Tourism space], *Turyzm [Tourism]*, 5 (2), 7–19.

Łukasiewicz, K. (2017). Standard jakości w turystyce wiejskiej i agroturystyce w Polsce i na Ukrainie w opinii potencjalnych klientów [Quality standards in rural tourism and agroturistry in Poland and Ukraine in opinion of potential customers]. *Roczniki [Annals]*, 19 (6), 147–152.

Lundberg, Ch. Gudmundson, A., & Andersson, T. D. (2009). Herzberg's Two-Factor Theory of work motivation tested empirically on seasonal workers in hospitality and tourism 2009. *Tourism Management*, 30, 890–899. Macbeth, J., Carson, D., & Northcote, J. (2004). Social capital, tourism and regional development: SPCC as a basic for innovation and sustainability, *Current Issues in Tourism*, 7 (6), 502–522.

Maciejewska, W. (2003). Edukacyjna rola gospodarstw ekologicznych w Wielkopolsce [Educational role of agrotouristic farms in region of Wielkopolska]. *Journal of Research and Applications in Agricultural Engineering*, 48 (1), 47–50.

Maćkowiak, M. (2010a). Stowarzyszenie agroturystyczne—fikcja czy skuteczna forma współpracy [Agrotourist association—fiction or effective form of cooperation]. *Roczniki Naukowe Stowarzyszenia Ekonomistów Rolnictwa i Agrobiznesu*, 12 (2), 191–201.

Maćkowiak, M. (2010b). Postawa mieszkańców jako czynnik warunkujący aktywizację turystyczną obszarów wiejskich. Studium przypadku wiosek tematycznych [The inhabitants' attitude as a tourist activation determinant of rural areas. The thematic villages case study]. *Ekonomiczne Problemy Usług*, 53, 191–201.

Maćkowiak, M., & Budych-Tomkowiak, E. (2012). Łowiectwo jako forma aktywności w turystyce wiejskiej–wybrane aspekty zachowań konsumentów [Hunting as a form of activity in rural tourism – selected aspects on consumer behavior]. *Ekonomiczne Problemy Usług*, 699, 53–64.

Maćkowiak, M., & Graja-Zwolińska, S. (2017). Znaczenie zaufania w budowaniu współpracy sieciowej w turystyce wiejskiej [The importance of trust in building collaborative rural tourism networks]. *Sieci współpracy w turystyce wiejskiej, stan obecny i wyzwania [Collaborative networks in rural tourism: current condition and challenges]*, Wojciechowska, J. (ed.), CDR Krakow, Krakow–Łódź, 57–70.

Maćkowiak, M., & Jęczmyk, A. (2013). Strategia hands-on activity w turystyce wiejskiej i jej wykorzystanie w tworzeniu edukacyjnych produktów turystycznych [The strategy of hands-on activity in rural tourism and its use in creating tourism educational products]. *Prace Naukowe Uniwersytetu Ekonomicznego we Wrocławiu*, 304, 134–143.

Majewska, J. (2008). *Samorząd terytorialny w kształtowaniu funkcji turystycznej gminy* [The role of local government in forming the tourism function of a commune]. Poznań University of Economics.

Majewski, J. (1994). Turystyka wiejska [Rural tourism]. *Poradnik Gospodarski [A Manager's Guide]*, 1, 1–33.

Majewski, J. (2003). Turystyka wiejska w programie PHARE TOURIN-kierunki strategiczne i ich realizacja [Rural tourism in PHARE TOURIN programmes: Strategic options and their implementation]. *Zeszyty Naukowe Akademii Rolniczej im. H. Kołłątaja w Krakowie* (90), 33–48.

Majewski, J. (2010). Wiejskość jako rdzeń produktu turystycznego–użyteczność podejść geograficznego i ekonomicznego [Rurality as the core of tourism product – usefulness of geographic and economic approaches]. *Acta Scientiarum Polonorum. Oeconomia*, 9 (4), 287–294.

Majewski, J. (2017). Forms of rural tourism in metropolitan areas. *Metropolitan Commuter Belt Tourism*, Sznajder, M. (ed.), Routledge, 168–179.

Majewski, J., & Zmyślony, P. (2014). Wiejski charakter–podmiejska lokalizacja. Turystyka wiejska na obszarze metropolitalnym Poznania [Rural character–suburban location. Rural tourism in Poznań metropolitan area]. *Turystyka i Rekreacja*, 11 (1), 120–126. Makała, H. (2014). Atrakcyjność dziedzictwa kulinarnego Podlasia [Attraction culinary heritage of Podlasie]. *Turystyka i Rekreacja*, 14 (2), 81–90.

Marciniak, G. (ed.) (2011). *Turystyka w 2010 roku* [*Tourism in 2010*]. Report, Central Statistical Office, Social Surveys Department, Warsaw.

Marczak, M., & Borzyszkowski, J. (2010). Analiza działań promocyjnych podejmowanych przez gospodarstwa agroturystyczne na Pomorzu Środkowym [Analysis of promotional actions taken by agrotouristic farms on Middle Pomerania]. *Acta Scientiarum Polonorum. Oeconomia*, 9 (4), 295–303.

Marković, S., Perić, M., Mijatov, M., Doljak, D., & Žolna, M. (2017). Application of tourist function indicators in tourism development, *Journal of the Geographical Institute Jovan Cvijić SASA*, 67 (2), 163–178.

Marks, E., Polucha, I., Jaszczak, A., & Marks, M. (2009). Agritourism in sustainable development: Case of Mazury in North-eastern Poland. *Proceedings of the International Scientific Conference: Rural Development*, 4 (1), 90–94.

Marks-Bielska, R., & Zielińska-Szczepkowska, J. (2011). Dotychczasowe wykorzystanie środków z Programu Rozwoju Obszarów Wiejskich (2007–2013) na rozwój agroturystyki w woj. warmińsko-mazurskim [Current utilization of funds within the Rural Development Programme 2007–2013 for agritourism development in Warmia-Mazuria voivodeship]. *Polityka ekonomiczna*, Sokołowski, J., & Sosnowski, M. (eds.), Wydawnictwo Uniwersytetu Ekonomicznego we Wrocławiu, 446–457.

Marks-Bielska, R., Babuchowska, K., & Lizinska, W. (2014). Agritourism as a form of business activity in rural areas. *Acta Scientiarum Polonorum. Oeconomia*, 13 (3), 69–79. Marzejon-Frycz, I. (2012). Specyfika agroturystyki Polski na tle wybranych krajów Unii Europejskiej [Specific of Polish agritourism in the context of chosen EU countries]. *Ekonomia*, 29, 66–85.

Maton, A. (1999). Agroturystyka w Belgii [Agrotourism in Belgium]. *Nowoczesne Rolnictwo. Nauka, Doradztwo, Praktyka*, 6 (11), 1–45.

Matusek-Psiuk, R. (1999). Osobliwości środowiska przyrodniczego Ziemi Białostockiej i ich wykorzystanie w turystyce [The natural peculiarities of the Białystok Region environment and their application in tourism]. *Zeszyty Naukowe Górnośląska Wyższa Szkoła Handlowa w Katowicach*, 6, 31–47.

Matuszewska, M., Skóra, J., Darul, D., Gajda, M., Górczewska, K., Kistowska, P., & Graja-Zwolińska, S. (2016). Promocja wielkopolskiej turystyki wiejskiej w mediach społecznościowych [Promoting rural tourism of the Greater Poland region in social media]. *Turystyka wiejska. Zagadnienia ekonomiczne i marketingowe* [*Rural tourism. Economic and marketing aspects*], Jęczmyk, A., Uglis, J., & Maćkowiak, M. (eds.), Wieś Jutra Publishing House, 172–182.

Matysiak, I., & Michalska, S. (2016). Social farming: a new model of dealing with ageing in rural areas in Poland? *Sociologia e Politiche Sociali*, 19 (3), 65–82.

Mazurek-Kusiak, A., & Golian, S. (2011). Ocena współpracy właścicieli nowo powstających gospodarstw agroturystycznych z samorządami, ODR-ami i jednostkami naukowymi w zakresie rozwoju promocji usług agroturystycznych [Assessment of the cooperation of owners of start-ups agritourism farm with local governments, agricultural advisory centre and scientific institutions in the development and promotion of agriturism service]. *Ekonomika i Organizacja Gospodarki Żywnościowej*, 90, 109–118.

McMahon, F. (1996). Rural and agri-tourism in Central and Eastern Europe. *Tourism in Central and Eastern Europe: Educating for Quality*, Richards, G. (ed.), Tilburg University, 175–182.

Mencwel, J., Milczewska, K., & Wiśniewski, J. (2014). *Koła Gospodyń wiejskich nie tylko od kuchni* [*Farmer's Wives' Associations: in the kitchen and beyond*]. Report from the study by the "Stocznia" Foundation for Social Research and Innovation, Warsaw.

Mickiewicz, B. (2009). Rola władz gmin wiejskich położonych na terenie Euroregionu Niemen w rozwoju społeczno-gospodarczym (w opinii mieszkańców)–badania własne [Role of rural districts self-government located on Euroregion Niemen in socio-economic development (in inhabitants' opinions) – own studies]. *Acta Scientiarum Polonorum. Oeconomia*, 8 (2), 83–92.

Miczyńska-Kowalska, M. (2017). Innovation in agritourism as perceived by students of University of Life Sciences in Lublin. *Acta Innovations*, (24), 53–64.

Midura, F. (2008). *Dziedzictwo kulturowe elementem ożywienia ruchu turystycznego na wsi*[*Cultural heritage as a stimulus of rural tourism traffic*]. V Konferencja Naukowo-Techniczna' Błękitny San' [5th Science and Technology Conference "Blue San River], Jabłonka, 33–42.

Mika, M. (2013). Postawy społeczności lokalnych wobec turystów i rozwoju turystyki–przykład gmin Beskidu Śląskiego [Attitudes of local communities towards tourists and tourism development – the example of municipalities in the Silesian Beskidy Mountains]. *Prace Geograficzne*, (134).

Milewski, D. (2005). Determinanty rozwoju funkcji turystycznej gmin nadmorskich województwa zachodniopomorskiego [Determinants of the development of the tourism function in coastal communes of the Zachodniopomorskie voivodeship]. *Ekonomiczne Problemy Turystyki*, 402, 213–225.

Ministry of Rural Development (2018). *Monitoring Rozwoju Obszarów Wiejskich* [*Rural Development Monitoring*], 3rd Stage, Stanny, Monika (project coordinator), European Fund for the Development of Polish Villages, Institute of Rural and Agricultural Development, Polish Academy of Sciences, Warsaw.

Ministry of Sports and Tourism (2014). *Opracowanie – Losy absolwentów szkół i uczelni kształcących kadry dla turystyki* [*What happened to graduates from tourism schools and universities: a report*].

Mitura, T., Bury, R., Begeni, P., & Kudzej, I. (2017). Astro-tourism in the area of the polish-slovak borderland as an innovative form of rural tourism. *European Journal of Service Management*, 23, 45–51.

Mokras-Grabowska, J. (2009). Możliwości rozwoju turystyki kulturowej obszarów wiejskich w Polsce [Possibilities of cultural tourism development in rural areas of Poland]. *Turystyka kulturowa*, 1, 14–31. Musiał, J. (2007). Rola dziedzictwa kulturowego i jego znaczenie edukacyjne dla rozwoju turystyki wiejskiej na przykładzie subregionu limanowskiego [Role and educational importance of cultural heritage in the development of rural tourism: a case study of the Limanowa sub-region]. *Turystyka wiejska a edukacja, różne poziomy, różne wymiary* [*Rural tourism vs. education. Different levels, different dimensions*]. Sikora, J. (ed.), Publishing House of the Poznań University of Life Sciences, 209–217.

Nawrocka, E. (2001). Możliwości oddziaływania władz lokalnych na rozwój turystyki wiejskiej na terenie gminy [The possibilities of local authorities to influence the development of rural torism in a municipality]. *Prace Naukowe Akademii Ekonomicznej we Wrocławiu*, 924, 43–56.

Niedziółka, A. (2008). Działalność stowarzyszeń agroturystycznych jako determinanta rozwoju agroturystyki w województwie małopolskim [Activity of agritouristic associations as a determinant in the development of agritourism in Malopolska Voivodship]. *Folia Universitatis Agriculturae Stetinensis. Oeconomica*, 51, 71–78. Niedziółka, A. (2009). Uwarunkowania rozwoju agroekoturystyki w województwie małopolskim

[Determinants of agriecotourism development in Małopolska voivodeship]. *Problemy Rolnictwa Światowego*, 8 (23), 142–151.

Niedziółka, A. (2010). Rola środków finansowych z unii europejskiej w rozwoju różnych form turystyki wiejskiej w województwie małopolskim [The role of financial resources from the European Union in different forms of rural tourism development in Malopolska Voivodeship]. *Ekonomiczne Problemy Usług*, 53721–734.

Niedziółka, A. (2017). 27 Tourism segmentation in rural tourism and agritourism in metropolitan areas. *Metropolitan Commuter Belt Tourism*, Sznajder, M. (Ed.), Routledge, 283–294.

Niedziółka, A., & Lis, M. (2008). Organic food as the factor of the development of agritourism in Poland. *Наукёвий вісник Львівськёгё націёнальнёгё університету ветеринарнёї медицини та біётехнёлёгій імені СЗ Гжицькёгё*, 10 (2–3),21–37.

Niedzwiecka-Filipiak, I., & Potyrala, J. (1996). Infrastructure as an important factor in the agrotourism development. *Zeszyty Naukowe Akademii Rolniczej we Wrocławiu. Melioracja*, 188–194.

Nowogródzka, T., & Pieniak-Lendzion, K. (2014). Propozycje ofert gospodarstw agroturystycznych a oczekiwania konsumentów [Proposals offers agritourism farms and consumer expectations], *Zeszyty Naukowe Uniwersytetu Przyrodniczo-Humanistycznego w Siedlcach Seria: Administracja i Zarządzanie*, 100, 97–108.

Oleksiuk, I., & Werenowska, A. (2019). Promotion of regional and traditional products, *Środkowoeuropejskie Studia Polityczne*, (2), 135–149.

Oleśniewicz, P., Widawski, K., & Markiewicz-Patkowska, J. (2016). Atrakcyjność oferty zagród edukacyjnych w kontekście rozwoju agroturystyki [Pedagogical farm offer attractiveness in the context of agritourism development]. *Gospodarka, Rynek, Edukacja*, 17 (2), 25–31.

Orłowski, D., & Woźniczko, M. (2007). Edukacyjne znaczenie skansenów w propagowaniu tradycji kulinarnych w turystyce wiejskiej [Educational importance of open-air museums in promoting culinary traditions in rural tourism], *Turystyka wiejska a edukacja, różne poziomy, różne wymiary [Rural tourism vs. education. Different levels, different dimensions]*, Sikora, J. (ed.), Publishing House of the Poznań University of Life Sciences, 228–241.

Orłowski, D., & Woźniczko, M. (2015). Turystyka kulinarna na wiejskim rynku turystycznym [Culinary tourism in the rural tourism market], *Innowacyjność w turystyce wiejskiej a nowe możliwości zatrudnienia na obszarach wiejskich [Innovativeness in rural tourism vs. new rural employment opportunities]*, Kamińska, W. (ed.), Polish Academy of Sciences, Vol. CLXIII, Warsaw.

Otłowska, A., Buks, J., Chmieliński, P. (2006). *Przedsiębiorczość na obszarach wiejskich – stan i perspektywy rozwoju [Rural enterprise: current condition and development prospects]*. Institute of Agricultural and Food Economics – the National Research Institute, Warsaw.

Owsianowska, S., & Winiarski, R. (2017). *Antropologia turystyki [Anthropology of tourism]*. Jubilee publication on the 40th anniversary of the Faculty of Tourism and Recreation of the University of Physical Culture in Krakow. Krakow, Bronisław Czech University of Physical Culture.

Pałka, E. (2015). Innowacje w gospodarstwach agroturystycznych Polski Południowo-Wschodniej [Innovations in agrotourist farms in South-East Poland]. *Studia i Materiały Wydziału Zarządzania i Administracji Wyższej Szkoły Pedagogicznej im. Jana Kochanowskiego w Kielcach*, 19 (4–1), 73–88.

Pastuszek, A. (1994). *An agritourism guide to Bornholm.* Bonin Agricultural Consultancy Center, 9, 1–10.

Patoła, G. (2016). Wsparcie agroturystyki w Małopolsce w ramach PROW 2007–2013 [Support for agrotourism in Małopolska Province from Rural Development Programme 2007–2013]. *Turystyka i Rozwój Regionalny,* 5, 95–106.

Patura, T. (1995). Agroturystyka-kiedy wolna od podatków. *Serwis Informacji Rolniczych. Ośrodek Doradztwa Rolniczego w Piotrkowie Trybunalskim,* 3, 1–15.

Pawlikowska-Łagód, K., Dąbska, O., Wołoszynek, E., & Wójcik, M. (2017). Uniwersytety Trzeciego Wieku jako jedna z form aktywizacji seniorów wybrane aspekty [Universities of the Third Age as way to mobilize the elderly], *Starość niejedną ma twarz – badanie interdyscyplinarne and starością [The many faces of being old: an interdisciplinary study on old age],* Libor, G. (ed.), 254–267.

Pawlikowska-Piechotka, A., Gołębieska, K., Łukasik, N., Ostrowska–Tryzno, A., & Sawicka, K. (2016). Rural sanctuaries as "smart destinations'–sustainability concerns (Mazovia region, Poland). *European Countryside,* 8 (3), 304–321.

Pawlusiński, R., & Piziak, B. (2009). Tourism potential and possibilities of its exploitation in the Subcarpathian (Podkarpackie) voivedeship. *Folia Geographica,* 49 (14), 223–237.

Pearce, D. G. (1995). *Tourism today: A geographical analysis.* Harlow: Longman.

Perepeczko, B. (2003). Popytowe uwarunkowania rozwoju turystyki wiejskiej w perspektywie bliższej i dalszej [Short-term and long-term demand conditioning of the development of rural tourism]. *Zeszyty Naukowe Akademii Rolniczej im. H. Kołłątaja w Krakowie,* (90), 185–192.

Perepeczko, B. (2009). Przyroda jako rdzeń wiejskiego produktu turystycznego [Nature as the core of a rural tourism product]. *Marka wiejskiego produktu turystycznego [Brand of a rural tourism product],* Palich, P. (ed.), Publishing House of the Gdynia Maritime University, 23–29.

Pisarek, M. (2007). Wykorzystanie Internetu w doskonaleniu zawodowym właścicieli gospodarstw agroturystycznych województwa podkarpackiego [Using the Internet in professional training for agritourism farm owners in the Podkarpackie voivodeship], *Turystyka wiejska a edukacja. Różne poziomy, różne wymiary [Rural tourism vs. education. Different levels, different dimensions],* Sikora, J. (ed.), Publishing House of the Poznań University of Life Sciences, 102–109.

Pisarek, M., Bienia, B., Brągiel, E., & Dykiel, M. (2013). Wykorzystanie Internetu w promocji wiejskiej bazy noclegowej w woj. Podkarpackim [Using the Internet as a medium to promote of farm tourism in the region Podkarpacie]. *Komunikowanie i doradztwo w turystyce wiejskiej,* Krzyżanowska, K. (ed.), Wydawnictwo SGGW, 50–59.

Pisarek, M., Gargała-Polar, M., & Czerniakowski, Z. (2017). Zioła jako atrakcja produktów sieciowych w agroturystyce [Herbs as an attraction in agritourism network products]. *Sieci współpracy w turystyce wiejskiej – stan obecny i nowe wyzwania [Collaborative networks in rural tourism: current condition and new challenges],* Wojciechowska, J. (ed.), Łódź University, 165–180.

Poczta, J., & Malchrowicz-Mośko, E. (2016). Turystyka sportowa jako czynnik promocji zdrowia i aktywności fizycznej na obszarach wiejskich w Polsce [Port tourism as a factor promoting health and physical activity in rural areas in Poland], *Handel Wewnętrzny,* 6 (365), 373–388.

Poczta-Wajda, A., & Poczta, J. (2016). The role of natural conditions in qualified agritourism–case of Poland. *Agricultural Economics,* 62 (4), 167–180.

Podhorodecka, K., & Dudek, A. (2019). Disadvantages connected with the development of tourism in the contemporary world and the concept of sustainable tourism, *Problems of Sustainable Development*, 14 (2), 45–55.

Podolec, B. (2000). *Analiza kształtowania się dochodów i wydatków ludności w okresie transformacji gospodarczej w Polsce [Analysis of the evolution of the population's income and expenditure during the economic transformation in Poland]*. PWN Scientific Publishing House.

Pomianek, I. (2019). Spatial diversity of counties of the Warsaw capital region in terms of the development degree of the tourism function in 2005–2017. *Annals*, 4, 384–393.

Popiel, M. (2016). Znaczenie turystyki wiejskiej w życiu osób niepełnosprawnych [The importance of rural tourism in the lives of the disabled]. *Turystyka Wiejska. Zagadnienia ekonomiczne i marketingowe [Rural tourism. Economic and marketing aspects]*, Jęczmyk, A., Uglis, J., & Maćkowiak, M. (eds.), Wieś Jutra Publishing House194–202.

Popławski, Ł. (2009). Gospodarstwa agroturystyczne, ekoagroturystyczne i ekologiczne jako czynnik rozwoju turystyki na obszarach chronionych województwa świętokrzyskiego Eco-tourist and ecological farms as a factor of tourism development within the protected areas of the Świętokrzyskie Voivodeship]. *Zeszyty Naukowe Małopolskiej Wyższej Szkoły Ekonomicznej w Tarnowie*, 3 (14), 139–152.

Potočnik-Slavič, I., & Schmitz, S. (2013). Farm tourism across Europe. *European Countryside*, 5 (4), 265–274.

Prochorowicz, M. (2007). Formy edukacji w zakresie świadczenia i poprawy jakości usług turystycznych na wsi [Forms of education on how to deliver and improve the quality of tourism services in rural areas]. *Turystyka wiejska a edukacja, różne poziomy, różne wymiary [Rural tourism vs. education. Different levels, different dimensions]*, Sikora, J. (ed.), Publishing House of the Poznań University of Life Sciences, 50–60.

Przeorek, R. (2014). Możliwości finansowania rozwoju turystyki wiejskiej w Polsce [Funding opportunities of development of rural tourism in Poland]. *Ekologia i Technika*, 22 (3), 129–139.

Przezbórska, L. (2000). Miejsce i rola turystyki wiejskiej i agroturystyki w zrównoważonym rozwoju wsi i rolnictwa wielkopolski [Place and role of rural tourism and agritourism in sustainable development of rural areas and agriculture of the Wielkopolska region]. *Roczniki Naukowe Stowarzyszenia Ekonomistów Rolnictwa i Agrobiznesu*, 2 (2), 140–146.

Przezbórska, L. (2002). Agroturystyka i turystyka wiejska w programach rozwoju terenów wiejskich Unii Europejskiej [Agritourism and rural tourism in rural development programmes of the European Union]. *Roczniki Naukowe Stowarzyszenia Ekonomistów Rolnictwa i Agrobiznesu*, 4 (6), 150–155.

Przezbórska, L. (2005a). Przemiany gospodarstw agroturystycznych Wielkopolski w latach 1990–2003 [Transformation of agritourism farms of the wielkopolska region between 1990 and 2003]. *Roczniki Naukowe Stowarzyszenia Ekonomistów Rolnictwa i Agrobiznesu*, 1 (07), 109–203.

Przezbórska, L. (2005b). *Classification of agri-tourism/rural tourism SMEs in Poland (on the example of the Wielkopolska Region)*. XIth International Congress of EAAE (European Association of Agricultural Economists), The Future of Rural Europe in the Global Agri-Food System, August 24–27, Copenhagen, Denmark.

Przezbórska, L. (2006). Waloryzacja turystyczna obszarów wiejskich województwa wielkopolskiego [Tourism valorization of rural areas of the Wielkopolska Region]. *Prace Naukowe Akademii Ekonomicznej we Wrocławiu*, 1118 (2), 254–260.

Przezbórska, L. (2007). Determinanty rozwoju agroturystyki w Polsce na przykładzie wybranych regionów [Determinants of agri-tourism development in Poland (on the example of chosen regions]. *Acta Scientiarum Polonorum Oeconomia*, 62, 113–121.

Przybyla, K., & Kulczyk-Dynowska, A. (2017). Transformations of tourist functions in urban areas of the Karkonosze Mountains, Drusa, M., Yilmaz, I., Marschalko, M., Coisson, E., Rybak, J., & Segalini, A. (eds.), *IOP Conference Series-Materials Science and Engineering*, 245, IOP Publishing, Bristol, 1–7.

Raciborski, J. (1997). Zakres doradztwa prawnego na rzecz podmiotów podejmujących świadczenie usług agroturystycznych [Range of law advisory for subjects which start to render touristic services]. *Zagadnienia Doradztwa Rolniczego*, 4, 103–109.

Rapacz, A. (2000). Możliwości i sposoby oceny atrakcyjności turystycznej i inwestycyjnej regionów turystycznych [Possible ways of assessing the attractiveness of tourism regions to tourists and investors], *Polityka samorządu terytorialnego w dziedzinie turystyki* [*Local government's tourism policy*], Boruszczak, M. (ed.), Academy of Tourism and Hotel Management in Sopot, 155–165.

Rasoolimanesh, S. M., Roldán, J. L., Jaafar, M., & Ramayah, T. (2017). Factors influencing residents' perceptions toward tourism development: Differences across rural and urban world heritage sites. *Journal of Travel Research*, 56(6), 760–775.

Reid, D. G., Mair, H., & Taylor, J. (2011). Community participation in rural tourism development, *World Leisure Journal*, 42, 20–27.

Rodacka, M. (1997). Wykorzystanie turystyczne dziedzictwa kulturowego Kaszubów [Tourism uses of the cultural heritage of Kashubians]. *Turyzm* [*Tourism*], 7 (1), 57–71.

Roman, M. (2009). Inicjatywy klastrowe w agroturystyce na przykładzie Okopskiej Organizacji Turystycznej [Okopska Tourism Organization as an example of cluster initiative within agrotourism]. *Infrastruktura i ekologia terenów wiejskich*, 06, 187–195.

Roman, M. (2010). Szanse i bariery rozwoju agroturystyki w gminie Suchowola w opinii społeczności lokalnej [Chances and barriers of the development of the farm tourism in the Suchowola commune in the opinion of the local community]. *Roczniki Naukowe Stowarzyszenia Ekonomistów Rolnictwa i Agrobiznesu*, 5 (12), 204–207.

Roman, M., & Niedziółka, A. (2017). *Agroturystyka jako forma przedsiębiorczości na obszarach wiejskich* [*Agritourism as a form of rural entrepreneurship*]. Publishing House of the Warsaw University of Life Sciences.

Roman, M., & Roman, A. (2019). Wykorzystanie dziedzictwa kulturowego w działaniach terapeutycznych realizowanych w turystyce wiejskiej [The use of cultural heritage in activities therapeutic implemented in rural tourism]. *Zagadnienia Doradztwa Rolniczego*, 3, 72–83.

Roman, M., & Wojcieszak, M. (2018). Znaczenie social farmingu w wybranych krajach Unii Europejskiej jako przykład przedsiębiorczości w turystyce na obszarach wiejskich [The importance of social farming in selected European Union countries as an example of entrepreneurship in tourism in rural areas]. *Prace Naukowe Uniwersytetu Ekonomicznego we Wrocławiu*, 535, 161–171.

Roman, M., Zaburanna, L., & Krzyżanowska, K. (2018). Status and Development Trends of Agritourism in Poland and the Ukraine. *Folia Turistica*, 46, 115–129.

Romanowski, A. (2005). Stan bazy noclegowej gospodarstw agroturystycznych Śląska Opolskiego [Accommodation resources of agritourism farms in Opolian Silesia], *Śląsk Opolski*, 1 (2), 81–91.

Rosner, A., & Stanny, M. (2016). *Monitoring obszarów wiejskich. Etap II [Monitoring of rural areas. Stage II]*. Institute of Rural and Agricultural Development, Polish Academy of Sciences (IRWIR PAN), Warsaw.

Rosner, A., & Stanny, M. (2017). *Socio-economic development of rural areas in Poland*, The European Fund for the Development of Polish Villages Foundation (EFRWP), Institute of Rural and Agricultural Development, Polish Academy of Sciences (IRWIR PAN), Warsaw, p. 166. Sala, K. (2016). Wioski tematyczne jako przykład innowacyjności w turystyce wiejskiej [Thematic villages as an example of innovativeness in rural tourism]. *Zeszyty Naukowe Małopolskiej Wyższej Szkoły Ekonomicznej w Tarnowie*, 2 (30), 117–126.

Sala, J. (2016). Zmiany w uwarunkowaniach uczestnictwa Polaków w turystyce w latach 2006–2015 [Changes in the conditions for the Polish population's participation in tourism in 2006–2015)]. *Zmiany zachowań turystycznych Polaków i ich uwarunkowań w latach 2006–2015 [Changes in the Polish population's tourism behavior and its conditions in 2006–2015]*, Berbeka, J. (ed.), Krakow University of Economics, 72–88.

Sala, J. (2017). Tendencje rozwojowe turystycznej bazy noclegowej w Polsce w latach 1990–2014 [Development trends in tourism accommodation resources in Poland between 1990 and 2014], *Trendy w Turystyce [Tourism trends]*, Biernat, E., & Dziedzic, E. (eds.), Warsaw School of Economics, 13–30.

Sammel, A. (2010). Znaczenie Internetu w rozwoju turystyki wiejskiej i agroturystyki w województwie zachodniopomorskim [The meaning of internet in development of the rural tourism and the agri-tourism in Zachodniopomorskie Province]. *Ekonomiczne Problemy Usług*, 53, 563–570.

Sammel, A., & Dańczak, A. (2002). Zwierzęta w gospodarstwach agroturystycznych województwa zachodniopomorskiego-stan obecny i perspektywy [Farm animals on agri-tourism farms in Zachodniopomorskie Province – current situation and prospects]. *Journal of Research and Applications in Agricultural Engineering*, 47 (1), 55–57.

Sammel, A., & Prochorowicz, M. (2013). Wykorzystanie środków unijnych na potrzeby turystyki wiejskiej w województwie zachodniopomorskim w ramach SPO restrukturyzacja i modernizacja sektora żywnościowego oraz rozwój obszarów wiejskich w latach 2004–2006 [The destination of European Fund on need of rural tourism in Zachodniopomorskie Province in from SOP restructuring and modernization of food sector as well as the development of rural areas in years 2004 2006]. *Ekonomiczne Problemy Turystyki*, (2), 141–153.

Sawicki, B. (2007). Perspektywy rozwoju elementów edukacji i wychowania młodzieży licealnej w agroturystyce i sylwanoturystyce [Development prospects of high school-level educational components of agritourism and silvitourism]. *Turystyka wiejska a edukacja, różne poziomy, różne wymiary [Rural tourism vs. education. Different levels, different dimensions]*, Sikora, J. (ed.), Publishing House of the Poznań University of Life Sciences, 345–346.

Sawicki, B. (2009). Kreowanie marki w agroturystyce i turystyce wiejskiej. *Marka wiejskiego produktu turystycznego*, Palich, P. (ed.), Wydawnictwo Akademii Morskiej w Gdyni, 11–17.

Seraphin, H., Sheeran, P., & Pilato, M. (2018). Over-tourism and the fall of Venice as a destination, *Journal of Destination Marketing & Management*, 9, 374–376.

Sharpley, R. (2014). Host perceptions of tourism: A review of the research. *Tourism Management*, 42, 37–49.

Sieczko, A. (2011a). Mazowieckie produkty lokalne jako potencjał budowy sieciowego produktu markowego turystyki wiejskiej [Mazovian local products as the potential construction of a network of branded products of rural tourism]. *Folia Pomeranae Universitatis Technologiae Stetinensis. Oeconomica*, 64, 185–192.

Sieczko, A. (2011b). Rekreacja w agroturystyce i turystyce wiejskiej ze szczególnym uwzględnieniem dawnych gier i zabaw [Recreation in agritourism and rural tourism with particular reference to the old games and amusements]. *Ekonomiczne Problemy Usług*, 79, 265–274.

Sieczko, A. (2014). Produkty regionalne i tradycyjne w promocji regionów [Traditional and regional products in the promotion of regions], *Turystyka i Rozwój Regionalny*, 2, 79–89.

Siedlecka, A., & Wielogórska, G. (2018). Czynniki determinujące prowadzenie działalności agroturystycznej przez wiejskie gospodarstwa domowe na przykładzie województwa lubelskiego [Factors determining the agritourism activity by rural households on the example of the Lublin Province], *Roczniki Naukowe Stowarzyszenia Ekonomistów Rolnictwa i Agrobiznesu*, 20 (1), 130–136.

Siekierski, J. (2009). Ekonomiczne uwarunkowania rozwoju turystyki wiejskiej w województwie małopolskim [Economic conditionings of rural tourism development in the Małopolska Province]. *Roczniki Naukowe Stowarzyszenia Ekonomistów Rolnictwa i Agrobiznesu*, 4 (11), 299–304.

Siekierski, J., & Popławski, Ł. (2009). Usługi turystyczne jako forma przedsiębiorczości na obszarach wiejskich [Tourist services as a form of enterprise in rural areas]. *Zeszyty Naukowe Małopolskiej Wyższej Szkoły Ekonomicznej w Tarnowie*, 14 (3), 153–164.

Siemiński, P., & Poczta, W. (2014). Możliwości rozwoju agroturystyki i turystyki wiejskiej w ramach PROW 2007–2013 i 2014–2020 [Possibilities of the development of agro-tourism and rural tourism under the RDP 2007–2013 and 2014–2020]. *Roczniki [Annals]*, 16 (6), 432–437. Sikora, J. (1998). Socjologiczne uwarunkowania rozwoju agroturystyki jako dodatkowego źródła dochodu w gospodarstwach rolnych [Sociological conditions for the development of agrotourism as an extra source of income in agricultural households]. *Zeszyty Naukowe. Seria 1.Akademia Ekonomiczna w Poznaniu*, 270, 96–114.

Sikora, J. (1999). *Organizacja ruchu turystycznego na wsi* (Organization of rural tourism). Warsaw, Wydawnictwa Szkolne.

Sikora, J. & Karczewska, M. (2004). Wpływy i wydatki środków pieniężnych w gospodarstwach agroturystycznych [Incomes and expenditures of agritourism farms], *Gospodarka turystyczna w XXI wieku [21st century tourism economy]. Szanse i bariery rozwoju w warunkach integracji międzynarodowej [Opportunities for and barriers to development in a context of international integration]*, Bosiacki, S. (ed.), University of Physical Culture in Poznań, 86–94. Sikora, J., & Wartecka-Ważyńska, A. (2007). Społeczność wiejska w kształtowaniu funkcji agroturystyki [The role of rural community in forming the agritourism function]. *Turystyka wiejska a edukacja, różne poziomy, różne wymiary [Rural tourism vs. education. Different levels, different dimensions]*, Sikora, J. (ed.), Publishing House of the Poznań University of Life Sciences, 22–32.

Sikora, J., & Wartecka-Ważyńska, A. (2009). Kapitał ludzki i kapitał społeczny czynnikiem rozwoju agroturystyki [Human and social capital as a key factor in agritourism development]. *Ekonomiczne Problemy Turystyki*, 325–339.

Sikora, J., & Wartecka-Ważyńska, A. (2013). Popyt na rynku turystyki wiejskiej w Polsce w świetle badań empirycznych [Demand on the market of rural tourism in Poland in the light of empirical studies]. *Prace Naukowe Uniwersytetu Ekonomicznego we Wrocławiu*, 304, 291–303.

Sikorska-Wolak, I. (2007). Turystyka jako system dydaktyczno-wychowawczy [Tourism as an educational and didactic system]. *Turystyka wiejska a edukacja, różne poziomy, różne wymiary [Rural tourism vs. education. Different levels, different dimensions]*. Sikora, J. (ed.), Publishing House of the Poznań University of Life Sciences, 12–21.

Sikorska-Wolak, I. (2009). Kształtowanie funkcji turystycznych obszarów wiejskich – potrzeby i możliwości, *Turystyczne funkcje obszarów wiejskich*, Sikorska-Wolak, I. (ed.), Wydawnictwo SGGW, Warszawa, 20–24.

Sikorska-Wolak, I. (2010). Tourism in a social-economic activation of country areas (Lubelszczyzna Case Study). *Acta Scientiarum Polonorum-Oeconomia*, 9 (4), 477–488.

Sikorska-Wolak, I., & Zawadka, J. (2011). Postawy społeczności lokalnej wobec rozwoju turystyki wiejskiej [Attitudes of the local community towards the development of rural tourism]. *Folia Pomeranae Universitatis Technologiae Stetinensis. Oeconomica*, 228 (64), 93–102.

Sikorska-Wolak, I., & Zawadka, J. (2016). Innowacyjne rozwiązania w turystyce wiejskiej [Innovative solutions in rural tourism]. *Roczniki Naukowe Stowarzyszenia Ekonomistów Rolnictwa i Agrobiznesu*, 18 (4), 207–212.

Silkes, C. (2012). Farmers' markets: A case for culinary tourism. *Journal of Culinary Science and Technology*, 10, 326–336.

Simancas Cruz, M., & Peñarrubia Zaragoza, M. P. (2019). Analysis of the accommodation density in coastal tourism areas of insular destinations from the perspective of overtourism, *Sustainability*, 11 (11), 1–19.

Smoleńska, O., & Machnik, A. (2013). Współczesne uwarunkowania funkcjonowania turystyki wiejskiej–rozwój gospodarstw agroturystycznych w kierunku specjalizacji w rekreacji i ekologizacji oferty [Contemporary conditions and determinants of rural tourism and the function of agritourism: agrifarm development specializing in ecology and recreation]. *Studia Periegetica*, 2 (10), 127–138.

Snopek, B. (1996). Agroturystyka w Anglii [Agritourism in England]. INO. Informacje, Nowości, Oferty. *Ośrodek Doradztwa Rolniczego w Płocku*, 3, 44–45. Sokół, J. L. (2009). Jeleniowate na wolności i w chowie fermowym jako atrakcja dla turystów [Animals of deer family living in the wild or on a farm as a tourist attraction]. *Ekonomia i Zarządzanie*, 1, 107–119.

Sokół, J. L., & Kołoszko-Chomentowska, Z. (2010). Produkty zwierzęce jako atrakcja w gospodarstwach agroturystycznych [Animal products as an attraction of the agritourist farms]. *Ekonomia i Zarządzanie*, 2, 137–146.

Sokołowicz, Z., & Krawczyk, Z. (2010). Znaczenie chowu drobiu w gospodarstwach agroturystycznych w opinii mieszkańców województwa podkarpackiego [Importance of poultry breeding in agrotourism farms in the opinion of inhabitants of Podkarpackie province]. *Roczniki Naukowe Stowarzyszenia Ekonomistów Rolnictwa i Agrobiznesu*, 12 (4), 314–316.

Spychała, A., Graja-Zwolińska, S., Tacu, G., & Păduraru, T. (2017). Perception of modern agritourism. Wielkopolskie Province (Poland) and the Northeast Region (Romania) case study. *European Journal of Service Management*, 23, 63–70.

Stanny, M., & Rosner, A. (2014). *Monitoring obszarów wiejskich. Etap I [Monitoring of rural areas. Stage I]*, The European Fund for the Development of Polish Villages

Foundation (EFRWP), Institute of Rural and Agricultural Development, Polish Academy of Sciences (IRWIR PAN), Warsaw.

Stanny, M., Rosner, A., & Komorowski, Ł. (2018). *Monitoring rozwoju obszarów wiejskich. Etap III. Struktury społeczno-gospodarcze, ich przestrzenne zróżnicowanie i dynamika* [*Monitoring the rural development. Stage 3. Socioeconomic structures, territorial differences between them and the pace of changes in them*], The European Fund for the Development of Polish Villages Foundation (EFRWP), Institute of Rural and Agricultural Development, Polish Academy of Sciences (IRWIR PAN), Warsaw.

Stec, A. (2015). Zastosowanie metody Hellwiga do określenia atrakcyjności turystycznej gmin na przykładzie województwa podkarpackiego [Application of Hellwig method to determine the tourist attractiveness of municipalities – Podkarpackie Voivodeship example]. *Metody ilościowe w badaniach ekonomicznych*, 16 (4), 117–126.

Stefko, R., Vasanicova, P., Litavcova, E., & Jencova, S. (2018). Tourism intensity in the NUTS III Regions of Slovakia, *Journal of Tourism and Services*, 9 (16), 45–59.

Stasiak, A. (2015). Rozwój turystyki kulinarnej w Polsce [Development of culinary tourism in Poland], *Kultura i turystyka–wokół wspólnego stołu* [*Culture and tourism: around a shared table*], Kowalczyk, A., & Stasiak, A. (eds.), Regional Tourism Organization of the Łódzkie voivodeship, 119–149.

Stepaniuk, M. (1999). Waloryzacja turystyczna środowiska przyrodniczego i kulturowego gminy Tykocin [The tourist valorization of natural and cultural environment of Tykocin district]. *Inżynieria Środowiska*, (11), 61–91.

Stepaniuk, K. (2009a). Ocena funkcjonalności wybranych witryn internetowych promujących działalność agroturystyczną w woj. podlaskim [The evaluation of functionality of selected websites promoting agrituorism in podlaskie district]. *Ekonomia i Zarządzanie*, 1, 128–132.

Stepaniuk, K. (2009b). Indywidualizm a masowość. Studium wybranych aspektów internetowych strategii promocyjnych gospodarstw agroturystycznych w woj. podlaskim [Individuality or mass. The selected aspects' study of Internet strategies of Agritourism farms' promotion in podlaskie district]. *Ekonomia i Zarządzanie*, 1 (1), 120–127.

Stepaniuk, K. (2010a). Google SideWiki jako nowe narzędzie potencjalnie zwiększające efektywność internetowej promocji działalności agroturystycznej w województwie podlaskim [Google SideWiki as a new tool potentially increasing the efficiency of Internet promotion of agritourist activity in podlaskie district]. *Ekonomia i Zarządzanie*, 2, 103–110.

Stepaniuk, K. (2010b). Wybrane koncepcje związane z projektowaniem, wdrożeniem i rozwojem działalności e-agroturystycznej na przykładzie województwa podlaskiego [Selected concepts of design, implementation and development of e-agritourist activity in Podlaskie Voivodeship]. *Acta Scientiarum Polonorum. Oeconomia*, 9 (4), 509–517.

Stępnik, K. (2017). *Koncepcja gospodarstw opiekuńczych w Polsce* [*The concept of care farms in Poland*]. Presentation at a voivodeship-level seminar organized by the Brwinów Agricultural Consultancy Center, Branch Office in Krakow.

Strzelecka, M., & Wicks, B. E. (2010). Engaging residents in planning for sustainable rural-nature tourism in post-communist Poland. *Community Development*, 41 (3), 370–384.

Strzelecka, M., Boley, B. B., & Strzelecka, C. (2017). Empowerment and resident support for tourism in rural Central and Eastern Europe (CEE): The case of Pomerania, Poland. *Journal of Sustainable Tourism*, 25 (4), 554–572.

Strzembicki, L. (1999). Zachowania nabywców usług turystyki wiejskiej w Polsce [Behaviour of rural tourism services purchasers in Poland]. *Zeszyty Naukowe Małopolskiej Wyższej Szkoły Ekonomicznej w Tarnowie*, (2), 223–235.

Strzembicki, L. (2002). Zachowania konsumentów na krajowym rynku turystyki wiejskiej [Consumers' behaviour in the domestic rural tourism market]. *Zeszyty Naukowe Akademii Ekonomicznej w Krakowie*, 612, 77–94.

Strzembicki, L. (2005). Zachowania konsumentów na krajowym rynku turystyki wiejskiej [Consumer behavior in the domestic market for rural tourism]. *Konsument na rynku turystycznym w warunkach społeczeństwa opartego na wiedzy i informacji [Consumers in the tourism market in the context of a knowledge and information society]*, Strzembicki, L. (ed.), Publishing House of the Katowice University of Economics, 60–69.

Strzembicki, L. (2006). Zachowania konsumpcyjne turystów wypoczywających na obszarach wiejskich Polski [Consumer behavior of tourists choosing a Polish rural destination]. *Scientific Journals of the Krakow University of Economics*, 716, 55–73.

Suchta, J., & Koczowski, F. (1992). Koncepcja ekorozwoju Warmii i Mazur [The eco-development concept for Warmia and Masuria]. *Zeszyty Problemowe Postępów Nauk Rolniczych [Advancements in Agricultural Sciences and related Problems, Scientific Journals]*, 401, 321–334.

Świeca, A., Krukowska, R., & Tucki, A. (2007). Possibilities for the development of tourism in the Lublin Region, *Tourism Theory – Conditions – Experiences*, Godlewski, G., & Bochenek, M. (eds.), Biała Podlaska, 69–98.

Świstak, E., Bilska, B., Stępień, A., & Tul-Krzyszczuk, A. (2013). Produkty regionalne jako element budowania konkurencyjności obszarów wiejskich [The role of regional products in building the competitiveness of rural areas], *Budowanie konkurencyjności obszarów wiejskich [Building the competitiveness of rural areas]*, Krzyżanowska, K. (ed.), Publishing House of the Warsaw University of Life Sciences, Warsaw, 139–148.

Szczepańska, B., & Szczepański, J. (2019). Współczesne role kół gospodyń wiejskich w społecznościach lokalnych (na przykładzie województwa dolnośląskiego) [Contemporary roles of farmer's wives' association in local communities (using the Lower Silesia Voivodeship)]. *Folia Sociologica*, 68, 67–79.

Sznajder, M. J. (ed.). (2017). *Metropolitan Commuter Belt Tourism*. Taylor & Francis.

Sznajder, M., & Przezbórska, L. (2004). Identification of rural and agri-tourism products and services. *Roczniki Akademii Rolniczej w Poznaniu – Seria Ekonomia*, 3, 165–177.

Sznajder, M., Przezbórska-Skobiej, L. & Scrimgeour, F. (2009). *Agritourism*, CABI.

Szromek, A., Kruczek, Z., & Walas, B. (2019). The attitude of tourist destination residents towards the effects of overtourism—Kraków case study. *Sustainability*, 12 (1), 1–17.

Szubert-Zarzeczny, E. (1996). *Turystyka w procesie przekształceń systemowych w Polsce [Tourism in the Polish economic transformation process]*. Publishing House of the Oskar Lange University of Economics in Wrocław, Wrocław.

Szwichtenberg, A. (1991). *Stymulatory i bariery rozwoju funkcji turystycznej w polskiej strefie nadbałtyckiej [Enablers of and barriers to the development of the tourism function in the Polish Baltic coast area]*, KONB Publishing House, Koszalin.

Szwichtenberg, A. (1993). Turystyka alternatywna i ekoturystyka-nowe pojęcia w geografii turyzmu [Alternative tourism and eco-tourism: new terms in tourism geography], *Turyzm [Tourism]*, 3 (2), 51–59.

Szwichtenberg, A. (1996). Nowe spojrzenie na zagospodarowanie turystyczne Półwyspu Helskiego [A new look at the tourism development of the Hel Peninsula], *Turyzm [Tourism]*, 6 (1), 77–88.

Szwichtenberg, A. (1998). Pojmowanie turystyki wiejskiej w Polsce i na świecie [The concept of rural tourism as seen in Poland and around the world]. *Turyzm [Tourism]*, 8 (1), 29–37.

Szyguła, A. (2014). Agritourism as the best form of entrepreneurship in the Polish countryside. *Вісник Львівськёгё університету. Серія: Міжнарёдні віднёсини*, (34), 125–132.

Szymańska, J. (1996). Agroturystyka formą kreowania przedsiębiorczości wiejskiej [Agritourism as a rural enterprise enabler]. *Prace Naukowe Akademii Ekonomicznej we Wrocławiu*, (741), 129–134.

Szymoniuk, B. (2003). *Rural clusters in the Lublin Region (Eastern Poland)– good solutions for a young democracy*, Contribution to the 43rd Congress of the European Regional Science Association, Finland, August 27–31.

Targosz, R. (1997). Agroturystyczne atrakcje w Wielkiej Brytanii [Agritourism attractions of the United Kingdom]. Rolniczy Rynek [Agricultural market]. *Voivodeship Agricultural Consultancy Center in Wrocław*, 10, 1–12.

Tobiasz-Lis, P., Wójcik, M., Dmochowska-Dudek, K., & Jeziorska-Biel, P. (2019). Thematic village as the new anchor for local development: a lesson from Masłomęcz, Poland. *Europa Regional*, 26 (2), 29–42.

Tomaszewska, M., Bilska, B., Grzesińska, W., & Szymańska-Radecka, M. (2014). Rozpoznawalność oznaczeń produktów tradycyjnych i regionalnych wśród konsumentów województwa mazowieckiego [Recognizability of traditional and regional products designations among consumers from Mazowieckie Voivodeship]. *Marketing i Rynek*, (6), 757–773.

Tucki, A. (2009). Propozycja regionalizacji turystycznej województwa lubelskiego [The proposal of the tourist regionalization of Lubelskie Voivodeship]. *Folia Turistica*, 21, 145–164.

Tucki, A., & Świeca, A. (2008). Rola samorządów lokalnych w rozwoju turystyki na przykładzie regionu lubelskiego [Role of local government units in tourism development: a case study of the Lublin region]. *Turystyka w środowisku geograficznym [Tourism in a geographic environment]*, Wyrzykowski, J. (ed.), Publishing House of the University of Wrocław.

Tyran, E. (2005). Rozwój turystyki wiejskiej a fundusze strukturalne Unii Europejskiej [Development of rural tourism and structural funds of the European Union]. *Agrobiznes*, 1070 (2), 370–374.

Tyran, E. (2006). Competitiveness of rural tourism. *Agrobiznes*, 1118 (2), 449–454.

Tyran, E. (2007). Regional and traditional products as an important part of rural tourism offer. *Acta Scientiarum Polonorum. Oeconomia*, 6 (3), 121–128.

Tyran, E. (2008). Specialization as a driving force of competitiveness in agritourism. *Roczniki Naukowe Stowarzyszenia Ekonomistów Rolnictwa i Agrobiznesu*, 5 (10), 166–170.

Tyran, E. (2009). Pojęcie produktu markowego w turystyce wiejskiej [Concept of branded product in rural tourism]. *EPISTEME*, 8, 11–21.

Uglis, J., & Jęczmyk, A. (2009). Agroturystyka szansą ożywienia obszarów wiejskich [Agritourism As a Chance for Rural Areas Boom]. *Roczniki Naukowe Stowarzyszenia Ekonomistów Rolnictwa i Agrobiznesu*, 4 (11), 341–346.

Uglis, J., & Kosmaczewska, J. (2017). Traditional and modern tour operators in agritourism. *Metropolitan Commuter Belt Tourism*, Sznajder, M. (ed.), Routledge, 234–245.

Uglis, J., & Kozera-Kowalska, M. (2016). Kształcenie kadr dla turystyki wiejskiej [Education for rural tourism staff], *Turystyka Wiejska – zagadnienia ekonomiczne i marketingowe [Rural tourism. Economic and marketing aspects]*, Jęczmyk, A., Uglis, J., & Maćkowiak, M. (eds.), Wieś Jutra Publishing House, 32–42.

Uglis, J., & Krysińska, B. (2012). Próba zdefiniowania profilu agroturysty [The Attempt to Define a Profile of Agritourist]. *Ekonomiczne Problemy Usług*, 669 (84), 155–166.

Utzig, M. (2018). Wydatki konsumpcyjne w wiejskich i miejskich gospodarstwach domowych jako miara ich poziomu życia [Urban and rural households' consumption expenditures as a measure of their level of living]. *Roczniki Naukowe Stowarzyszenia Ekonomistów Rolnictwa i Agrobiznesu*, 20 (4), 195–199.

Vojnovic, N. (2018). Tourist intensity in Croatia's leading tourist towns and municipalities, *Geoadria*, 23 (1), 29–50.

Wawrzyniak, B., & Wojtasik, B. (2004). Zmiany wykształcenia mieszkańców wsi w latach 1996–2002 [Changes of education level of rural population since 1996 to 2002]. *Acta Scientiarum Polonorum Oeconomia*, 3 (2), 139–147.

Werenowska, A. (2014). Possibility of applying modern forms of communication of agritourism farms with environment groups. *Economic Science for Rural Development*, 35, 80–86.

Wiatrak, A. P. (1996). Wpływ agroturystyki na zagospodarowanie obszarów wiejskich [Impact of agritourism on rural development]. *Zagadnienia Ekonomiki Rolnej [Issues of Agricultural Economics]*, 1, 34–46.

Wiatrak, A. (1997). Możliwości finansowego wsparcia rozwoju turystyki wiejskiej [Possible options of financial support for the development of rural tourism]. *Problemy Turystyki [Problems of Tourism]*, 20 (1–4),81–92.

Wiatrak, A. (2003). Baza agroturystyczna w Polsce i uwarunkowania jej rozwoju [Agri-tourism base in Poland and its Conditions of development]. *Zeszyty Naukowe Akademii Rolniczej im. H. Kołłątaja w Krakowie*, 402, 9–18.

Wiatrak, A. (2008). Kierunki rozwoju turystyki wiejskiej w świetle planów rozwojowych Polski [The directions of development the rural tourism in the light of developmental plans the Poland]. *Wieś i Rolnictwo*, 138 (1), 74–87.

Wiatrak, A. (2017). Istota i cele krajowych i regionalnych inteligentnych specjalizacji w sektorze rolnym [The nature and objectives of national and regional smart specialisations in the agricultural sector]. *Roczniki Naukowe Stowarzyszenia Ekonomistów Rolnictwa i Agrobiznesu*, 19 (5), 210–216. Wilk, I., & Keck-Wilk, M. (2013). Oczekiwania turystów dotyczące oferty gospodarstw agroturystycznych. [Tourists' expectations towards the agritourism farms' offer]. *Journal of Agribusiness and Rural Development*, 02 (28), 243–250.

Wilkin, J. (2011). Wielofunkcyjność wsi i rolnictwa a rozwój zrównoważony [Sustainable development and the multifunctionality of rural areas and agriculture]. *Wieś i Rolnictwo*, 4 (153), 27–39.

Wiśniewska, B. (1998). Rola doradztwa w rozwijaniu przedsięwzięć agroturystycznych [Role of consultancy in the development of agritourism projects]. *Wieś i Rolnictwo [Rural Areas and Agriculture]*, 4, 136–139.

Wojciechowska, J. (2002). Uwarunkowania rozwoju agroturystyki w Polsce w świetle zmian ilościowych, Stan i perspektywy rozwoju agroturystyki w województwie pomorskim [Conditions for agritourism development in Poland in the light of quantitative changes. Current state and development prospects of agritourism in the Pomorskie voivodeship]. *Conference proceedings*, Academy of Tourism and Hotel Management, Gdańsk, 69–80.

Wojciechowska, J. (2006). Geneza oraz ewolucja turystyki na obszarach wiejskich w Polsce [Tourism in rural areas in Poland – the origins and evolution]. *Folia Turistica*, 17, 99–119.

Wojciechowska, J. (2007a). Formy i mechanizmy działań podmiotów wspierających rozwój agroturystyki w Polsce [Forms and mechanisms of action of operators acting as enablers of agritourism development in Poland]. *Turystyka wiejska a edukacja, różne poziomy, różne wymiary* [*Rural tourism vs. education. Different levels, different dimensions*], Sikora, J. (ed.), Publishing House of the Poznań University of Life Sciences, 30–38.

Wojciechowska, J. (2007b). Typy gospodarstw agroturystycznych w Polsce i sylwetka ich właścicieli [Types of Polish agritourism farm and owner profiles]. *Turyzm*, 17 (1–2),159–171. Wojciechowska, J. (2009). *Procesy i uwarunkowania rozwoju agroturystyki w Polsce* [*Processes of and conditions for the development of Polish agritourism*]. Postdoctoral theses of the Łódź University, Łódź. Publishing House of the Łódź University.

Wojciechowska, J. (2011). Twenty Years of Polish Agritourism: the Past and the Future, *Tourism*, 21 (1–2),67–72.

Wojciechowska, J. (2018). *Agroturystyka. Signum turystyki i obszarów wiejskich* [*Agritourism. A symbol of tourism and rural areas*], Polskie Wydawnictwo Ekonomiczne, Warsaw. Wojcieszak, M. (2017). Analiza ofert wybranych gospodarstw agroturystycznych powiatu gnieźnieńskiego oraz ich wpływ na zainteresowanie konsumentów produktami turystycznymi [The analysis of offers of agritourist farms in Gniezno County and their impact on consumer interest in tourist products]. *Journal of Tourism and Regional Development*, (7), 109–119.

Wojcieszak, A., & Wojcieszak, M. (2018). Uwarunkowania funkcjonowania gospodarstw opiekuńczych na terenach wiejskich [Conditions of functioning of care farms in rural areas]. *Zagadnienia Doradztwa Rolniczego*, 3 (93), 20–31.

Wolańska, T., & Lisowska, J. (1998). Outdoor recreation in rural tourism. *World Leisure & Recreation*, 40 (1), 39–41.

Wos, B. (2009). Ekoturystyka szansą zrównoważonego rozwoju terenów wiejskich [Ecotourism as a chance of sustainable development of rural areas]. *Infrastruktura i Ekologia Terenów Wiejskich*, 05, 115–122.

Wos, B. (2014). Api-tourism in Europe. *Journal of Environmental and Tourism Analyses*, 2 (1), 66–74.

Woźniak, M. (2008). Czynniki kształtujące konkurencyjność gmin wiejskich na przykładzie województwa podkarpackiego [Factors influencing competitiveness of rural communities on example of Podkarpackie Province]. *Roczniki Naukowe Stowarzyszenia Ekonomistów Rolnictwa i Agrobiznesu*, 2 (10), 285–289.

Woźniak, M. (2014). Przedsiębiorczość turystyczna kierunkiem rozwoju atrakcyjnych krajobrazowo gmin wiejskich [Tourist entrepreneurship as the direction of the development of landscape attractive rural communities]. *Prace Naukowe Uniwersytetu Ekonomicznego we Wrocławiu*, (366), 605–617.

Woźniak, M., & Batyk, I. M. (2017). Szlaki kulinarne jako forma konkurencyjności oferty turystycznej [Culinary trails as a form of competitiveness tourist offers]. *Zeszyty Naukowe Uczelni Vistula Turystyka*, 54 (3), 98–111.

Woźniak, M., Szara, K., & Wojciechowski, A. (1998). Działalność Ośrodka Doradztwa Rolniczego w zakresie rozwoju agroturystyki (na przykładzie Rejonowego Zespołu Doradztwa Rolniczego w Gorlicach) [Agricultural extension service center activity on agritourism (e.g. regional agricultural extension service set in Gorlice)]. *Zagadnienia Doradztwa Rolniczego*, (3), 116–126.

Woźniczko, M. (2014). Infrastruktura okołoturystyczna w turystyce wiejskiej na przykładzie wioski tematycznej „Kraina Rumianku" w Hołownie. jako element wzbogacający ofertę obszarów wiejskich [Tourism-related infrastructure in rural tourism: a case study of the thematic village "Land of chamomile" in Hołowno as an enhancement to the offer of rural areas], *Infrastruktura okołoturystyczna jako element wzbogacający ofertę obszarów wiejskich [Tourism-related infrastructure as an enhancement to the offer of rural areas]*, Jastrzębski, C. (ed.), Publishing House of the School of Economics and Law in Kielce, Kielce, 69–84.

Woźniczko, M., & Orłowski, D. (2011). Szlaki kulinarne komponentem wiejskiego produktu turystycznego [Culinary trails as a component of the rural tourism product], *Turystyka wiejska na drodze do komercjalizacji [Rural tourism's path to commercialization]*, Jastrzębski, C. (ed.), Publishing House of the School of Economics and Law in Kielce, Kielce, 101–123.

Wrzochalska, A. (2005). Wybrane cechy społeczno-ekonomiczne ludności wiejskiej a rozwój wsi i rolnictwa [Selected socioeconomic characteristics of the rural population vs. rural and agricultural development], *Institute of Agricultural and Food Economics – the National Research Institute*, 3, 1–44. Zajda, K., Kołomycew, A., Sykała, Ł., & Janas, K. (2017). *LEADER and Community-Led Local Development Approach. Polish Experiences*. Łódź University Press.

Zamelska, M., & Kaczor, B. (2015). Realizacja procesu dydaktycznego — zmiany programowe w latach 1974–2014 [Implementing the educational process: evolution of the curriculum in 1974–2014], *40 lat kształcenia na Wydziale Turystyki i Rekreacji Akademii Wychowania Fizycznego im. Eugeniusza Piaseckiego w Poznaniu [40 years of education at the Faculty of Tourism and Recreation of the Eugeniusz Piasecki University of Physical Education in Poznań]*, Bosiacki, S., & Stuczyński, M. (eds.), University of Physical Education, Faculty of Tourism and Recreation, 30–47.

Zarski, T., Klimaszewski, K., & Dworakowski, R. (2005). *Agri-Environmental Measures, Animal Welfare and Agrotourism – PolishFarmers' View*. Warsaw, ISAH, 2, 157–160.

Zawadka, J. (2010a). *Ekonomiczno-społeczne determinanty rozwoju agroturystyki na Lubelszczyźnie (na przykładzie wybranych gmin wiejskich) [Socioeconomic determinants of agritourism development in the Lublin region (a case study of selected rural communes)]*, Publishing House of the Warsaw University of Life Sciences, Warsaw.

Zawadka, J. (2010b). Ewolucja działalności agroturystycznej w Polsce i typologia wiejskich gospodarstw turystycznych [Evolution of agritouristic activity in Poland and typology of rural tourism farms]. *Acta Scientiarum Polonorum. Oeconomia*, 9 (4), 627–638.

Zawadka, J. (2012). Preferencje turystów dotyczące wypoczynku w gospodarstwach agroturystycznych na Lubelszczyźnie [Preferences of tourists on holiday in agritourism farms in the Lublin Province]. *Ekonomiczne Problemy Usług*, 669 (84), 167–173.

Zawadka, J. (2015). Możliwości finansowania przedsięwzięć z zakresu turystyki wiejskiej w perspektywie 2014–2020 [Potential Sources of financial support for the implementation of projects in the field of rural tourism, in the current term financing – 2014–2020]. *Innowacyjność w turystyce wiejskiej a nowe możliwości zatrudnienia na obszarach wiejskich [Innovation of Rural Tourism and New Opportunities of Employment on Rural Areas]*, Kamińska, W. (ed.), PAN KPZK Studia, 163, 279–289.

Zawadka, J. (2016). "Hity turystyki wiejskiej' w świadomości Polaków ["Hits of rural tourism' in the consciousness of Poles]. *Roczniki Naukowe Stowarzyszenia Ekonomistów Rolnictwa i Agrobiznesu*, 18 (3), 398–401.

Zawadka, J. (2017). Sieciowy produkt turystyczny i perspektywy jego rozwoju na przykładzie szlaku kulinarnego Mazowiecka Micha Szlachecka [A networked tourism product and its development prospects: a case study of the Noble Bowl of Mazovia culinary trail], *Sieci współpracy w turystyce wiejskiej. Stan obecny i nowe wyzwania [Collaborative networks in rural tourism: current condition and new challenges]*, Wojciechowska, J. (ed.), KSOW, 149–164.

Zbikowski, J., & Kasprzyk, A. (2010). Krajobraz kulturowy wsi podlaskiej jako istotny element atrakcyjności turystycznej regionu [The culture landscape of the Podlasie village as the element of the tourist atractiveness of the area]. *Problemy Ekologii Krajobrazu*, 27, 163–169.

Żelazna, K., & Woźniczko, M. (2005). Uwarunkowania wyboru oferty turystyki wiejskiej w kontekście konkurencji na rynku usług turystycznych [The conditioning of choice of offer rural tourism in context of competition on the market of tourist services]. *Turystyka i Rekreacja*, 1, 89–91.

Ziernicka-Wojtaszek, A., & Zawora, T. (2010). Dziedzictwo kulturowe obszarów wiejskich małopolski jako atrakcja agroturystyczna [Cultural heritage of the Malopolskie province rural areas as a form of agro-tourism attraction]. *Infrastruktura i ekologia terenów wiejskich*, 02, 119–130.

Ziółkowski, B. (2006). Rolnictwo ekologiczne a turystyka wiejska-próba modelowego ujęcia wzajemnych zależności [Ecological farming and agritourism - the attempt of depiction mutual dependences]. *Journal of Research and Applications in Agricultural Engineering*, 51 (2), 224–229.

Zrobek, J. (2007). Marketing terytorialny w tworzeniu przewagi konkurencyjnej na terenach wiejskich [Territorial marketing in creating competitive advantage in rural areas]. *Studia Prawno-Ekonomiczne*, 75, 239–247.

Zwoliński, Ł. (2007). *Wybrane cechy demograficzne ludności wiejskiej w latach 2000–2005 [Selected demographic characteristics of the rural population in 2000–2005]*. Institute of Agricultural and Food Economics – the National Research Institute, 58, 1–47.

5 General trends and their potential impact on the development of tourism and leisure activities in Poland

Monika Wojcieszak-Zbierska

This chapter presents a selection of factors that set the global tourism trends, taking account of development drivers of the tourism industry. It shows the specialization of tourism providers based on how agritourism farms evolve towards recreation, greening, and educative dimensions in their offer. Also, this chapter identifies the role and importance of social farming in the sustainable development of rural tourism. It describes the characteristics of social farming, taking both enablers and inhibitors into account, and some examples of social farming in Poland and in selected European Union countries are presented. In addition to explaining how the assumptions of the experience economy affect the competitiveness of rural tourism, it indicates some opportunities for using state-of-the-art technologies in promoting rural tourism products. Further, this chapter discusses the technological capacity that will be tapped into by agritourism farmers and consumers in the tourism market. Another interesting aspect consists in presenting the consumer journey from FOMO (Fear of Missing Out) to JOMO (Joy of Missing Out). These processes are described in the context of the tourism market. This chapter also presents some examples of JOMO tourism products and services in Poland. The last section shows the importance of smart villages in creating the image of Polish rural areas. It describes the characteristics of that concept and presents some examples of how it operates in Poland.

5.1 Factors affecting global tourism trends

Tourism is among the fastest-developing industries of the global economy (WTTC, 2019; UNWTO, 2019; OECD, 2018a; Alejziak, 2011; Ninemeier & Perdue, 2008; Cooper & Hall, 2008). A worldwide phenomenon, it accounts for a considerable part of the service sector and has a strong economic impact. It affects economic performance measured as the share of GDP or the number of people employed in entities active in creating tourism products. The total contribution of the tourism sector to GDP in 2018 was USD 8,811.0 billion (10.4% of GDP). In 2019, it grew by 3.6% to reach nearly USD 9,126.7 billion (10.4% of GDP) (WTTC, 2019). According to a WTTC report (2019), over 123 million people were directly employed in the tourism sector in 2018 (3.8% of total employment). Further growth is expected by forecasters: 125 million jobs by the end of 2020 (3.9% of total employment, a growth rate of 2.2%) and over 144 million jobs by 2029 (WTTC, 2019). This is supposed to include employment in travel agencies, hotels, motels, tourism facilities, airlines, other passenger transport services, and the catering and leisure industry supported directly by tourists. However, the COVID-19 pandemics have profoundly changed the situation in the tourism market. According to a report (UNTWO, 2020), tourism is among the sectors directly affected by the current pandemic crisis. As the epidemic hazard has arisen in countries around the globe, restrictions have been imposed on a number of activities, including those related to accommodation and tourism facilities, short-stay accommodation services, spa resorts, pilgrimages, etc. The restrictions on human traffic resulted in a decline in the number of users of accommodation services in all kinds of tourism facilities, including hotels, agritourism farms, hostels, etc. (UNWTO, 2020). As a consequence of the tourism crisis engendered by the COVID-19 pandemic, Poland saw a sharp decline (by more than 22%) in the number of incoming international tourists in 1Q 2020. According to plausible scenarios for 2020, the annual decline could vary in the range of 60% to as much as 80% compared to 2019 figures, depending on how fast will the traveling restrictions be eased in each part of the world (UNWTO, 2020). In a political note, OECD (2020a) emphasizes that the crisis has a clear impact on the whole tourism industry, and the reopening and reconstruction of destinations will require a common approach because it takes more effort to open a sector than to close it. Most importantly, the reopening must follow a sustainable approach. The consequences of the pandemics will depend not only on its duration (which will have an impact on the entire tourism sector and its survival) but also on potential long-term changes in the tourists' traveling behaviors caused by the crisis. The crisis will have a permanent effect on traveling behaviors; a greater emphasis will be placed on a hygienic, healthy lifestyle and on the use of cashless, contactless payment methods. Currently, being secured against health risks by tourism operators will become an important aspect to consumers (Folinas & Metaxas, 2020; Hall et al., 2017). In addition to accessing financial support allocated to the tourism sector, service providers will find it important to develop a strategy designed to make the destinations safe to consumers (Folinas & Metaxas, 2020). A broad international debate which is currently ongoing has identified the key priorities and challenges for the tourism sector (OECD, 2020a):

- *Re-analyzing the tourism sector:* the crisis is an incentive to build a tourism system which will ensure greater sustainability and robustness to better respond to any reoccurrence of such a situation in the future. According to the assumptions, it will be necessary to provide political assistance in solving the sector's structural problems, avoiding the reemergence of tourism management issues (e.g. overtourism) and pursuing the key priorities, such as encouraging new business models, embracing digitization and promoting communications. In their quest for less crowded destinations, the tourists will look for agritourism farms and organic farms.

(OECD, 2020b)

- *Reconstructing the destinations and ensuring safety:* support and repair measures should be implemented on a comprehensive basis across all sub-sectors and aspects which include accessibility, communications, transport, accommodation, resorts, restaurants, events, tourism associations, tourism technology providers, tour operators, tourism facilities, and agritourism farms. Rebuilding a secure system is a tremendous challenge for local economies.
- *Tourism innovation and investments:* each country, including Poland, should make sure the sector is ready to resume its activity and implement innovation. It will be necessary to implement a series of investments that result in structural and physical changes to address the requirements for health protection and meet the visitors' expectations in the first bounce-back period. The Polish government has come up with several solutions to prevent the tourism industry crisis. The first one, referred to as the "Polish Tourist Voucher," is designed to financially support both Polish families and the tourism sector weakened by the effects of the COVID-19 pandemics. It is a form of one-off support for Polish families which consists in paying PLN 500 for each child aged below 18 and providing one supplementary benefit in the amount of PLN 500 for children with a medical certificate of disability. The voucher can be used to pay for hotel and accommodation services or tourism events organized by a tourism undertaking or a public benefit purpose foundation in the Polish territory. Currently, work is in progress on establishing a directory of tourism resources. These efforts are coordinated by the Polish Tourism Organization.[1] Another proposition is the "Have a safe holiday in Poland" program. Prepared by the Polish Tourism Organization, it consists in certifying tourism accommodation facilities. The purpose of the project is to strengthen the confidence the tourists have in their destination by implementing pandemic prevention and inspection measures, and to provide the Polish population with information on facilities that are ready do deliver services as per the guidelines of the Chief Sanitary Inspectorate. The program is intended for all providers of accommodation services in the Polish territory.[2] "Your rural holidays" is the third major project. It was initiated in November 2016 to strengthen rural tourism. However, due to the pandemic context, it has been broadly promoted during the 2020 holiday season. Its main purpose is to make rural areas viewed as a tourism market that offers diverse attractions all year round.[3]

Furthermore, a report by the OECD (2020a), identifies digitization as one of the areas where pandemics provide a stimulus for potential development of the sector. Indeed, the current crisis has accelerated the digital transformation of tourism. Digital solutions are a response to demand-side requirements because they are developed to provide the tourists with a "live" remote experience (visiting virtual museums, mines, etc.). In the long run, consumers are likely to demand self-service, contactless, customized experiences which may encourage the implementation of solutions such as automatic check-in/check-out kiosks, self-service transport endpoints, etc.

Recently, many papers have been published that analyze the development factors and trends of today's tourism sector (WTTC, 2019; OECD, 2018a; Ban & Ramsaran, 2016; Angelkova et al., 2012; Nowak et al., 2012; Andereck & Nyaupane, 2011; Mathew, 2009). These include empirically informed studies and popular science papers resulting from many years of observation of the tourism market. The broad range of factors affecting tourism includes enablers and inhibitors, i.e. those which either promote or restrict its development and which are general or specific to rural tourism. Enablers of rural tourism include:

- *environmental factors:* natural and landscape values; geographic location; cultural and historic heritage; traditional hospitality; motivation to improve the standards of living (Niedziółka & Roman, 2017);
- *investment factors:* aligning the accommodation facilities with applicable standards; making the farms and their surroundings suitable for tourists; good marking of tourist trails; building the environmental protection infrastructure (reservoirs, lookout towers); building, upgrading and adapting agritourism facilities (Panasiuk, 2014);
- *organizational factors:* self-organization of farmers active as accommodation and service providers, e.g. agritourism associations, local action groups (LGAs), local tourism organizations (LTOs); cooperation with local government units on creating a development strategy for sustainable tourism; educative measures taken to encourage people to engage in a tourism activity (Niedziółka & Roman, 2017).

In turn, factors that inhibit the development of rural tourism include:

- troubled family life of the farmer (fear);
- greater workload on women;
- psychological barriers related to the farmers' language proficiency (some of them cannot speak any foreign language);
- poor sanitary standards of residential buildings intended to be used for agritourism purposes;
- poor environmental awareness among tourists;
- an inadequate system for the domestic and international promotion and distribution of the rural tourism offer;

- poor knowledge of marketing tools among agritourism service providers and institutions;
- lack of coordination between different rural tourism institutions (Niedziółka & Roman, 2017, pp. 81–85).

Aramberri (2010), Buckley et al. (2015), Grigaliūnaitᐧ et al. (2015) indicated that the inhibitors can include the following: fiscal crises in key economies; disasters affecting financial institutions; epidemics and diseases of affluence; political and social instability; food crises; wars and natural disasters. In turn, reports by the World Tourism Organization (2019) point out the following enablers:

- increased awareness of traveling opportunities;
- shortening of working time;
- physical activity during holidays;
- shift towards a healthy lifestyle;
- aging of the population;
- the trend to start a family at an older age;
- the evolving 2+1 family model;
- the growing number of childless couples;
- traveling security.

The importance of each factor varies over time and is continuously modified by external aspects related, for instance, to the social, economic or political situation (Çalışkan, 2014; Crouch, 2010; Jovanović & Ilić, 2016; Muganda et al., 2013). Also, the development of tourism is impacted by cooperation enablers such as globalization, technological advancements, computerization, political and trade agreements, and national policies around the world. In turn, a paper by Buckley et al. (2015) indicates such factors as social, economic and environmental conditions; social fears related to masses of people traveling over long distances; the use of new technologies, i.e. social media; consequences of armed conflicts; the use of tourism as a tool in making geopolitical deals; and the growing links between tourism and nature protection in many parts of the world. Authors of the extensive international literature (Buckley et al., 2011; Buhalis & Darcy, 2011; Hoang, 2015; Dickinson & Lumsdon, 2010) identify eight key factors that have an impact on tourism (Table 5.1).

The development of transport modes is another important factor that makes the society more mobile. Growing tourist expectations, diverse motivations behind traveling, increased competition between tourist regions, lifestyle trends, and weather are aspects that definitely affect the choice of a tourist destination. The global tourism system should respond to the growing number of interrelated challenges, including the current uncertainty related to the financial crisis, pandemic, tourist preferences and expectations, mobility, terrorism threats, and demographic change. Hence, there is need for crisis-

Table 5.1 Key factors impacting the tourism sector

Factor type	Description
Economic	Deemed to be of utmost importance in the international literature. They have the greatest impact on the development and form of today's tourism, and include: a real increase in incomes; an adequate distribution of incomes; a stable financial situation; favorable socioeconomic conditions. They also mean an appropriate use of available funds allocated to tourism development
Socio-demographic	Mostly related to the growing world population and the increasing lifespan. Changes in the employment structure, general availability of free and holiday time (e.g. the establishment of public holidays), cultural changes, and the evolution of mindsets taking place in some communities are the main factors that impact the global trends in tourism
Sociological and cultural	These are factors related to aspects such as the changing family model, aging societies, and problems of the disabled. They affect the development and adjustment of products and services available to older or disabled people. The existence of electronic media has an effect on changing the cultural model. Today, easily accessible virtual tours (e.g. museum tours) offer unlimited possibilities to people without the need for them to actually enter the facility
Political	The government's policy regulates all spheres of economic life through specific policies, e.g.: the social policy (through its national legislation, a country can support or restrict the tourism activity of its nationals and foreigners, for instance based on the healthcare or public security system); the monetary and fiscal policy (e.g. taxes, duties, and stability of the national currency in global markets); the economic policy (which has an effect on how fast the economy grows, on real incomes, and state-supported investments in, or related to, the tourism sector); the international policy (e.g. international agreements, changes to border crossing rules, such as the Schengen Agreement, and liberalization of tourism markets)
Technological (technical)	Mainly related to technical and technological advancements in the transport infrastructure. The development of tourism and related facilities resulted in increased tourist traffic, thus stimulating growth and quality improvements of the tourism infrastructure. The 21st century is marked by profound changes not only in how fast but also in how far and how comfortably people travel. The expansion of inland, air and sea transport has contributed to making the distances smaller. Moreover, the fact that tourists watch TV, access the Internet and rely on specialized tourist applications, computerized booking systems (e.g. Booking.com), travel marketing platforms and social media (Twitter, Facebook, etc.) has had a considerable impact on tourism

(Continued)

Table 5.1 (Cont.)

Factor type	Description
Environmental (ecological)	Because of its large scale, tourism undoubtedly had, and continues to have, an impact on the natural environment. The tourism sector totally depends on natural resources. Hence, it is important to preserve the natural environment in a sound condition and reduce its degradation. This means implementing the principles of sustainable development. Therefore, the purpose of tourism is to protect the natural and cultural heritage. Environmental factors also include the existence of natural disasters (floods, earthquakes, tsunamis, hurricanes) and pandemics (COVID-19, SARS, Ebola, Zika)
Security	More and more attention is paid to aspects related to the security of traveling, a factor which certainly has an impact on the geography of traveling. Problems that have a political or cultural background affect the tourism traffic in the region concerned. Locations hit by wars or disasters (e.g. the 2004 tsunami in the Indian Ocean which hit the Asian, Indonesian, African, and Oceanian coasts) are less attractive to tourists. Because of such events, the number of incoming tourists drops and so does the amount of income. However, as can be seen today, there is a new trend which makes some people travel to such regions (e.g. Irak, Pakistan, Syria); this is more about observing or participating in some disturbing events than being a direct part of them
Other	Other factors that impact the global tourism include: regional promotion measures; commitment to improve service quality; taking care of the range of experiences lived in attractive locations; improving the skills of tourism staff in attractive locations (including their ability to speak multiple languages)

Source: Compilation based on Scott et al. (2012); Yeoman (2012); Buckley et al. (2015); Kruczek (2014); Panasiuk (2014); UNWTO (2019); OECD (2018a); Weaver (2010); Pickering (2011); Buckley (2011).

Note: The impact of the factors differs between time periods, countries, and regions (disaggregated by the urban/rural dimension).

averting strategies which can contribute to fighting these adverse aspects. As an economic industry, tourism is not exempt from crises. Importantly, such situations are reflected in global tourism trends. It can be assumed that tourism trends are affected by a multitude of diverse factors, primarily including the tourists, destinations, competitive processes in the market, and restrictions resulting from the principles of sustainable tourism development. Recently, the UNWTO (2019) has identified six major tourism trends (Figure 5.1).

The demand- and supply-side transformation of today's tourism market drives considerable changes in the purchasing behavior of tourists. There are changes in the way the travels are organized, in the distance traveled, in the quality and frequency of trips, and in the ways people spend their free time.

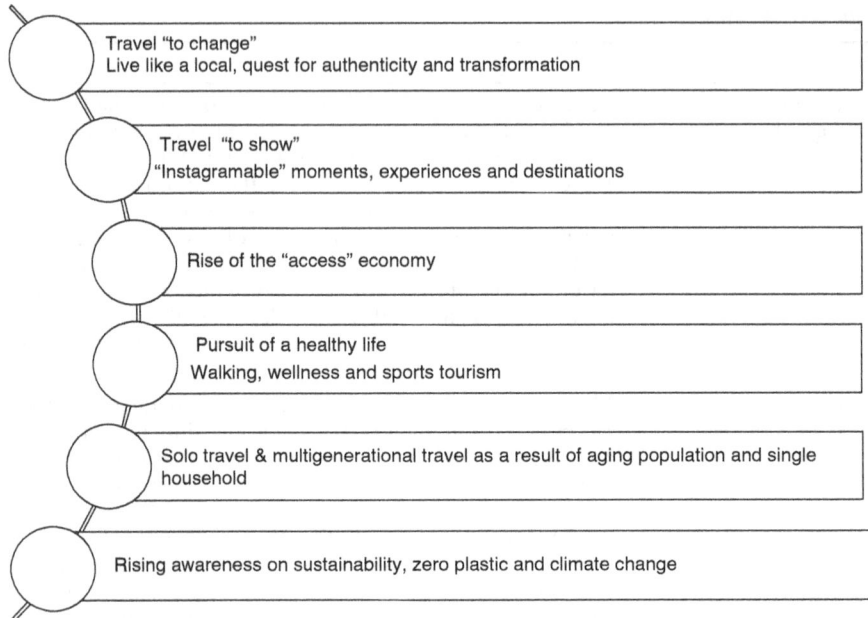

Figure 5.1 Global trends in tourism travels.
Source: UNWTO (2019, p. 5).

However, the surroundings and advantages of a location continue to be the main selection criteria when choosing a tourist destination. The first trend indicated by the UNWTO (2019) is that people travel to change the surroundings they live in. As an important aspect, the travelers look for authentic places where they can relax (UNWTO, 2019). In the international literature, many authors (Delener, 2010; Panasiuk, 2011; Wang et al., 2012; Butler, 2012) emphasize that the growing mobility makes tourists increasingly often move between towns instead of staying in one location. Another trend is that people travel to meet their need for something new. In this case, experiences, emotions, and feelings are what make tourists travel to a specific destination. Increasingly often, it can be seen that tourists find it important to look for locations which provide them with an authentic experience and true emotions. Rather than products and services, travelers buy the accompanying experience. Value is no longer viewed only in a utilitarian context (price and utility of the offer to the customer); instead, the spotlight is on its relationships with a category of experiences that allow the buyers to make their dreams come true and shift to a desired lifestyle (for more information, see Section 5.2). Another trend is related to the availability and openness of the economy. Today's technology allows the tourists to travel far away using state-of-the art solutions. The development and use of technology, the impact of IT, and the growing accessibility of the Internet are the key components which should be skillfully used by tourism operators. Modern tourists expect to be offered a virtual museum or city tour and to participate in traditional ceremonies and holidays which are staged shows carefully designed for them. The commitment to a healthy and active leisure and the development of wellness and sports tourism testify to the shift towards focusing on mental and physical health. Today, it can be observed that tourists believe wellness to be "a life philosophy, a state of mind, balance, harmony and wellbeing which goes beyond the physical to encompass the mindsets and emotions" (Bajdalska & Knefel, 2018). Another important part of wellness is the broadly defined physical activity which includes a myriad of interesting forms of action. This means not only training in fitness clubs but also any kind of physical activity of humans, from forest walks through to boxing classes. Note also that outdoor workout has nowadays become highly popular; it is supposed to combine fitness classes with accommodation in an attractive destination away from large urban agglomerations (especially in rural areas). Even in large urban agglomerations, fitness clubs increasingly often organize outdoor workout to make their offer more attractive to customers while leveraging the natural landscape assets (Bajdalska & Knefel, 2018).

Population aging, as is witnessed today (especially in Europe, including Poland), results from the extended life expectancy, on the one hand, and from low fertility rates, on the other. The elderly are estimated to account for around one-fifth of the population in developed countries (EUROSTAT, 2019). According to EUROSTAT forecasts, they will represent nearly one-third of the population in 2050 (EUROSTAT, 2019). This has multidimensional socioeconomic consequences, including in the tourism market. The growing group of seniors demand tourism services as they have plenty of free time, fixed incomes, and increasingly higher education levels (Janiszewska, 2017). Today, seniors are viewed as an important segment of the tourism market. According to UNWTO forecasts (2019), the number of individual trips is predicted to grow ca. 17% faster than that of organized package tours. There will be greater interest in more frequent, shorter (weekend) explorative and leisure trips.

The last trend is related to sustainable development aspects, including sustainable tourism. Today, rural areas have been rediscovered as a place where, in addition to passive leisure, people can also actively spend their time (Roman, 2018). Tourists increasingly often opt for rural holidays. Hence, rural tourism—due to its particularities—is competitive to large, modern urban resorts (Przezbórska-Skobiej, 2015; Sikora, 2012; Wojcieszak, 2017). Specific kinds of tourism attractive to visitors are agritourism, eco-tourism, and eco-agritourism, which are highly valued by Polish tourists (Roman et al., 2020). The basic development path for agritourism consists in improving and extending the leisure offer of agritourism farms intended for various groups of visitors, e.g. families with children, young people or seniors. Also, the importance of ecology in tourism is on the rise. Tourists increasingly often prefer environmental leisure and opt for green tourism (eco-tourism). The overarching goal of green tourism is the commitment to preserve a durable and sustainable development of tourism resources and values. The whole concept is underpinned by the integration of tourism activities with environmental protection goals, socioeconomic life, and the formation of new attitudes and behaviors among tourists and tourism organizers (Roman, 2018). Eco-tourism can considerably contribute to local development in regions which capitalize on natural assets, attractive landscape, and cultural heritage. Furthermore, it is related to similar forms of economic activity, namely agricultural processing, handicraft, and other kinds of tourism which are adapted to regional conditions: agritourism, active tourism, culinary tourism, health-oriented tourism, and spa tourism. Note that what eco-tourism enthusiasts view as an advantage is not only the contact with nature but also the opportunity to discover regional cultures and traditions and explore the local cuisine based on organic products. Hence, it can be concluded that eco-tourism certainly helps the tourists enhance their education. Travelers increasingly often want to access and expand knowledge and improve their language skills. Educational tourism is based on individuals'

> self-development capacity defined as a set of factors which make them autonomous from social pressures and allow them to take many actions to make them feel in control of their lives. Humans need to develop their creative capacity and to express their creativity. This is the basis for their creative activity.
>
> (Dhiman, 2012)

In Poland, educational farms, educational excursions, and questing are a perfect example of how education is promoted as part of tourism activities (for more information, see Section 5.3).

Today, there is a noticeable evolution in how tourists view tourism. Indeed, tourists increasingly often opt for a 3E (Entertainment, Excitement, Education) formula which has largely replaced the traditional 3S (Sun, Sea, Sand) pattern. The classic 3E model can also be enhanced with a fourth E which stands for Engagement (Stasiak, 2015; WTTC, 2019). Currently, individual tourism can be observed to turn into mass tourism characterized by diverse, specific individual needs resulting from the tourists' commitment to stay fit and mentally balanced, pursue their hobbies, and experience strong emotions. This shift gives rise to new products and services designed to excite and educate by prompting the tourists to think and reflect. These days also witness the emergence of other kinds of tourist traffic, such as: moon tourism (Chang & Chern, 2016), geo-tourism (Allan, 2015), eco-tourism (Roxana, 2012), agritourism (Bansal & Kumar, 2011), healthcare tourism (Kogiso, 2015), business tourism (Nicula & Popsa, 2014), congress and conference tourism (Ayaz & Şamata, 2017), cultural tourism (Han & Imran, 2018), pilgrimage tourism (Štefko et al., 2014), and educational tourism (McGladdery & Lubbe, 2017).

Twenty-first-century tourists pay great attention to security aspects, especially when it comes to health. The travelers' perception of security and wellbeing can be seen to grow in importance, too. There is increased demand for diverse interesting forms of qualified and specialized tourism. The structure of tourism products offered undergoes changes because of the shift in motives behind traveling. The number of tourists has been on the rise until 2020. This is both the cause and the consequence of changing trends in the market for tourism services. However, the situation caused by the COVID-19 pandemic has affected the whole tourism industry, resulting in a decline of ca. 43% in tourism traffic (OECD, 2020b). The reopening and reconstruction of destinations will require a shared effort from the government and tourism organizations and operators to guarantee a safe experience to the travelers. It will be possible to overcome these problems by implementing adequate solutions and tapping into experiences of different countries. Sustainable development and competitiveness are two significant elements affecting the global condition of a tourism destination. Tourism companies can become competitive by making an appropriate and efficient use of resources while promoting the protection of biodiversity and preventing climate change (UNWTO, 2019). The creation of global trends in the tourism market is a complex, multifaceted process which takes place at multiple levels (WTTC, 2019; OECD, 2018a; Ramjit, 2015; Ritchie & Crouch, 2010; Spencer & Nsiah, 2013; Schiff & Becken, 2011). Based on the analysis of the ongoing evolution of tourism, several key development trends can be identified, including: development of alternative tourism; growing popularity of exploration trips; tourists becoming more and more physically active; growing popularity of healthcare trips; and the development of rural tourism. The trends followed by the tourism industry, as listed above, are true at a global, national, and regional level, though they may differ in intensity.

5.2 Experience economy in shaping the competitiveness of rural tourism

Today, there is a clear interest in the natural environment and in active forms of contact with nature (Yeoman & McMahon-Beattie; 2019, Engen & Mehmetoglu; 2011, Poczta, 2013; Sadowski & Wojcieszak, 2019; Balińska, 2019). Tourists increasingly prefer individual trips that address their diverse interests and specific needs related to their commitment to stay fit, pursue a hobby and live exciting experiences. There is growing demand for various forms of adventure and specialized tourism, e.g. health resorts or rambling. The structure of tourism products and services offered changes, too, due to a shift in motives behind traveling and the quest for adventure. For the providers, experiences sought by tourists are an incentive to develop products or services that meet the travelers' expectations. Hence, the tourism environment becomes more and more competitive and dynamic, and evolves under the impact of various global trends (Aşan & Emeksiz, 2018; Bakas et al., 2019). The experience-gaining process plays a major role as a factor that motivates people to make a tourist trip (Chen & Chen, 2010; Aşan & Emeksiz, 2018;

Bakas et al., 2019; MacCannell, 2002; Bakas et al., 2018; Gartner, 2004; Jepson & Sharpley, 2015; Kastenholz et al., 2018; Łukaziewicz, 2017). That process has a great impact on tourism activity as it enables increasing tourism demand. Undoubtedly, this is an important aspect for potential service providers and allows them to professionally orchestrate an adequate tourism offer for specific customers. The 21st century is marked by a rapid development of the experience economy which uses feelings, emotions, and experience—rather than products—as basic commodities. Many authors (Andersson, 2007; Quadri-Felitti & Fiore, 2015; Pine & Gilmore, 2013; Lee et al., 2011; Kim et al., 2020; Galvagno & Dalli, 2014; Ritchie et al., 2011) emphasize that today's tourism is a specific "experience industry." The currently growing importance attached to experiences could be viewed as a megatrend in the tourism industry. Also, as Mikos von Rohrscheidt (2011) remarked, the tourism market now deals with "new tourists," a group which grows steadily. A "new tourist" is someone more experienced, flexible, mobile, independent, guided by another system of values than his/her predecessors, and, most importantly, someone with other lifestyle preferences. "New tourists" want not only to "see" and "learn" but also to "understand," "live," "experience," and enter into a direct contact with the place they visit. This is why a discussion is still ongoing around the world on the importance and meaning of experiencing in the economy in the context of tourism. Marciszewska (2010) indicated several characteristics of the experience economy which refer to the tourism industry. She believes that the widespread creation of experiences by state-of-the-art technologies, the need for active participation (immersion) from product users, customization and personalization (personal nature) of experiences (the need to tailor them each time), and the fact that experience is what people remember (including all of its upsides and downsides) are aspects of vital importance. The creation of emotional tourism products has become even more important as it can be noticed that tourism operators make continuous efforts to intensify the tourism experience (Tan et al., 2013; Tarssanen & Kylänen, 2005; Santos et al., 2018; Alexiou, 2020). Transforming the tourism and related infrastructure into unique tourist attractions, enhancing traditional services or bundles with elements that make the experience more emotional, and using state-of-the-art technologies are certainly the key measures taken to meet the objective of providing a greater satisfaction and meeting tourist expectations. The theories that explain this phenomenon include the "economy of experiences" developed by Pine and Gilmore (1998). According to the authors, an economic experience exists in the context of tourism. This means that potential tourists expect they can derive a series of experiences from the tourism offer of service providers. According to Pine and Gilmore (1998), a modern person lives in an era of development, globalization, and an advanced service-based economy and technology designed to provide the customers (tourists) with a highly emotional experience. The experience, feelings, emotions, and moods become a product and are regarded as a new source of value for tourists. As regards tourism, that approach is not as new as it seems because tourism means selling emotions, experiences, dreams, feelings, and memories related to a journey. However, the new part is the conscious, professional, and intentional development and creation of tourism products filled with emotions. Numerous examples of this approach can be found around the globe (space flights, snake massage, sleeping in the desert, etc.). Importantly, many authors (Niezgoda, 2013; Roman, 2018) note that experience emerges in the consumption process and is a source of satisfaction to the consumer:

> In services of a largely intangible structure (which include tourism), experience is a value in itself. If positive, it enhances consumer satisfaction. In addition to being a part of a service product offered in the market, experiences and feelings strongly determine its quality.
>
> (Niezgoda, 2013, p. 24)

This is mainly because of the characteristics of services which are unique and intangible. Tourism services are not basic services; hence, consumers have time to consider the way they want their need to be satisfied, and can analyze their own experiences. In modern market conditions, innovative products are the ones that set the standards and are believed to be of highest quality. To stand apart in the market, service providers should make efforts to ensure a high-quality emotional experience. Traveling has always involved strong internal experiences and feelings. However, they were considered a side effect of traveling rather than an important goal of tourism activity. Increasingly often, the tourists seem to attach greater importance to their internal feelings, emotions, and satisfaction from being in a specific location than to the service level. This finding shows that experiences are a foundation for a comprehensive strategy of a tourism brand, and underpin the goals and purposes of all actions of service providers. Adopting that paradigm of tourism has required the development of methods and tools for professional development of tourism products, bundles, and services strongly associated with emotions and feelings, especially when it comes to rural tourism. Niezgoda (2013) noted that "it is necessary to shift from a tourism service bundle to a tourism experience bundle" (Niezgoda, 2013; p. 26) in order to turn an ordinary product to an extraordinary experience (LaSalle & Britton, 2003). The experience delivered to the customer becomes a new, highly important battlefield for competing service providers.

> In the context of tourism, the key objective of the economy of experiences is believed to consist in shaping a positive customer experience in the cognitive, affective and behavioral (conative) dimension, leading to desired consumer attitudes expressed as satisfaction and loyalty.
>
> (Skowronek, 2011, p. 11).

Time compression, which means putting the largest possible amount of offers and experiences in the shortest possible time frames, is characteristic of today's tourism (Mroczkowska, 2008). Leisure is not enough for modern tourists; they want to be part of something special, be surprised, and live a unique, fascinating adventure. They expect to stay in an extraordinary, fairy-tale-like place that will provide them with true emotions and feelings, engage all of their senses, and ensure a non-trivial personal experience and (most importantly) memories. In an effort to address that demand, service providers and tourism companies prepare increasingly surprising, shocking, and fancy products and services. Tourism experience has long been investigated and analyzed by tourism researchers, such as Cohen (1979) and Urry (2007), who come from all over the world and represent different fields of science, e.g. sociology or psychology. An interesting approach was presented by Ooi (2003) who indicated and highlighted the most frequently tackled research problems related to this topic:

• psychology of tourism experiences;
• outcomes of tourism experiences;
• level (depth) of engagement and experience;
• phenomenology of tourism experiences;
• differences between tourism experiences and day-to-day living.

Moreover, multifaceted analyses of tourism experiences were carried out as part of sociological research by authors such as Cohen (1979), Urry (2007), and Wieczorkiewicz, (2008). Tourism experience was also discussed in the ample marketing literature, for instance in the context of defining a tourism product, considered to be a "set of utilities related to tourist trips, i.e. tourism goods and services available in the market which allow to plan, make and live the trips and pool the related experiences" (Stasiak, 2015, p. 23). Yeoman and McMahon-Beattie (2019; p. 13), identified seven microtrends related to the economy of experiences, namely:

• "once is not enough,
• an experience of luxury,
• having a better holiday time,
• escaping from the modern world in a quest for authenticity,
• identity,
• exceptional day-to-day living; experience as the highest priority,
• genuinity / value."

The first trend refers to the category of one's own values. This means that some people will be able to live a specific experience (e.g. a wedding ceremony on a volcano) only once. Due to having access to state-of-the-art technologies, young people will have a greater pool of experiences than those aged over 65. Interestingly, older people decide to make their dreams come true and live specific experiences because they think this might be their only chance to do so (Souza et al., 2019; Weinberger et al., 2017). In Poland, examples include attending the midnight mass as the Wieliczka salt mine or the Midsummer Night.[4] The second trend identified by the authors is associated with luxury. As they emphasize, the definition of luxury has evolved over the years (Yeoman & McMahon-Beattie, 2019; Hennigs et al., 2015). Today, generations X, Y, and Z[5] believe luxury (i.e. access to unique forms of leisure and attractions) to be something obvious and natural. However, note that the world currently struggles with a great economic and financial crisis resulting from the worldwide epidemic caused by COVID-19, and therefore luxury can become slightly less accessible. A similar conclusion was presented by Foresight Factory (2017), and Seo and Buchanan-Oliver (2015), after the 2008–2009 crisis. Poland has many unique and luxurious attractions to offer, e.g. the Sopot Pier, the Royal Baths Park in Warsaw, the Oscypek Museum in Zakopane or the Old Market Square in Krakow.[6] In turn, examples of unique rural attractions include Gałkowo, a village at the very heart of the Masurian Lake District which is best known for Ferenstein's stud farm. Each year, it hosts periodic events, e.g. equestrian exhibitions and horse riding shows (Gałkowo Cup). It is surrounded by beautiful forests and lakes; the nearby river Krutynia, together with 10–20 lakes, forms a popular canoeing trail.[7] The third trend is associated with the concept of improving the way people relax. This means the consumers are more and more aware and value their free time. While some of them prefer luxury holidays, others need to live a thrilling experience derived from attractions which are supposed to be ensured by service providers and which, above all, will satisfy their need for self-improvement (Yeoman, 2012). Examples found in Polish rural areas include what is referred to as the Polish Provence (Lavender Fields in Warmia–Masuria). The Lavender Fields agritourism farm based in Nowy Kawków, Warmia, runs the Living Museum of Lavender and offers workshops where people can learn to make natural cosmetics by themselves, including creams, aromatic waters, and essential oils. This is also an opportunity to learn more about other uses of lavender (liqueurs, mead, vinegar, wine).[8]

In turn, the fourth approach emphasizes the role of consumers (tourists) who seek alternative and, as a consequence, authentic forms of leisure (Yeoman et al., 2014). Tourists look for authentic products, services, and bundles which are true and unique because only such an authenticity is viewed as value added by the buyers (Yeoman & McMahon-Beattie, 2019). They can find it in a regional meal, culture, folklore, tradition or rites (Polish examples include the Dunajec River Gorge rafting, common carol singing, harvest festivals or Corpus Christi celebrations). Another approach involves the aspect of tourist mobility. Today, tourists use extended forms of communication (whether they travel by plane or are on a remote trip) and rely on Internet companies (e.g. Booking.com) to book accommodation. All of this makes them free to build their own tourist identity which is not limited by their origin or geographic location. In the era of globalization, people form a networking society. The availability of state-of-the-art technology, messengers, and innovations is the reason why communication channels use a broad access to products and services (Yeoman, 2016). A study by Boztug et al. (2015) also confirms that 21st century tourists are people open to novelty who seek experiences, emotions, and feelings, and who value creating their own identity (for instance, in Poland, tourists can use the *Agroturystyka* application to easily find a facility matching their criteria and check the local attractions, prices, and availability of accommodation). Another trend is the celebration and holidaymaking. As noted by (Yeoman & McMahon-Beattie, 2019), consumers (tourists) keep looking for reasons to break their routine and enjoy unplanned celebrations. This means that holidays are a reason to travel for some of them; in Poland, examples include traveling to Częstochowa to celebrate Our Lady of Częstochowa (August 26); traveling to Warsaw to celebrate the National Independence Day (November 11); the Saint Adalbert indulgence feast in Gniezno; traveling to Jerusalem at Christmas; Saint Patrick's Day; or Oktoberfest. Thanks to social media, the tourists can choose the event they want to participate in. According to the authors of "Generations X, Y and Z," in the context of the philosophy of tourism, the last trend of the economy of experiences shows that consumers look for increasingly exciting and innovative solutions (Yeoman & McMahon-Beattie 2019; Hennigs et al., 2015). Provided with more and more possibilities to personalize their travels, tourists can create a broader range of unique experiences. Increasingly often, consumers want to enjoy a quiet holiday away from the Internet (Collins & Weiss, 2015; MacLaren et al., 2013). Importantly, Penn (2007), Penn and Fineman (2018) emphasize that changes which take place in the economy of experiences are mostly consumer-oriented. Instead of spending their time doing nothing, the tourists want to live something special, and therefore make both theoretical and practical preparations to their trips. Today, people clearly renew their interest in history, culture, folklore, and day-to-day lives of the population of a region or country (Cohen, 2010; Żemła, 2017). In the era of globalization, DMCs (Destination Management Companies)[9] identify more and more new methods for the destinations to improve their competitiveness and be an attractive market player. When analyzing the core product, it can be concluded that competition is poorly efficient at that level since it is hard to change the natural resources of a geographic area. Hence, the only thing to do is to compete with value added which, in the context of a regional tourism product, can mean providing tourists with an adequate experience (in addition to the region's image and brand). Hence, this is consonant with the commitment to a 4E tourism model. Tourists increasingly often prefer environmental leisure which allows them to have a quiet and calm getaway from the urban lifestyle. Rural tourism is a perfect platform for an emotional and spiritual experience (Jepson & Sharpley, 2015) while also being an interesting topic of research. It can be analyzed in a multitude of contexts, such as: diversification of rural functions; changes in demand for tourism services; non-agricultural activity; trends in the tourism market (Balińska, 2019). The development of rural tourism is important in stimulating the development of emotional experiences, and may affect tourism preferences (Kurczewski, 2016; Lacher et al., 2013; Martinez et al., 2019; Poczta, 2013; Shu-Yi Chi, 2019; Stasiak, 2013). Rural tourism is driven by people seeking attractive, unique, unforgettable experiences from specific locations. Tourism experiences can be measured on different scales. The tourists differ in what they find attractive (products, local food, folklore, culture, rites, habits, etc.). One of the distinctive features of today's tourism, including rural tourism, is the quest for experiences and *genius loci* (a unique climate and spirit of a destination) (Duda, 2016; Niezgoda & Markiewicz, 2014). Tired, often stressed by their hectic lifestyle, being in a hurry, and living in a monotonous landscape, tourists start to look for calm, quiet places and new experiences in a new environment. Hence, they pick rural areas (Niezgoda & Markiewicz, 2014). The authenticity of rural areas is viewed through the prism of characteristics that meet a tourist's personal needs. This demand is addressed by the growing number of services related to what is defined in the literature as the slow life movement (Duda, 2016). Such measures fuel the search for newer methods and places of leisure and recreation; this, in turn, stimulates the development of new tourist destinations, including rural areas (Roman et al., 2020). Compared to what has been seen so far, slow tourism is a different and innovative philosophy of pursuing and organizing tourism. The differences include: avoiding traveling by car; using alternative forms of transport; long stays instead of short trips; contacting the local community and its culture; and (most of all) experiencing the natural environment (Dall'Aglio, 2011). Rather than being focused on luxury, slow tourism adepts are ready to contact the local population, make slow tours, or participate in the work and day-to-day living of owners (farmers) which may be both an entertaining and educative activity (Dickinson & Lumsdon, 2010). Rural development and functional changes (resulting from the customization of their offer) make rural areas aligned with the growing interest in slow tourism which complies with the principles of sustainable development. This is what makes more and more Polish villages embark on the economic and social development path. Recent years have witnessed growing demand from tourists for creative and active ways of spending their free time. Meanwhile, rural areas have started to perform new functions (Balińska & Wojcieszak-Zbierska, 2020). The traditional agricultural and tourism functions are now supplemented with the educative (cognitive) function. Research (Chaminuka et al., 2012; Chrysolite, 2014; Dykas & Tokarski, 2013; Fiedora & Kociszewski, 2010; Galvagno & Giaccone, 2019; Kastenholz et al., 2012) found that the rural tourism experience also has a noticeable educative and aesthetical dimension. In line with the rapid development of the Internet and of tourism portals, rural tourism has entered the era of the sharing economy. Developing shared platforms and tourism bundles and enhancing rural tourism products and services certainly are good ways to put rural tourism on a sustainable development path in today's economy. The rural tourism experience is considered to be a new value and the main source of consumer satisfaction. While these measures must not necessarily be part of an economic activity pursued by tourism operators, they are supposed to play an essential role in providing the customers with emotions, feelings, and a good mood (Marciszewska, 2010). In Poland, rural tourism products are intentionally designed to trigger strong emotions. Some trends and processes already known in Western Europe are present in Poland, too (e.g. LARPs[10]). Capitello et al. (2013), Bravi and Gasca (2014), and Benckendorff and Zehrer (2013) indicated that the experience derived from the consumption of different characteristics of a destination or good could play a major role in shaping buyer preferences. The creation of value based on the way the resources are delivered to buyers (rather than on the presence itself of resources) is of great importance to the provision of tourism services in the experience-based economic model, especially when it comes to rural tourism (Frochot & Batat, 2013). Complexity and multidimensionality are the key characteristics of the tourism product. In line with the "economy of experiences" concept, each tourism product, service or bundle should be enhanced with some additional benefits in the form of emotions, feelings, experiences, and satisfaction generated in the tourists' minds before, during and after the trip. Polish examples of such products include sensory gardens, i.e. gardens that encourage the stimulation of different senses (Dudkiewicz et al., 2014), and therapeutic gardens (Latkowska & Miernik, 2012). The purpose of the first type of gardens is to make the visitor open and sensitive to colors and smells. The general concept is that sight should be dominated by other senses (e.g. an evening walk around the park, at a time when the eyes fail in the darkness while other senses become stronger). The human body must rely on hearing and smell, and this is exactly what the garden is supposed to do: to stimulate the perception of surroundings. The gardens should be designed so as to activate all senses: smell, sight, touch, hearing, and taste. While they are established to address the needs of blind and disabled people, they are also intended for children, youth, and adults. Polish examples include the Botanical Garden in Kielce, the Muszyna garden near Krynica, and the "Ostoja" Sensory Garden in Wrocław.[11] The second type are therapeutic gardens, defined by Eckerling (1996, p. 23) as

> space designed primarily to make people feel better. The role of such a garden is to provide the users with a sense of security, comfort, and relaxation. Therapeutic gardens are established near healthcare facilities (hospitals, rehabilitation centers, hospices) and social institutions (nursing homes, retirement homes).

A specific group of therapeutic garden users are sick people who stay in a hospital, care home or care farm. Gardens are among the key components of what is referred to as the therapeutic environment. In Poland, such gardens are mostly established in the vicinity of sanatoriums, rehabilitation centers, and hospitals.

Tourism practices which: are based on natural values; take care of sustainable development at regional level; are enhanced with educative aspects and adequate infrastructure that regulates tourist traffic (at a reasonable level adequate to local values), especially in naturally valuable areas, become increasingly popular, year by year, both in Poland and around the world (WTTC, 2019). Traveling to places which offer attractive landscape and natural assets and allow visitors to watch unique species of fauna and flora is on an upward trend; this should be leveraged by rural tourism undertakings, also in a context of local, regional, and networking projects that enable creating and building educational trails, information and promotion components and similar resources focused on addressing tourist needs. Today's rural tourism sector absolutely needs, and will need in the near future, to focus its efforts on building partnerships and networks between stakeholders. This will result in delivering innovative products and an integrated, comprehensive offer for tourists that will provide them with emotions, feelings, and experiences. Hence, it will be necessary to support integrated products and local brands of key regional and supra-regional importance as they will become a perfect showcase for a town or region.

5.3 New directions in rural tourism

The new millennium witnesses changes taking place in all areas of life at an unprecedented rate. Globalization and human progress experienced in recent years all around the globe are driven by the growing role of science, technology, and education in economic processes. In the 21st century, economic growth is mainly fueled by knowledge and related innovativeness which is value added in the tourism offer. The international literature (Hall et al., 2017; Butler, 2014; Tranchenko, 2015; Seken et al., 2019; WTTC, 2019; UNWTO, 2019; OECD, 2018a) emphasizes that tourism is an extremely complex social process. It is an element and function of cultures, and a reflection of cultural change. Usually, it is associated with travel, recreation, sports, and experiencing nature. However, what is a form of leisure for some is an important revenue stream for others. One of the forms of tourism is rural tourism (including agritourism), a true alternative which enables the creation of new jobs and revenue streams while improving the standards and ways of rural living (Roman, 2015a; for more information, see Section 3.1). Importantly, it also helps promoting a region and preserving historical traditions and customs (Bogusz & Kmita-Dziasek, 2015 and Section 4.4). The current policy of the European Union is focused on supporting projects designed to combine farming with rural tourism by promoting sustainable and responsible rural tourism (for more information, see Section 3.6). According to numerous studies (Hall et al., 2017; Bernard, 2012; Moric, 2013), rural tourism is a developing sector which offers interesting, innovative solutions and development options resulting from its capacity to respond to some of the emerging trends that affect tourism demand (Belletti, 2010).

Unlike urban tourism, rural tourism addresses the changes in the level and scope of demand. Today's tourists expect security and freedom which, in the context of recent events related to COVID-19, will be a priority aspect when choosing a town or accommodation facility.

The new dimension of "rural tourism" requires that—in addition to their basic function, which is to produce food—the farms be specialized while ensuring adequate access to leisure and education by providing an appropriate pool of experiences and emotions and caring for the environment (Bernard, 2012; Mastronardi et al., 2015). The European Union's 2014–2020 Rural Development reports (European Commission, 2017)[12] emphasize that countries such as France, Italy, Austria, and Switzerland pioneered that trend, and now have a broad and rich portfolio of rural tourism products and services. In the future, an important trend will consist in promoting agritourism farms, eco-agritourism farms, and other rural facilities. By nature, rural tourism prompts people to cooperate and embrace shared projects and investments. It is a major driver of value added across a whole series of actions, including social, economic, and cultural measures, taken in rural areas where tourism is developed in a sustainable manner. Compared to other tourist destinations, the essential competitive advantage of rural areas is a harmonious combination of culture and nature and their mutual interrelation caused by historical events which create a mix of attractive tourism products. Today's image of rural tourism is often equated with organic products, fresh air, healthy foods, and contact with people (Zrakić et al., 2017; Kosmaczewska, 2009; Poczta, 2013).

> The potential for rural tourism development is determined, on the one hand, by unused farm resources and rich natural and cultural resources of rural areas and, on the other, by the growing attractiveness of environmental leisure away from urban rush, the commitment to be active when on holiday and the need to reduce accommodation costs.
>
> (Karnafel-Wyka, 2011; p. 4)

The forms of rural tourism include agritourism. In the multipurpose and sustainable rural development model, which has been put in place in the last couple of years, an important role is played by a tourism activity which enables tapping into the endogenous potential of rural areas, including their ample natural, cultural, and human resources (Guzal-Dec et al., 2015). The natural environment is part of the cultural and economic heritage. Opening rural areas to tourism is equivalent to making them accessible to tourists (Ilbery & Saxena, 2011). The dynamic development of life, large urban agglomerations and metropolises, and environmental pollution resulted in changing peoples' lifestyles and made them want to escape the urban noise and turmoil. For the urban population, having a farm holiday is an attractive form of leisure. Getting to know the lives, customs, and day-to-day routines of rural dwellers, and being in direct contact with nature and rich regional resources allows the tourists to truly enjoy their holiday away from modern-day nuisances. Today's agritourism should be part of a knowledge- and inno-vation-based economy (Machnik, 2010; Perechuda & Hołodnik, 2012). Farm owners should first know the local customs, attractions, and traditions, and be able to use them in developing a professional offer. When implemented properly, the tourism development strategy can contribute to increasing the number of tourists using different kinds of services. Both the Polish and international literature has addressed the specialization of the agritourism farms' offer as a *sine qua non* for their survival in the agritourism market (Zawadka, 2010; Barbu, 2013; Keane, 2013). Agritourism farms can embark on many different specialization paths, such as greening. Should they do so, the activity of eco-agritourism farms can become a perfect rural tourism attraction. As a consequence, both farms themselves and entire rural regions where they are located can become attractive to tourists because they would encourage visits from naturalists, ornithologists, entomologists, photo-graphers, and people who look for healthy, unprocessed food.

Nowadays, providers can be observed to differentiate and enhance their offers with innovative solutions. Undoubtedly, this is the consequence of addressing the growing and emerging expectations, desires, and needs of tourists. The attractiveness of rural tourism largely depends on innovative solutions. Innovations in rural tourism mean creating a new tourism product from the scratch as well as establishing a professional marketing environment for existing local natural and cultural assets (Krzyżanowska, 2014; Majewski, 2015). Innovative solutions also mean improving, enhancing, and differ-entiating previously offered products and services, e.g. organizing handicraft workshops to make the tourists' stay more attractive (Sikorska-Wolak & Zawadka, 2015). Initially, this process was related to farm specialization (an offer designed for disabled people, families with children, angling fans or mushroom pickers). Particularly valuable innovative solutions are those cre-ated on the initiative, and with the involvement, of the local community and many local and other operators (Sikorska-Wolak et al., 2014; Klimek, 2010), including: entering into cooperation and developing integrated rural tourism products; establishing networks of specialized tourism service farms; estab-lishing tourism clusters, educational farms, thematic villages, tourism bundles, organic farms, etc. There are numerous examples of rural tourism specializa-tion, and therefore this chapter will only present those which enjoy the greatest popularity in Poland.

Educational farms are the first example of specialization and innovation in tourism. The concept behind them is mostly about promoting popular culture and tradition, making the visitors familiar with regional food, and fostering the widespread adoption of farm education projects. These efforts are mostly undertaken in rural areas by the local population. Currently in Poland, there are 259 active educational farms which together offer a total of 867 educa-tional bundles within 12 thematic areas.[13] In Poland, in order to be recog-nized as an educational establishment, the farm should be a member of the

Educational Farms Network and pursue no less than two educational goals from among those listed below:

- vegetable production education;
- animal production education;
- education on the processing of agricultural products;
- environmental and consumer awareness education;
- education on the material culture of rural areas, traditional professions, handicraft, and folk art.

As another important criterion, the farm should keep livestock or cultivate crops which are presented to groups of children and youth received under school programs or used as a tourist attraction for families with children or single adults (Bogusz & Wojcieszak, 2018; Bogusz & Kmita-Dziasek, 2015). The next interesting example of innovation in rural tourism is thematic villages. That concept emerged in Europe after 1990 as part of programs such as the Community Initiative EQUAL, LEADER or the Rural Renewal Program. The first thematic villages were established in Lower Austria.[14] In Poland, the first initiative to create a thematic village was taken in 1998 (Zawadka, 2010). Thematic villages, i.e. villages with a leitmotiv, are an innovative method for reviving the rural economy. What the thematic villages offer is usually based on existing natural, cultural, and historical resources (for more information, see Section 4.4). Due to changing lifestyles and emphasis being placed on ecology, staying in an organic farm could be an attractive, emotionally loaded experience for the tourists. Modern consumers pay attention to healthcare which is related to environmentally friendly behaviors. The growing ecological awareness gives rise to a new "eco-tourist" who wants to stay in an organic farm and buy ecological products and services while taking care of natural environment. An active, modern, highly environmentally aware tourist could encourage environmentally friendly behaviors and make the operators (tourism farms) realize the need for developing an offer that meets environmental protection requirements (Niezgoda & Markiewicz, 2011). Eco-tourists become increasingly aware of sustainable development and environmental protection principles. As a consequence, they are more willing to spend their holiday at an eco-farm, and increasingly often choose locations where local residents are open to visitors. International authors (Vrsaljko et al., 2017; Cigale et al., 2013; Horlings & Marsden, 2014) also point that due to growing interest in natural environment and tourism, these two components can be combined into a new product, eco-agritourism, which is increasingly identified in the worldwide literature. Eco-agritourism promotes biodiversity and encourages the promotion and sales of organic products while also protecting ecosystem diversity and rural areas together with their architecture and culture. This is certainly a new trend which is willingly followed by Italians, Austrians, and French and Polish people. Eco-agritourism provides an opportunity to offer permanent employment to residents of less favored areas, such as isles, mountains, and foothills. Another interesting rural tourism trend is referred to as WWOOF.[15] This is a world-wide movement linking volunteers with organic farmers and growers to promote cultural and educational experiences. Based on a barter process, it helps build a sustainable global community. WWOOF farmers invite tourists to work (4 to 6 hours a day) in return for food, lodging, and the experience of sharing their day-to-day lives with people who live and work at the farm. Both short- and longer-term stays are possible.[16] The Polish WWOOF organization comprises ca. 30 organic farms, including the Giże resort (the most popular one), the Amanda farm (the most specialized one) and the Indianki[17] farm (the one in greatest need of volunteers, i.e. WWOOFers). Tourism becomes a more and more personal experience and consumers are increasingly aware of sustainable development principles. This is also reflected in the development of what is referred to as slow tourism: a movement that largely enables deep, authentic relationships with other people, culture, places, food, heritage, customs, and the natural environment. The concept of slow tourism is consistent with assumptions behind the gap year,[18] which is often believed to be a continuation of the Grand Tour.[19] Many authors emphasize that slow tourism is part of rural tourism (Gralak, 2012) as it is based on avoiding noise and modern-day rush. In rural areas, slow tourism ensures a better satisfaction of tourists' needs related to their sense of individuality, feelings, emotions, an in-depth exploration of places they visit, and establishing durable relationships with the local community (Fullagar et al., 2012; Lowry & Missoon 2016). Another innovation in rural tourism is network products: geographically dispersed tourism attractions and services which share a common coherent concept and a leading differentiator (value or service). Network products require cooperation from local producers, service providers, and agritourism farm owners. Examples include the Noble Bowl of Mazovia, which is part of the "Mazovian Folklore and Taste Trails." Agritourism farms willingly join such projects because it allows them to extend their product and service portfolio which will undoubtedly attract potential agritourists. Tourism clusters are another example (Moric, 2013; Roman, 2009), with rural tourism micro-clusters being an interesting concept. The micro-cluster model is highly popular in less developed areas, and can be used, for instance, by countries such as Montenegro, Moldova, Poland (e.g. the "Equestrian Tourism" Łódź cluster, "Wisłoka Valley Tourism Cluster," the North-East Innovative Tourism Cluster "Europe's Crystal," the Northern Greater Poland Tourism Cluster "Noteć Valley"), Slovakia or Lithuania (Moric, 2013; Sikora, 2010). Whether agritourism farms can improve their competitiveness depends on aspects such as mutual cooperation within cluster structures. Thanks to joint efforts, clustered agritourism farms can be competitive in the market and, importantly, can stimulate innovativeness. The development of a food cluster (for instance) can bring benefits to rural communities and farmers by attracting tourists who will spend money, by increasing the awareness of local identity and image, and by promoting local products.[20]. Another interesting trend is the establishment of "rural tourism centers," a project which consists in using the region's own brand in combination with such trends as the economy of experiences or slow tourism. For rural areas capable of creating their own brand, it is particularly interesting to tap into the culinary assets of tourism centers. Tourists increasingly often rely on web pages, smartphone applications, and social media to find a surprising (or sometimes shocking) place and taste the regional products. Therefore, it is important that farmers and rural producers make skillful use of tools to attract consumers or boost online sales (through Facebook, Pinterest, etc.). Examples of such tourism centers include the project implemented in Estonia under the LEADER initiative. It has many stakeholders based in the south-eastern part of the country. Inspired by a program deployed in Groningen (an eastern province of the Netherlands) in cooperation with the National Geographic, the project consisted in installing 21 yellow rectangles in the territories of six collaborative local action groups. The rectangles were placed strategically to capture the regional particularities and attract attention from tourists, making them feel like they were looking through an open window. Of the 125 locations proposed by local communities, project partners picked 21 places of historical, cultural, and natural value in south Estonia. Signage with local information and presentation of small enterprises[21] is available along the route.

Another trend which received recognition from tourists is the culinary trail put in place by the Scottish government. The purpose of that initiative was to familiarize the tourists with how chocolate is produced. The trail runs between Scottish artisanal chocolate manufacturers.[22] They discovered the opportunity to collaborate in order to encourage tourists to visit their facilities, often located in remote rural areas, and developed accompanying presentations, trips, and classes. In turn, in Slovakia, agritourists enjoy spending their free time in apiculture and mini-spa farms. These are the two key trends which have become popular among families with children and the elderly. A new concept which has been developed both in European Union countries and around the world is an offer dedicated to seniors, i.e. silver agritourism. Usually, the beneficiaries of that concept are Germans, Italians, and French people. Essentially, it consists in that agritourism farms dedicate and offer specific products and services to this customer group. As an important aspect, they provide a quiet and calm experience which is certainly what the elderly look for (Roman, 2015b; Ciani, 2012; Lee et al., 2015, Brandth & Haugen, 2010; Wojciechowska, 2009, Kłoczko-Gajewska, 2013).

Rural tourism allows tapping into rural and agricultural resources and making use of the available infrastructure. The transformation of rural landscape also involves the reconstruction, adaptation, and maintenance of residential facilities and of regional architecture, which means preserving the local characteristics, customs, and folklore. However, preserving the region's cultural identity does not mean isolating the rural community from the civilization. The innovative nature of a product, service or bundle could be an advantage that fosters interest from tourism. The development of the tourism industry should be fueled by consistent measures taken to adjust an offer to what the tourists expect, attract new target groups and leverage state-of-the-art technologies. In the context of rural tourism, innovativeness means introducing extensive modifications to products offered while creating new and implementing improved solutions for customer support and quality management. Tourism innovations include developing an original tourism product, such as a location (area, facility, trail), a professional marketing environment associated with natural and cultural assets (e.g. festivals, cultural events), and services or service bundles (e.g. organized canoeing or trekking trips). Tourists mostly want to experience the natural landscape and stay in a genuine rural setting. They want their experience to be filled with emotions and unforgettable impressions. In that context, the quality of products or services offered is an authentic experience which tourists look for.

5.4 Social farming in the promotion of social/ecological sustainability in rural areas

Sustainable development is a topic addressed in international literature by representatives of the scientific community (Senni, 2013; Hassink et al., 2014; Ferwerda-van Zonneveld et al., 2012; Leck et al., 2014; González et al., 2014; Czyżewski & Staniszewski, 2018; Wilkin, 2010; Rosner, 2012; Pawlak & Poczta, 2010; Wojcieszak & Wojcieszak, 2018). However, no unequivocal definition exists for sustainable development (Woś, 1998; Wilkin, 2011; Kłodziński, 2008; Matuszczak, 2013; Rosner, 2012; Rudnicki, 2010; Stanny, 2013). Conversely, a broad range of definitions exist which share some common features; accordingly, sustainable rural development means an appropriate use of natural resources in a way that enables their self-renewal, growth in agricultural output (attained through improvements in input productivity rather than an increased consumption of inputs), making farming less vulnerable to fluctuations, and ensuring symbiosis between agricultural goals (Czyżewski & Staniszewski, 2018; Wilkin, 2011; Stanny, 2011; Kołodziejczak, 2010; Halfacree, 2009). In view of the above, it can be concluded that sustainable development is viewed as a path of economic development together with the associated social development. It allows the preservation, or even restitution, of the natural environment and ensures the absence (or a considerable reduction) of irreversible adverse phenomena and consequences affecting the environment. At the same time, it enables the exploitation of natural resources, implementation of investments, and creation of modern, innovative technologies designed to maximize the economic, natural, and social resources which provide a basis for addressing the current and future needs of humanity.

Today, the sustainable development concept is the focal point for social and political discussions in most highly developed countries, in many developing countries and in international social and political organizations (Wesołowska, 2011; Woods, 2009; Wójcik, 2012; Zegar, 2012; Tittonell, 2014). The multifunctional nature of rural areas is also related to the development of many functions, including non-agricultural ones (Rizov, 2005). That concept is underpinned by the assumption that agriculture, as an economic activity, can provide the communities with multiple market and non-market benefits in addition to its basic function, which is to deliver food, resources, etc. (Wilkin, 2010). The international literature often views sustainable development in a context of building a sustainable, competitive economy which makes an efficient use of its resources and is oriented towards leveraging innovative processes and technologies (Pretty et al., 2011; Stoate et al., 2001).

As emphasized by a wide group of experts, rural areas become less and less involved in farming; that process is reflected in different aspects of rural living (Turþeková et al., 2015; Wilkin, 2010; Woś, 1998; Halamska, 2011; Czarnecki, 2011; Czapiewski, 2010; Stanny, 2011; Scuderi et al., 2014; Pretty et al., 2011; Król & Stępnik, 2018; Chmielewska, 2018; Bielińska et al., 2014; Karwat-Woźniak, 2013; Garnett et al., 2013). More and more often, the sustainable development concept is considered in its social context (Karwat-Woźniak, 2013). Examples of such projects include the creation of new revenue streams (jobs) for the farming and non-farming populations which cannot be employed on a full-time basis in a farm (Roman & Wojcieszak, 2018). The development of any initiatives between the rural population, farmers, and entrepreneurs, as well as the social integration of less favored social groups are complex challenges faced both at national and international levels in Europe and elsewhere around the world (Roman & Wojcieszak, 2018). Strategic EU documents indicate that agriculture, in addition to its traditional function (which is to manufacture marketable goods), is vested with new missions, which particularly include environmental protection and the social function (Figure 5.2) (Lanfranchi et al., 2015).

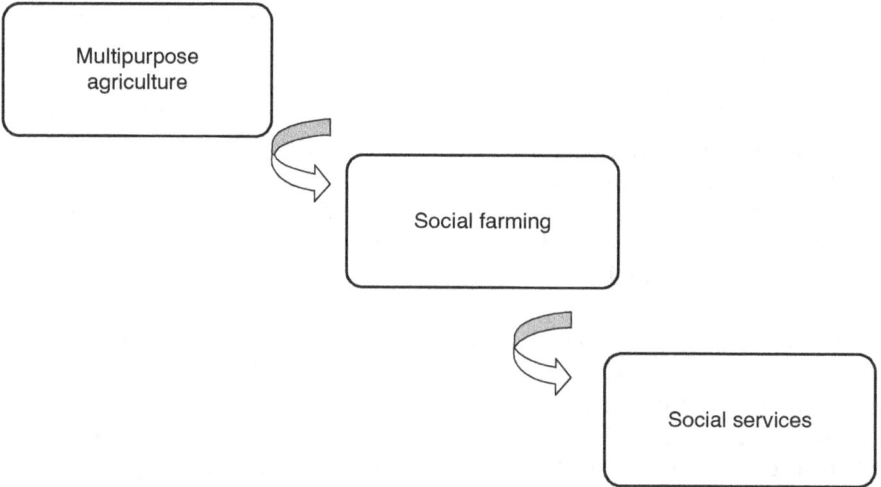

Figure 5.2 Position of social farming.
Source: Król (2017, p. 5).

Social farming is not a new concept, and is backed up by a long tradition. Already in the Middle Ages, prisons and monasteries used the natural environment for therapeutic purposes (Elings & Hassink, 2006). In Europe, the beginnings of social farming are estimated to date back to the second half of the 19th century. At that time, farms and institutions were established in rural and remote areas to take care of intellectually disabled people (Di Iacovo, 2014). Each country had a different reason to develop social farming. In Italy, it was driven by the mass closure of psychiatric hospitals; in Germany, by the closure of institutions in charge of supporting marginalized people; in the Netherlands and Ireland, by religious movements and communities. The international literature divides the development of social farming into four different periods (Figure 5.3), each being different in terms of social awareness, sectoral interests, etc.

In the first stage, only a few examples of social farming and voluntary action based on strong motivation can be cited. The first models of care farms were created mostly on the initiative of private agricultural holdings. That pioneering stage was experienced in Austria, Finland, and Sweden (Chmielewski et al., 2017). The next stage was multipurpose farming, with a growing role of social farming which started to be financed with agriculture and rural development funds. Examples include the UK and Italy (Subocz, 2019). The third stage consisted in recognizing agriculture as being part of the social assistance system (Bombach et al., 2015), and is witnessed in countries such as Germany and Ireland. The last development stage is based on the assumption that social farming is viewed as an integration model (Dessein & Bock, 2010) and is part of the agricultural sector and of the social assistance system. The Netherlands and France are examples of countries where the integration model is already in place (Subocz, 2019).

While social farming measures taken by European countries are similar, many differences exist between them due to such aspects as historical events,

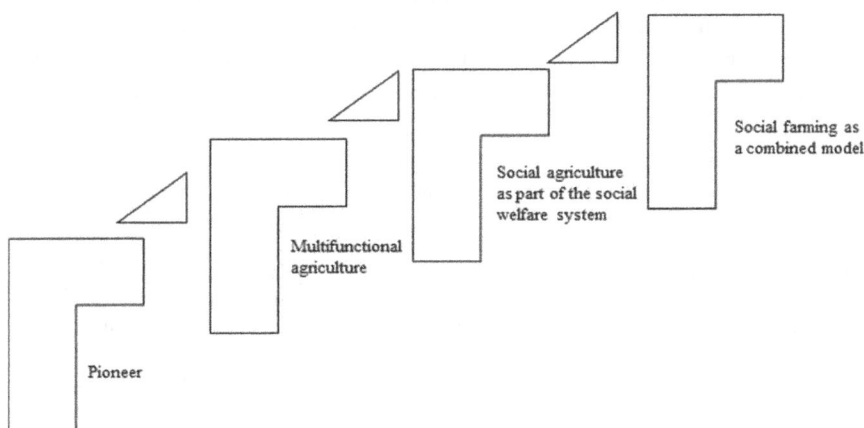

Figure 5.3 Development stages of social farming.
Source: Lanfranchi et al. (2015).

approaches, and targets set for the activities. Germany, Ireland, France, and Slovenia follow an institutional approach with a prevailing role of public institutions. A private approach based on therapeutic farms is popular in the Flemish part of Belgium and in the Netherlands. In turn, a mixed approach based on social cooperatives and private farms is widespread in Italy (Subocz, 2019). Interestingly, social farming can be mostly associated with the care and social sectors (as is the case in Italy and France) or with the healthcare sector (in the Netherlands), or be located somewhere between the social and the healthcare sector (in Germany, UK, and Ireland) (Opinia... [Opinion...], 2013). Social farming has attained the highest development level in the Netherlands (Hine et al., 2008 after: Bombach et al., 2015). The Dutch social farming model is an example of good practices; indeed, of all the European countries, the Netherlands has the fastest-growing population of farms offering social services because several ministries (the Ministry of Agriculture, Nature and Food Quality; the Ministry of Health; and the Ministry of Sports) financially support such activities (Subocz, 2019).

In Poland, social farming is at a pioneering stage of development (Opinia..., 2013; Subocz, 2019; Krzyżanowska, 2014; Roman & Wojcieszak, 2018). This is mostly because of the lack of an appropriate public policy and of financial support for such activities in rural areas. Bombach et al. (2015) emphasize that all initiatives are taken on a bottom-up basis as a private activity of farmers financed with private funds available under NGO schemes. However, the development of social farming has a number of positive impacts which can be viewed from three perspectives: that of the guest, of the farmer, and of the local community (Table 5.2).

The relevant literature does not provide a single definition of social farming because it is a multidimensional phenomenon spanning over a broad range of different practices in countries around the world. Research carried out by Di Iacovo (2014) clearly found that an urgent—if not critical—need exists to develop knowledge on social farming as a significant component of rural multifunctionality. He believes social farming means rural development which takes social and environmental aspects into consideration as per the principles of

Table 5.2 Benefits from the development of social farming in Poland

For the guest	For the farmer	For the local community
Improved wellbeing; becoming more active and engaged; delivery of new services	New revenue streams or increase in revenue; self-fulfillment; expansion/diversification of activity; promoting a new image of agriculture	Increased availability of social/healthcare services; local development; social activation; making rural areas more attractive; fighting against social exclusion

Source: Own elaboration based on Król (2019).

sustainable development. Therefore, it is important to put the social farming concept in economic practice as a response to the disadvantageous demographic situation in Europe. According to the European Economic and Social Committee, social farming is "an innovative activity that combines two concepts. The first one is about multipurpose agriculture whereas the second one refers to social services in the area of healthcare at local and national level" (Opinia..., 2013). Social farming is also referred to as "farming for health," "care farming," "green care" or "green therapies" (Opinia..., 2013). These terms refer to social reintegration or care activities which can contribute to improving the health status and to social inclusion of disadvantaged people. As Di Iacovo (2014) emphasize, social farming is both a traditional and innovative approach which relies on vegetable and animal resources to promote (or generate) therapy, rehabilitation, social integration, education, and social services in rural areas. However, this is strictly related to agricultural activities where (small) groups of people can stay and work together with native farmers (Di Iacovo, O'Connor, 2009). Hence, social farming spans over all rural farms, agritourism farms, and agricultural holdings whose basic activity consists in integrating physically, mentally or emotionally disabled people (Roman and Wojcieszak, 2018) and caring for the elderly, children, and youth. In the relevant literature, many authors (Dessein & Bock, 2010; Farming for Health, 2018, Scuderi et al., 2014; FAO, 2014; Di Iacovo et al., 2014; Elings & Hassink, 2006; European Network for Rural Development 2010) indicate that the delivery of social farming services brings numerous benefits to local operators, stakeholders, farm owners, and the local community. Social farming also means aspects related to rural development, such as:

- an appropriate organization of local services;
- farmers' reputation and evolution of farmers' attitudes in their relationships with local communities;
- reorganization of the local economy and introduction of new components of solidarity and mutuality (Dessein & Bock, 2010).

Today, social farming has many common characteristics across Europe. First, there is a strict relationship with the traditional activity of rural economy, i.e. social farming takes place in farms (agricultural holdings, agritourism farms, rural farms, etc.). Second, social farming demonstrates great flexibility and a multifunctional nature which means the scope of services differs between the countries and depends on traditions, methods and guidelines put in place (Lanfranchi et al., 2015; Elings, 2012). Currently, there are four activity sectors in social farming:

a   rehabilitation and individual and group therapy;
b   professional and social integration;
c   education;
d   care services for the elderly and children (Dessein and Bock, 2010).

It seems that care farms provide a development opportunity for rural areas in Europe, and especially in Poland. They enhance the offer of agricultural holdings and agritourism farms with new functions. Offering a contact with nature, combined with the values and cultural resources of a rural environment, is the starting point for creating high-quality services which will truly affect the users' quality of living. The economic aspect is important, too, because care farms will support the development of local economies and will provide the farmers with an opportunity to diversify and increase their incomes (Matysiak & Michalska, 2016). Thus, a care farm is an attractive form of a tourist product which consists in combining farming and agritourism activities with taking care of those in need for support, mainly including the elderly (Brandth & Haugen, 2010). Service providers who run a farming or agritourism activity are in a position to deliver diversified care services (Di Iacovo et al., 2016). It all depends on who is the potential customer because the offer can be targeted at specific groups, e.g. physically or mentally disabled people; reintegration of addicted people (e.g. drugs, alcohol); the elderly; reintegration of ex-convicts; problematic youth; people with brain damage due to accidents or diseases such as dementia (Alzheimer's disease); autistic people; child care (before and after school hours); people with occupational burnout; long-term unemployed; asylum seekers. Usually, care farm owners dedicate their services to one or several target groups. Care service buyers may be provided with a temporary or permanent on-farm accommodation. Both short and long stays are possible (Table 5.3).

Such services are often accessed by single persons (who can play team sports or card games; embroider; pick mushrooms; prepare meals for the whole community; watch TV; hike; discuss their problems). These advancements in rural tourism can be considered an innovation (Toby, 2014). Another important aspect is the approach of the farm owner who should be an open person eager to help.

In recent years, Poland has witnessed a growing number of social farming initiatives. Social farming activities span over a large number of target groups based both in rural and urban areas (the elderly, youth, migrants, the disabled, convicts, etc.). As noted earlier, Poland is at a pioneering stage of social farming services. One of the first measures consistent with that concept is the establishment of care farms. In Poland, the relevant procedure was developed

Table 5.3 Forms of service offered by care farms

Specification	Without on-farm accommodation	With on-farm accommodation
Temporary stay	Short stay, day care	Short stay, 24h care
Permanent stay	Long stay, day care	Long stay, 24h care

Source: Manintveld (2014).

by the Krakow Branch Office of the Brwinów Agricultural Consultancy Center, one of the major institutions in charge of bottom-up actions. The concept was the outcome of implementing the project "Care farms: building a collaborative network" co-financed by the National Rural Network under the 2014–2020 Rural Development Program.[23] The objective of the project was to

> promote, foster the widespread adoption and support the development of the social farming concept, especially including the establishment of what is referred to as care farms, as a way to diversify rural incomes and as an enabler of sustainable rural development, multipurpose agriculture and activation of the rural population.
>
> (Dąbrowski et al., 2017; pp. 9–10)

In Poland, only 30 care farms are active. This is a new, innovative idea which so far has been willingly used by agritourism farms. In 2017, a pilot project was implemented in Poland by the Kuyavia–Pomerania Agricultural Consultancy Center based in Minikowo and by the "Tuchola Forest" Local Action Group. The European project was co-financed with PLN 2,937,855 from the European Fund "Green care: care farms in the Kujawsko-Pomorskie voivodeship." The participants were 15 farm owners and their families from five districts located in one of the Polish 16 voivodeships. The project consisted in that the farms were supposed to provide day care services (8 hours per day) to 75 elderly people. The objective was to increase the availability of care and activation services for dependent persons offered in rural day centers and to improve care service competencies (Wojcieszak & Wojcieszak, 2018). Currently, another project ("Care farm 2018–2020") is being implemented with the participation of 15 more farms. Just as in the previous edition, the guests will be offered care services 5 days a week, 8 hours a day for 6 months in groups of four to eight persons (a total of 225 people will access care and activation services).[24]

In Poland, the current legal framework restricts the ability to combine farming activities with care services; this is mostly due to the absence of appropriate legal regulations. Care farms can be operated as an economic activity in the form of a social cooperative, association or foundation. This makes them eligible for co-financing from various sources, e.g. social economy support funds, and central and local government funds allocated to the establishment and development of care facilities (Dąbrowski et al., 2017). Polish creators of the care farm concept (i.e. the Krakow Branch Office of the Brwinów Agricultural Consultancy Center) propose that services offered by care farms be mostly targeted at economically inactive or dependent elderly (60+) people (Dąbrowski et al., 2017; Chmielewski et al., 2017; Subocz, 2019). They also believe that Polish care farms should be organized in a manner which is characteristic of the social assistance system, i.e. be operated as: day care centers, family assistance centers or sheltered institutions (Dąbrowski et al., 2017; Chmielewski et al., 2017). Polish care farms are regularly inspected by psychologists to make sure the guests are provided with an adequate service level and their needs are addressed (based on what is referred to as a community interview). There is also organized help from the volunteers who, just like farms owners who decide to provide these services, must attend dedicated training sessions.

5.4.1 Examples of care farms

Germany, a country with a high level of care services, saw the establishment of care farms in the 1800s (Chmielewski et al., 2017). Initially, their mission was to care for prisoners and people excluded from society. Most of them were located in rural areas and hence could rely on local farmland to address their own needs. Their core activity consisted in activating their guests by engaging them in day-to-day operations of the facility. Three types of social farming emerged in Germany in line with the development of care activities (Chmielewski et al., 2017):

- sheltered workshops, whose main goal was to ensure jobs for the disabled;
- educational farms, whose mission was to disseminate farming and food processing knowledge among children and youth;
- private farms who offered care services for different target groups.

Another care farm, Meves-Hof, is located in north Germany, 80 km away from Hamburg, in a region with a strongly developed agriculture. In addition to offering care services, the Maves family keep suckler cows with calves for fattening. Their farm has been active since late 1800s, and started to offer care services in 2013. Four persons are dedicated to this activity. The farm can accommodate disabled and able-bodied children, dementia sufferers and people with mental health problems. Care services are provided to children aged 5 to 12, men and women affected by dementia aged 65 to 100, and men and women with mental disorders aged 20 to 60. The guests are offered with many classes, including handcrafting with natural materials, processing agricultural products (fruits, vegetables, cereals), special games for dementia sufferers, natural experiments, and stroking and feeding animals. How often social farm guests attend the classes depends on the target group:

- children visit the farm once in a month,
- dementia sufferers visit the farm once in a month or once a week,
- persons with mental disorders visit the farm once in a month.

The UK has over 250 care farms, including more than 200 in England, eight in Wales, 20 in Scotland and 15 in Northern Ireland (Chmielewski et al., 2017). Examples include Butterlope Farm, located 2 miles from Plumbridge, a village in the Glenelly valley (Butterlope Farm, 2020). It has 180 acres of meadows and upland areas. The owners have 350 sheep, 15 hens and two shepherd dogs. They also have a kitchen garden, and a plastic film hothouse where they cultivate diverse vegetables. The owners were trained under the INTERREG IVA Social Farming Across Borders (SoFAB) project. They started their farming activity in 2006. Established in 2015, their social farm provides care services to men and women with learning difficulties and people recovering from mental disorders. The participants assist in day-to-day activities, e.g. caring for animals, cultivating fruits and vegetables, preparing meals, carrying out dry-stone masonry work or doing some minor cleaning. The guests can obtain accredited qualifications during farm internships (Chmielewski et al., 2017).

A good example of a Polish care farm is the establishment run by Małgorzata and Grzegorz Oparka in Lubiewice. They learned of the opportunity to set up a care farm from a local paper and web pages of the Agricultural Consultancy Center in Minikowo. Farm owners accessed a pilot project implemented in the Kujawsko-Pomorskie voivodeship by two institutions, namely the Toruń Marshall's Office (in charge of providing financing under the project) and the Agricultural Consultancy Center in Minikowo (in charge of substantive and practical preparation of farm owners). The first step was a qualification survey designed to retrieve information on the owners and the facility where care services were supposed to be provided. For the applicants, the next stage was the completion of a training course for carers of the elderly and disabled. Upon completing the training session and meeting the required formalities, each accredited farm was provided with funds from the Toruń Marshall's Office (a minimum wage of PLN 2250 per month + earnings from rental of rooms). In accordance with relevant guidelines, the farm was required to provide one day-care room, one relaxation room, access to the kitchen, a bathroom, and a toilet. In turn, substantive assistance was provided by the Agricultural Consultancy Center in Minikowo. The first visitors completed their session in January 2017, and the last group left in June 2018. A total of three care programs were delivered:

- three groups of five people (each enrolled for a 6-month program);
- mixed groups;
- groups where the youngest and the oldest participants were 62 and 88 years old, respectively.

As part of the program, visitors could participate in painting, bakery, and other workshops. Also, they prepared jam, made embroidery, and painted Christmas tree decorations together. They demonstrated great interest in handicraft and decoupage workshops, manufacturing cosmetics, and making felt shopping bags.

Another example of a care farm is the establishment run by Iwona Cybulska in Stary Sumin. Just as described above, the owner took part in the pilot program "Green care." She obtained the same remuneration and met the requirements for the number of rooms and access to relevant premises.

The owner has a 15-hectare vegetable farm. Just like at the farm of Mr. and Mrs. Oparka, the first guests of Ms. Cybulska started their program in January 2017, and the last group ended their session in June 2018. Ms. Cybulska received a total of three groups of five people each (with each group staying 6 months at the farm). These were mixed groups: the first and third were composed of four women and one man; the second comprised five women. The youngest and the oldest guests were 62 and 88 years old, respectively. The main classes attended by the visitors were: embroidery and sewing workshops; preparing Christmas decorations; cooking and making jam together; and picking forest fruits and mushrooms.

In recent years, more and more operators interested in rural areas have noticed the importance of social farming; numerous examples of this activity can be found across EU-27 member countries. This is the consequence of an ever greater interest in, and understanding of, the role rural resources can play in improving the physical and mental condition of the population. At the same time, social farming is an opportunity for the farmers to provide new, innovative products and services, which means enhancing and diversifying their activity and multifunctional role in the society. Such an integration of agricultural and social activities can also ensure additional revenue streams for the farmers and could improve the image of agriculture.

While social farming has been developing for many years in Europe, its development level differs between the countries. In Poland, the social farming concept is at a pioneering stage, and so are the initiatives which have only started to be implemented. For instance, the first forms of care farms have emerged, and are mostly focused on seniors or sick people. However, it seems important to urgently lay a legal framework for their development and to secure funds for their functioning.

5.5 Technological changes and their impact on products and consumers

5.5.1 Using technology in promoting rural tourism products

According to a 2018 DESI[25] report, over 83% households in the European Union have access to broadband Internet. The lowest rates of digital exclusion are found in the Nordics, i.e. Finland, Iceland, Sweden, Estonia, Norway, and Denmark, whereas the highest rates are recorded in new Union members, i.e. Poland, Bulgaria, and Romania.[26] According to the DESI report on the progress made by European Union countries along the digitization path, as referred to above, Poland is ranked 23rd. This means there are spots in the Polish territory which either lack access to or have considerable difficulties in readily accessing a high-quality Internet link. Therefore, the European Union takes extensive measures to equalize Internet access within and between countries and regions. To do that, it develops strategies which include the implementation of actions (e.g. the report of Union telecommunication regulations, including the creation of investment incentives to build broadband Internet networks) intended to establish a single digital market which would create appropriate conditions for the provision of digital and innovative services (Rutka, 2017).

Today, changes can be observed which result from the intensive development of information technologies (Brych et al., 2013; Edelman, 2018). The ease, speed, and scale of information exchange between market players fuel the evolution of consumers' purchasing behaviors (Perenco & Rosa, 2011). Twenty-first-century consumers are educated and aware; without leaving their homes, they use mobile devices to find information on interesting offers (delivered to them based on their location), compare prices and read feedback (Bostrom & Klintman, 2011). Modern mobile technology creates unlimited possibilities for accessing goods and services. However, an important advantage consists in developing adequate customer relationships which have an ever greater impact on the way the consumers behave in the market. Consumer expectations for the market offer presented to them can change under the impact of different factors; these changes differ in pace and scope between product categories. Multidirectional changes in consumer behavior are the result of modern technologies (high-tech) defined as sophisticated technological solutions. Modern consumers are dynamic buyers strongly active in the virtual world; they share information in web forums and buy online, thus becoming increasingly aware, critical, and independent purchasers (Krzepicka, 2015). They rely on technology in their relentless quest for new experiences, feelings, and emotions. Increasingly often, they buy products not only to address their basic needs; instead, they want to express themselves and showcase their social status (or the fact of belonging to a specific group) in their environment (Dybka, 2013). Consumer purchasing behavior is a complex process composed of stages which are supposed to result in choosing and buying a product or service. The literature identifies five phases of that process, each of a different duration: realizing a need; looking for options; assessing the options; making the purchasing decision; impressions after buying (Rudnicki, 2012). State-of-the-art technologies contribute to how consumers behave when both buying and consuming (using) products and services. They include a broad range of IT tools, especially the Internet and mobile phones, and are among the factors that make the customer realize his/ her needs because they

> allow to access information on new products, foster the emergence of new needs (as the customer sees a new product in the Internet), affect the change in expectations regarding a product (through the impacts of advertising and information on social changes which correspond to consumer values and attitudes and are available in modern media).
>
> (Olejniczuk-Merta, 2010; p. 138)

The impact of state-of-the-art technologies is also noticeable in the market for tourism services because the global tourism traffic is integrally linked to tourism information, in a broad sense (ITUR).[27] Information is emitted and received by tourists, hotels, catering facilities, travel agencies, travel agents and intermediaries, agritourism farms, tour guides and leaders, commercial centers, rental companies, carriers, scientists, and educational centers.

The use of information and communication technologies (ICT) in choosing travel options and addressing tourism needs becomes widespread and includes more and more aspects of tourism consumption. When planning their trips, today's tourists increasingly use the capabilities offered by the Internet and modern digital technologies. ICT-driven transformations provide operators, agritourism farms, and customers (tourists) with a myriad of tools and capabilities they can use in playing their market roles. All of them are stimulated by information which enables continuous adjustment of organizational solutions to the potential of IT (Król, 2016; Peterson, 2015; Mohorovičić, 2013; Law et al., 2010). Indeed, ICT is an extensive and reliable source of information and knowledge delivered to providers of tourism offers. What also needs to be mentioned when considering the impact of ICT on the promotion of rural tourism products is the role and importance of marketing tools (Pawlicz, 2012; Bichler & Schmiderer, 2011). This is because tourism consumption is based on information, which means that consumers (tourists) demand great amounts of information throughout the consumption cycle (when preparing the travel, traveling, and coming back home). While rural tourism is nothing new, its development continues to face a number of obstacles, including poor promotion of rural areas and rural tourism operators, and lack of a coherent system for the booking and distribution of beds (Panasiuk, 2019). Tapping into the capabilities offered by the Internet and the global technological transformation could be decisive for gaining a competitive edge in the tourism market (Król, 2015; Król & Wojewodzic, 2006). Due to the rapid development of the Internet and to access to network services, tourism operators and agritourism farms active as service providers faced the need to rethink their business strategies related to formulating and implementing a business model based on new technologies and on what is referred to as network presence (Kemp, 2016). The use of the Internet in promoting, booking, and selling tourism services mirrors the global trends and has undergone profound structural changes over the last few years, especially as regards technology (Marek, 2015). The changes in how selected web services are provided are mostly the consequence of technical and technological development and of the philosophy of how the user is perceived. Having a website enables reaching millions of users in the global network. An adequately prepared graphical layout and reliable information on the activity contribute to building a professional image of accommodation providers and helps promote confidence in the market. Increasingly often, today's technological advancements prove to be an accelerator of development of mobile technologies which initiated the era of modern mobile communications. The ample literature identifies three types of applications for mobile devices: responsive sites (Responsive Web Design, RWD; Adaptive Web Design,

AWD), mobile sites (mobile web), hybrid apps, and native apps (Król, 2016). The first types of application for mobile devices are responsive sites (RWD, AWD). RWD is a technology of developing responsive pages which fit the resolution of the screen they displayed on (Król, 2016). As an advantage, they are highly ranked by search engines (currently, RWD is a standard supported and preferred by web search engines; RWD sites are ranked higher in search results than non-responsive pages or sites which use dedicated mobile pages). Their strengths also include coherence (responsive pages present the same content on all devices, whether fixed or mobile) and maintenance procedures. Indeed, responsive pages are cheaper to maintain because instead of two template sets (one for fixed computers, one for mobile devices) only one set is developed. The next type of mobile applications are mobile web pages, developed for mobile devices only. Increasingly often, mobile devices (mobile phones, tablets, etc.) prove to have a considerable share in generating the global Internet traffic. For several years now, changes can be observed in how data is catalogued (indexed) by Google. When implemented in web search engines, these changes force service providers to improve the technical design of their sites and the quality of the content they present. As the world's most popular search engine, Google builds the SERP (search engine results page) ranking, putting particular emphasis on original content. This means that accommodation providers dedicate particular efforts to care for their site's content, quality and visiting path. As Król and Bedla (2016) emphasize, the tourism industry makes often use of geo-IT techniques and tools. Moreover, geospatial services play a growing role in how information is transferred (Coleman et al., 2009; Lestari et al., 2014; Knight, 2011; Kemp, 2016; Buglass et al., 2017; Perenco & Rosa, 2011). The name and address of the tourist facility can be found on each agritourism farm website. This is spatial data which mostly refers to a specific location or geographic area (Kuna & Rzuciło, 2015). In Poland, nearly 33% of agritourism farms have a website.[28] According to research, such as the study carried out by Kosmaczewska in 2010, only 21% of agritourism farms in the Wielkopolskie voivodeship had their own website (Kosmaczewska, 2010). Another study, carried out by Dziechciarz in the Lubelskie voivodeship in 2010 (with a sample of 134 agritourism farms) confirmed that the main mission of websites was to provide information. He also indicated that the scope and quality of information available on the websites were not fully satisfactory (Dziechciarz, 2011). Similar conclusions were drawn by Krzyżanowska and Wojtkowski (2012) who found that most agritourism farm websites failed to meet the minimum criteria for site content and design, technical parameters and the offer itself. A 2015 study carried out with 300 websites (100 from each of the Warmińsko–Mazurskie, Małopolskie, and Podkarpackie voivodeships) found that the websites mostly had an information, marketing, and contact function. Only one website in five had a booking feature (Król, 2015). The most recent research by Gralak (2016) showed that most Polish agritourism farms published information on their activity in countrywide and regional agritourism portals, whereas only a small number (12.3%) had their own websites. Websites covered by this analysis were static and their sole function was to provide information. Compared to other forms of promotion, Polish agritourism farms make insufficient use of the Internet (Gralak, 2016).

To encourage tourists, service providers increasingly often rely on digital maps used as a component of different kinds of applications and websites which combine thematic content with an underlying map (Peterson, 2015). How fast all site components are loaded is an aspect of extreme importance for the tourists; this is the time the user spends waiting for access to the content. The site's performance largely determines how the site is perceived. Another interesting solution put in place is mobile applications (mobile software), i.e. software handled with a dedicated interface which enhances the functionality of mobile devices. It can be divided into corporate, location-specific, perishable, personal, transaction-oriented, and entertainment software, or into customized (dedicated, tailored) apps, standalone software (which can run offline), and mobile games. There are also applications preinstalled on mobile devices and applications downloaded by users themselves. Mobile apps have a tremendous impact on today's tourism consumption. First, they make the company more prestigious and change they way it is viewed. Second, for the operators, this is an easy way of communicating with the customer. As emphasized many times in the literature, mobile applications are a modern solution and a manifestation of the operator's competitive edge (Peterson, 2015, Mohorovičić, 2013, Natda, 2013). Examples of these solutions include the *e-turystyka* domain. It is "mostly viewed through the prism of programs, applications and devices which make it much easier to engage in tourism," and allows the tourist to access a bundle of information on accommodation, interesting events, monuments, etc. (Kalecińska, 2013, p. 17). The rapidly developing e-tourism market made it technically feasible to prepare mobile multimedia directories and guides, including for the purposes of agritourism. The tourists willingly use mobile applications as they allow to browse the offers and make a booking. Polish examples include the application titled *Agroturystyka. Wieś polska zaprasza* (Agritourism. Welcome to Polish Rural Areas) which provides a directory of over a thousand agritourism facilities recommended by the "Hospitable Farms" Polish Federation of Rural Tourism.[29] Also, the application is a multimedia guide for the tourists. Another example is the application titled "Małopolska Wieś dla seniorów" (Lesser Poland Rural Areas for Seniors), "Małopolska Wieś dla dzieci" (Lesser Poland Rural Areas for Children) which enables finding farms with dedicated products and services for specific customers. European examples of interesting applications include "Agriturismo On Line" and "Italia—Agriturismo," which offers a collection of Italian agritourism farms, "Falusi turizmus szálláshelyek" for people interested in Hungarian rural tourism and agritourism, and "Turismo Rural Murcia," which promotes agritourism in Spanish rural areas. In addition to enabling access to accommodation information etc., applications are also a way to express one's opinion on a location, region, interesting events, etc. That functionality is offered, for instance, by

Tripadvisor, which currently is the world's largest tourism platform, available in 28 languages and used by 463 million people each month.[30] Users from all over the world access the Tripadvisor website or application to browse over 859 million reviews and opinions on 8.6 million accommodation facilities, restaurants, attractions, airlines, and cruises.[31] Whether when planning their travel or after reaching a target destination, tourists can use the application to check the prices of accommodation, flights, and cruises, or buy tickets for popular trips and attractions. Another platform which is popular with travelers is Booking.com, established in 1996 in Amsterdam. By investing in state-of-the-art technologies which help avoid traveling difficulties, Booking.com allows tourists to access extraordinary places to stay, from agritourism farms, through to houses, boarding houses, and hotels.[32].

As one of the world's largest platforms, Booking.com allows facilities from all over the world global access to customers.[33] Travelers can use the website to find over 28 million accommodation options. Its application is available in 43 languages. Currently, 11,447 agritourism farms present their offer at Booking.com, most of which are based in Italy (5880), Poland (1198), Austria (667), and Portugal (605).[34] Airbnb, another platform, is available in 191 countries and 11 languages. The concept behind it is to build a global framework for all hosts and guests. Airbnb provides an online presentation of the host (as the owner of the accommodation facility) and experiences. Designed by local residents, Airbnb online experiences are interesting sessions which include trips, workshops, and participation in local events.[35] In order to propose an experience to his/her guests, the host must complete a defined procedure, i.e. design an experience, describe the related activities both in the application and on the page of the future experience, and send the concept to the Airbnb team. Once approved, the owner may prepare the session and welcome his/her first online guests.

In summary, mobile technology enabled further development of modern tourism. From the perspective of the tourism industry, the information technologies employed will allow more informed investment decisions to be made and a customized range of dedicated products and services to be offered, to address changing consumer preferences. Rural tourism takes on different forms, from agritourism through to commercialized (tourism) services, qualified, specialized, and targeted services, and cultural tourism. Undoubtedly, the particularities of agritourism and access to a broad range of technological capabilities provide an incentive for the development of accommodation resources while also ensuring a good promotion and distribution of the agritourism offer. In Poland, big cities continue to outperform villages when it comes to Internet access. Based on the analysis of "Information society in Poland in 2018," a study by the Central Statistical Office, it was found that during 2015–2018, the greatest improvement in network accessibility at household level was recorded in rural areas (by 12.6% compared to 2015) and in semi-urbanized areas (by 12.2% compared to 2015). Although the rural population has seen a considerable improvement in Internet access in recent years due to investments in digital infrastructure, high-speed

Internet links are not always available. Also, while the Internet becomes more and more accessible to rural residents, availability rates are relatively low compared to the urban population.

5.5.2 Consumer's path from FOMO to JOMO

Today, the tourism industry offers a broad range of products and services. Also, it witnesses some interesting trends, namely FOMO and JOMO (Kusnadi & Putra, 2019). FOMO stands for the Fear of Missing Out (Edelman, 2018; Elhai et al., 2018). Due to the rapid development of social media, consumers want to follow interesting events in real time. The fear of being disconnected from the virtual world often triggers frustration and negative emotions. Many web users around the world are concerned that they might miss something important whenever they go offline. FOMO can affect anyone, irrespective of his/her age, gender, and nationality. Currently, generation C (connecting)[36] can be observed to be very vulnerable to FOMO (Przybylski et al., 2013; Oulasvirta et al., 2012; Elhai et al., 2018; Kuss et al., 2018). FOMO is a term that was introduced in the 1990s by Dan Herman,[37] a marketing researcher and strategist. Initially, it was not related to the Internet. Later on, the intensive development of IT, social media, and consumer mobility made FOMO an interdisciplinary issue. The relevant literature (Przybylski et al., 2013; Buglass et al., 2017; Kusnadi & Putra, 2019) describes FOMO as:

- the fear that other people have more satisfactory experiences in their lives;
- feeling concerned when my friends have a great time without me, or if I do not know their plans; fearing that I might miss a planned or ad-hoc meeting;
- checking one's friends activity on a continuous basis; having a desire to know what they do at any moment;
- the need, or sometimes the constraint, to report how one spends his/her life (primarily including positive events) in the Internet (by posting blog, Instagram or Facebook entries);
- the need for having a phone always at the ready.

(Kusnadi & Putra, 2019, pp. 18–22)

FOMO also means a desire which results from the need to belong to a particular social group. People often satisfy that need by trying to become part of specific social groups (LaPlante et al., 2014; Abel et al., 2016; Marcotte, 2010). FOMO is very often combined with what is referred to as phonoholism, i.e. an addictive use of a phone. This is a situation where a person is no longer able to normally live his/her life without his/her phone. While phonoholism can affect people at different ages, the youth and children are particularly at risk, just as in the case of FOMO (Dębski, 2016). In tourism, FOMO means anxiety, if not fear, which strikes when an individual realizes he/she could miss a satisfactory experience, e.g. by not making a trip to exotic places or by being in a location without a phone service or Internet access.

The second concept is the Joy of Missing Out (JOMO), the opposite of FOMO, which means "enjoying being out of reach" (Kusnadi & Putra, 2019; Roberts & David, 2016; Scott & Woods, 2018). JOMO is a kind of travel where emphasis is placed on being isolated from modern technology (Table 5.4). For the tourist, the key is to escape from his/her day-to-day routine and responsibilities and to relax. Dickinson, Hibbert, and Filimonau (2016) found that viewing a tourist in the context of the dilemma posed by having a limited access to digital technologies when visiting tourism destinations could encourage tourism operators to propose a new product or service bundle. When on this kind of holiday, tourists experience extraordinary emotions by focusing on their tourism destination and forgetting about their day-to-day responsibilities. Furthermore, the interviewees noticed that "the dimensions of traveling experiences in places with a limited access to communication technologies make people break away from their routine for a moment and restore balance to their lives" (Dickinson et al., 2016; pp. 194–200). For the tourists, relaxing away from the technology is a way to have an extraordinary traveling experience which cannot be found elsewhere (Li et al., 2018).

Travel trends are related to many aspects, including the motivation to relax, experience nature, and contact friends and family members. Some services proposed by the tourism industry, such as special retreats, beauty treatments or agri-spas in remote parts of the world, are designed to isolate people from digitization and technology. That assumption is supposed to restrict tourists'

Table 5.4 JOMO vs. FOMO: basic information

Specification	Characteristics	
	JOMO	FOMO
Travel motives and behavior	Escape/relax Enhancement of kinship relationships Novelty seeking; isolating oneself; actualized autonomy	Digitization and continuous monitoring of the travel Posting information in real time Looking for novel tourism services
Traveler typologies	Venture Personality: singles, families Small groups, high-end tour packages	Social group Families Tour packages
Tourism products, activities	Singular and specific leisure (recreation, natural/outdoor sports), retreat, personal quest (wellness, SPA, yoga, spiritual/cultural experience), nature (private, remote, island)	Virtual tours Questing and LARPs A holiday full of emotions and experiences

Source: Compilation based on Kusnadi and Putra (2019).

activity in the digital world, which can affect the way they work and introduce greater mental pressure. The comparison of JOMO and FOMO suggests that tourism service consumers are guided by different needs. Some of them prefer being isolated from the digital realm when traveling, whereas others cannot think of traveling without online access. In Poland, this is a new trend among managers, corporation workers and big-city singles who spend their days trying to find a balance between intensive work and household chores. Examples of JOMO experiences include attending a calming retreat. The Benedictine Abbey in Tyniec near Krakow is a place where visitors can come and live. Tourists can take an active part in the cloister life and attend the Eucharist and the Liturgy of the Hours each day together with the monks. Men are allowed to access the refectory to enjoy a meal together with the whole Benedictine community. Rural areas, too, sometimes offer a service targeted at tourists who prefer a JOMO holiday. Examples include the "Złoty Raj" (Golden Paradise) boarding house located in the Kłodzko Valley, at the heart of the Golden Mountains.[38] As an advantage, it does not offer Internet access at all; the fascinating environment and proximity of various attractions (from beautiful natural assets to gold mines) is what encourages JOMO tourists to go there. Another example is "Siodło Budy," an agritourism farm in a rural setting near Białowieża. The charming local climate mixes with the wilderness of the forest. Farm owners provide a broad range of attractions, e.g. a Russian bath, tracing beavers and wisents with a hired guide, or eating authentic rural meals.[39] "Dom na łąkach" (the Meadow House) is an example of an agritourism farm based in the picturesque area of Low Beskids, at the foot of the Lackowa hill. The farm is located in ecological meadows, near a secluded road between two villages, Izby and Banica. The owners offer a broad range of services, primarily including a close contact with nature away from information technologies. This is a dream come true for those seeking a quiet and calm experience.

Today, broadband Internet access is among the basic advantages of human progress because many services have already been entirely digitized. Currently, before choosing a holiday destination, tourists increasingly often look for information in the Internet. They mostly use it to make detailed arrangements regarding accommodation booking, buying local attraction tickets, etc. However, some tourists prefer and choose a rural holiday so they can move away from the urban rush and other aspects of technological progress. As shown above, FOMO and JOMO are two extremely complex phenomena. FOMO is strongly related to reference groups, and primarily involves the need to build a team, contribute to a community of people who think alike and spend time together, and the willingness to stay in permanent contact. In turn, JOMO is a concept of traveling which places emphasis on being isolated from modern technologies, and being motivated to seek relaxing and quiet experiences and to escape from day-to-day routines and responsibilities. It therefore seems right that rural areas will become an adequate asylum for the development of JOMO-based services.

5.6 Smart village as a sustainable development target for rural tourism

Today's world witnesses territorial development, a process which may contribute to inequalities and boost inter-regional competition. Hence, there is a need for a policy focused on ensuring sustainability, cohesion, and competitiveness at the same time (Czupich et al., 2016; Szczech-Pietkiewicz, 2015; Nam & Pardo, 2011; Neirotti et al., 2014; Deller et al., 2001; Albino et al., 2015; Bell & Jayne, 2010; Teräs et al., 2015; Da Rosa Pires et al., 2014; Deller et al., 2001). The concepts of smart development referring to territorial systems are theoretically rooted in a series of theories and concepts of socioeconomic development, including: the theory of territorial competitiveness, social capital, innovative environment, innovation, territorial production systems, poles of development, and the core and periphery model (Dudek et al., 2016; Cocchia, 2014; Dej et al., 2014). European rural areas undergo changes which carry both a risk and a true opportunity for them to play a new, separate, and innovative role by proposing products and services. In that context, there is a great potential behind smart villages (SV) which have only recently appeared in European Union documents (European Commission, 2017). According to the first working definition,

> smart villages are communities in rural areas that use innovative solutions to improve their resilience, building on local strengths and opportunities. They rely on a participatory approach to develop and implement their strategy to improve their economic, social and/or environmental conditions, in particular by mobilizing solutions offered by digital technologies. Smart villages benefit from cooperation and alliances with other communities and actors in rural and urban areas. The initiation and the implementation of smart village strategies may build on existing initiatives and can be funded by a variety of public and private sources.[40]

"In smart villages, traditional and new networks and services are reinforced with digital and telecommunication technologies, innovations and a better use of knowledge which provides benefits to the population and enterprises" (Zwolińska-Ligaj et al., 2018). Smart village is a concept that refers to rural areas and to a community affected by factors which can play a relevant role in establishing SVs. In the European Union's rural development policy, the concept of smart villages emerged in response to the need for implementing the assumptions of the Europe 2020 strategy (The Smart Villages Initiative: Findings, 2014–2017, 2017; Da Rosa Pires et al., 2014). Its main priority was a sustainable, smart, and inclusive development. The smart and inclusive development concept is the answer to the quest for ways and actions to make the sustainable development concept a reality in the context of a series of growing problems affecting the development of Polish rural areas, such as: the depopulation of remote rural areas, the outmigration of the youth, adverse demographic conditions, and challenges related to climate change (Zwolińska-Ligaj et al., 2018; Stanny et al., 2016). Smart villages are also a new concept in the development of the European Union's policy (Naldi et al., 2015; Lombardi et al., 2012). It is of considerable importance to rural development, both because of the new opportunities for job creation, and from the perspective of the standards of rural living.

The demographic change in rural areas is among the key factors which affect the deployment and implementation of the action program for smart villages. According to estimations, the urban population of the European Union will grow by 24.1 million until 2050, whereas the population living in predominantly rural areas will decline by 7.9 million.[41] However, behind these global trends, some significant differences exist between particular parts of Europe and specific rural regions. Generally, rural areas operate in a symbiotic coexistence with smaller or larger towns. The Organization for Economic Co-operation and Development (OECD) analyzed the situation of rural areas as regards their relationship networks with cities, and found that rural areas located close, or having access, to cities are the ones which record the greatest growth in gross domestic product (GDP) (Komninos, 2011; Krievina et al., 2015). Another conclusion was that rural areas located close to cities operate on a sound basis. Indeed, small and medium cities play a major role in economic development of rural areas. Supporting the digital transformation of rural areas is an extremely important aspect which involves rural digitization and the ability for the rural population to use state-of-the-art tools and technologies. In Poland, rural households have recently improved their access to computers and the Internet. Indeed, the availability rate of computers in rural households increased from 67.1% to 75.0% during 2011–2018. However, that share continues to below the corresponding rate for cities (82.9%) and the nationwide average level (77.9%). The share of rural households with Internet access grew from 18.8% to 72.0% between 2005 and 2018, i.e. is only 5.7% below the average level for cities. In turn, the share of rural households with broadband Internet access increased from 15.2% in 2005 to 64.7% in 2018, i.e. 9.4% below the average level for cities (OECD, 2018b).

The smart villages concept is also associated with the potential and needs of a territory (especially including rural settlement) and with the rural landscape (European Commission, 2017; Bajer, 2012). In the international literature, many authors emphasize that this concept is largely underpinned by modern technology (Galdeano-Gómez et al., 2011; Markeson & Deller, 2012). Human capital (local community), human potential, and the development of a civic society are key priorities. Other aspects considered as part of the concept include: e-skills; skillful use of e-services; innovative environmental protection solutions; promotion of local products; implementation of a circular economy with respect to agricultural waste, supported with technology and ICT solutions; implementation of smart specializations in agri-food projects, tourism, cultural activity, etc. (European Commission, 2017; Renski, 2014; McGranahan et al., 2011).

The key assumption behind the smart villages concept is that technological progress, when effectively integrated with other rural development initiatives, can provide new opportunities for increasing incomes, providing services and reinforcing the social potential, which can significantly improve the quality of rural living.

(Van Gevelt & Holmes, 2015; p. 89)

In 2017, the European Parliament initiated an action at EU level to promote the creation of smart villages. The ENRD[42] Thematic Group on Smart Villages determined the way the European rural communities develop local, innovative, insightful solutions for and approaches to the revitalization of rural services in such areas as energy, transport, social care, healthcare, and education. A pilot project was implemented with focus placed on smart social villages. The outcome of that project was an assessment of what would be the future of smart village policies. The project was supposed to assess that issue in four thematic blocks. The first one was the mapping of opportunities and challenges faced by rural areas; the second involved presenting the definition of smart villages. The next stage consisted in identifying and presenting the best practices, whereas the last one showed the financing options for such projects. The report developed upon completing the project claimed that "the goal of smart villages is to make different policies interoperate to find better ways for promoting a holistic rural development" (Smart villages revitalizing rural services, EU, 2018, p. 14). This means leveraging the existing and emerging technologies and social innovations in an effort to add value to the population's lives with the use of adequate tools.[43] Currently, an important initiative of the European Union regarding smart villages is the preparatory action for 21st century smart rural areas, supported by the European Commission (DG AGRI), for 2020–2023. The purpose of that project is to promote and inspire the villages to develop and implement smart solutions and strategies across Europe, and to support future CAP interventions related to smart villages. Examples include Smart Rural, a project which is currently in progress, and is intended to support the villages in the implementation of smart solutions for the broad aspects of social and economic living.[44] However, as indicated by McCann and Ortega-Argilés (2013) and Mempel-Śnieżyk (2013), smart development in a rural context requires different initiatives to be placed within a broader, multi-level planning and management framework. Hence, it could be interesting to implement a smart specialization which is in place in most remote regions, and consists in building specialized links to urban markets. "The smart specialization policy not only needs to be adjusted to available local resources and potentials but also must be focused on changing the local social capital and its external links" (Visvizi & Lytras, 2018; Zavratnik et al., 2018). The "EU actions for smart villages" state that the key aspect of the smart village concept is that rural residents take the initiative and that national, regional and local authorities create an environment that encourages such projects and measures. Therefore, a specification was provided to show what the services provided in rural areas should look like (Table 5.5).

Table 5.5 New forms of service delivery in rural areas

Specification	Description
Integrated service delivery	Collaboration between service deliverers in terms of information, administration, training, etc.; cooperation between professional teams to provide more joined-up services; co-production between public, private, and community organizations, and particularly, community-based solutions; collocation of several services into one building or space
Alternative and more flexible delivery approaches	Hub and spoke models—where the services are provided regularly from a central location, but there are outreach services less regularly or at a lower level in more remote areas; new and improved services adapted to local needs (quality, marketing, the creation of totally new service approaches)
Digital solutions	Digitization can bring services closer to the customer, reduce cost, and have a major impact on quality of life in the countryside where structural change is rapid and distances to physical services, including health and social care, are increasing. Three potential gaps that need to be bridged on the road to creating smart villages are as follows: existing broadband network infrastructure, availability of digital networks, and digital literacy

Source: https://enrd.ec.europa.eu/sites/enrd/files/enrd_publications/publi-enrd-rr-26-2018-pl.pdf, online, April 28, 2020.

The EU's strategic documents on smart villages emphasize that rural areas[45] should attain a level of digital transformation that will enable a full use of the specific potential of a territory. It was also noted that digital technologies can considerably reduce rural problems caused by remoteness and low population density because they will enable online communications and access to a broad range of e-services. Also, smart villages open new paths to, for instance, improved mobility, development of rural enterprise, using the potential behind the bioeconomy, circular economy and more, ensuring high-quality education, health, and medical services or preventing social exclusion.

In Poland, the implementation of the smart villages concept has started only recently because it still looks for its own identity in the rural landscape of diverse ideas. Smart villages are mostly about projects designed to support and develop rural areas in accordance with the expectations and aspirations of rural residents. In Poland,

> smart villages refer to those rural areas (local communities) which make use of digital technologies and innovation in their daily lives; this is how they improve the quality of their living by enhancing the standard of public services and making a better use of local resources.
>
> (Kamiński & Leśniak, 2019, p. 17)

The areas of smart actions are identified in three domains of smart solutions:

1 *Public services:* e-care, e-health, remote education, transport (e.g. dial-a-ride buses),
2 energy (e.g. renewable energy systems, RES), and security (e.g. monitoring).
3 *Public management:* waste e-management (e.g. waste bin fill-level sensors), e-administration, land-use planning (e.g. digitization), and environmental e-monitoring (e.g. air quality sensors).
4 *Entrepreneurship:* precision farming, online trading (e.g. in local products), rural tourism (based on smart solutions) (Kamiński & Leśniak, 2019).

In Poland, the "Smart Village—Inteligentna Wieś" rural development concept was implemented thanks to co-financing with European Union funds disbursed under the 2nd Technical Assistance Scheme "National Rural Network" of the 2014–2020 Rural Development Program, as part of the "Moja SMART Wieś"[46] (My SMART Village) project. The implementation of the "Smart Village—Inteligentna Wieś" concept is the responsibility of the Institute of Rural and Agriculture Development of the Polish Academy of Sciences.

"Smart Village—Inteligentna Wieś" is a perfect example of a positive transformation which is possible in rural areas. The operation of smart villages can result in improved efficiency of service management and delivery. Rural areas become more competitive while respecting the economic, social, and environmental needs of the present and future generations. In Poland, Niedźwiady is an example of a sustainable, smart and innovative village which taps into state-of-the-art technologies to improve the quality and standards of living of its residents. It is located in the Kujawsko–Pomorskie voivodeship, in the Nakło district, Szubin commune. Niedźwiady established the Center of Astronomy, Culture and Education (*Centrum Astronomiczno-Kulturalno-Dydaktyczne*, abbreviated as CAKD) which makes use of modern technologies in the area of education, social and economic services, and more. CAKD is a project implemented under the 2014–2020 Regional Operating Program of the Kujawsko–Pomorskie voivodeship, with an investment amount of PLN 1,833,416.19 (the amount of co-financing with Union funds was PLN 1,255,187.53).[47] The Center implements social and economic projects focused on offering education and recovery courses to support pupils with specific development and educational needs. Two social projects are planned: the "Bliżej gwiazd" (Closer to the Stars) youth club and the "Kluczowe kompetencje—drogą do sukcesu" (Key Competences as the Road to Success).

The first project is dedicated to children and youth, including from families at risk of poverty or social exclusion, and is intended to improve the level of education. It will include: thematic classes and workshops designed to develop the pupils' passion for and interest in exact sciences, physics, astronomy, geography, mathematics or environmental sciences; active ways of spending free time; psychological and pedagogical counseling; learning support, etc. The second project is supposed to support primary school pupils in education processes in order to develop their key competences (in mathematics and natural sciences). The third key project implemented by the Center is "Kuźnia aktywności zawodowej" (Forge of Professional Activity). It is planned to include measures taken to promote the establishment of social economy operators (e.g. care or educational farms) and economic operators. It should be dedicated to building relationships and cooperation, including social integration that promotes the creation of new jobs.

The Center of Astronomy, Culture and Education established in Niedźwiady meets the assumptions of the "Smart Village—Inteligentna wieś" rural development concept. The creation of the Center has a positive effect on how rural areas are viewed, and contributes to improving the standards of rural living. It is an example of smart, innovative rural areas which tap into state-of-the-art technologies to improve the quality and standards of living of their residents. The smart village concept also includes the educational dimension as a crucial component. Adequate education will contribute to making the local community more competitive; but the most important thing is that human capital will start to use modern technological and scientific tools, e.g. with respect to time organization and management. The second example of a smart solution is the Sudety Educational Farm in Dobków, one of the two Lower Silesian educational centers operated by non-governmental organizations.[48] The farm was built at the center of the Land of Extinct Volcanoes, a geo-park in Dobków. The hills surrounding Dobków are the remains of historical volcanoes and a perfect place to find out how the forces inside the Earth formed the region of Kaczawskie Mountains and Foothills.[49] This is one of the most interesting geographic and tourism regions in Poland. The traces of three periods of volcano activity, marine inundations, deserts, and glaciations are visible and perfectly tell the history of Earth. The Sudety Educational Farm is a place were knowledge and skills are transferred in an interactive, dynamic way based on models, digital visualizations and experiments carried out by the participants themselves.[50]

Classes offered by the Sudety Educational Farm[51] are held in eight educational rooms, e.g. in the "Earth room" located in a former stable. Under splendid brick vaults, there is an interactive river model which allows the attendees to immerse their hands in water and trigger a flood by themselves. The room also includes a collection of regional rocks and minerals. Another innovation is the "volcano room" where the participants get to know different types of volcanoes found around the world. The volcano room also includes a geyser model which, just like a real one, ejects water at regular intervals. The "earthquake room" is another interesting feature where the visitors can experience Poland's largest earthquake simulation platform, and watch a volcano eruption or a 3D movie on how the Sudetes were formed. The room uses many interesting, innovative, and new technologies. It educates based on state-of-the-art interactive methods supported with technology, and uses non-conventional organizational solutions. Also, it has an impact on—and is strongly integrated and interoperates with—the local environment, and therefore Dobków can be concluded to be a Polish example of a smart village.

Energy independence of rural areas is one of the topics most frequently discussed within the EU. Poland, too, witnesses some investments focused on energy independence based on renewable sources of energy, and is home to several "bio-energy villages." A bio-energy village is an area capable of meeting its entire energy demand (in most cases, this means electricity) with local production based on renewables. Bio-energy villages can establish local heat networks or sell excess energy to a public network, thus making additional profits. The environmentally friendly nature of these investments is an important aspect, too. Bio-energy, i.e. energy generated through the conversion of biomass to fuels (usually made of waste which is anyway a byproduct of agricultural production and wood processing), does not involve any environmental burden. The Skierbieszów commune implements a project titled "Energia odnawialna w Gminie Skierbieszów" (Renewable Energy in the Skierbieszów Commune).[52] The program is co-financed by the European Regional Development Fund as part of the 2014–2020 Regional Operating Program of the Lubelskie voivodeship, Priority Axis 4 (Environmentally friendly energies), Measure 4.1[53] (support for the use of RES). The project consisted in buying and installing over 350 solar panels for the production of heat from solar radiation in order to heat household water. The overarching objective of the project was to increase the development potential of the Skierbieszów commune (while complying with sustainable development principles), to improve the population's standards of living, and enhance the natural environment. As an important aspect, the project will contribute to improving the quality (purity) of air in the commune by using renewable energies as an alternative source of heat. The project fits the smart village concept, although it spans over more than one village. However, its objective was to improve the quality of living for the residents, increase environmental awareness, and deploy state-of-the-art technologies to use RES.

Another interesting solution consistent with the smart village concept is the "Internet village." In Poland, the first Internet village was established in Grabowa, Łazy commune (Śląskie voivodeship). The project's headquarters are located in a local school which is open every day, can be accessed free of charge, and is aimed towards facilitating education of adults and children. "Internet village: remote education in rural areas" is a project financed by the European Union.[54] In line with the project's assumptions, 256 education centers were created in Polish rural areas. The first center in Grabowa is equipped with state-of-the-art computer devices and a broadband Internet connection. The classroom features a total of 15 computers with professional e-learning software, multi-function printers, and a multimedia projector. An online education platform was also developed which offers free e-learning courses on: how to use computers and software; the basics of entrepreneurship; how to look for a job; and language courses (Kamiński & Leśniak, 2019;

CDR, 2019; Rosner & Stanny, 2016). The Grabowa Internet village is also used by the employed and unemployed, youth, and children.

Smart villages are rural communities which rely on innovative solutions to improve their viability based on local assets and capabilities. Also, this is an approach in which the key role is believed to be played by the local community and their actions and undertakings. Smart villages are smart communities which know how to respond to challenges and changes taking place in rural areas. Instead of being focused on development planning, this approach promotes a line of action that enables the formation of adequate conditions for development which result from local particularities. Adequate conditions for development mean those which allow full use of the potential of a territory and its population. It is important to use immobile resources rooted in a location because they (but not only they) provide a framework for defining smart villages.

Notes

1 https://www.gov.pl/web/rozwoj/bonturystyczny, online access on September 26, 2020.
2 https://www.wot.org.pl/2020/06/koronawirus-informacje-dla-branzy-turystycznej/, online access on September 26, 2020.
3 https://odpoczywajnawsi.pl/wydawnictwa-turystyczne/, online access on July 8, 2020.
4 Also referred to as the Kupala Night (or the Feast of St. John the Baptist), it is a Slavic holiday related to the summer solstice, celebrated on the shortest night of the year. Often, it is called the feast of fire, water, sun and moon, harvest, joy, and love. It is commonly celebrated in territories where Slavic people live. However, a holiday of a similar nature is observed in territories inhabited by Baltic, Celtic, and Germanic nations, as well as by some Finno-Ugric peoples, e.g. Finns (it is one the most important holidays in the Finnish calendar) and Estonians. Called Līgo, it is a national holiday in Latvia (June 23 and 24) (Strzelczyk, 2007).
5 Delener (2010) indicated that the global trends in tourism are affected by how the society is distributed between generations X, Y, and Z. Today, generation X is the driving force behind multigenerational traveling. Generation Y prefers staying home instead of spending money on tourism. However, they increasingly often opt for diversity and more expensive trips, use state-of-the-art technology which requires a greater traveling experience, and choose more distant destinations than previous generations. The last generation (Z) are extremely mobile people for whom instant communications and virtual travels across continents, regions, and tourism attractions are something very common. The literature defines generation X as people born during 1965–1980 who grew up during the 1970s economic crisis. In turn, generation Y are those born during 1981–1994, raised in the era of globalization and common access to the Internet. Finally, generation Z means people born after 1995 (Hardey, 2011).
6 Based on Tripadvisor (2020a).
7 https://forsal.pl/galeria/1210411,najbardziej-luksusowe-miejsca-w-polsce-i-na-swiecie-zdaniem-polakow.html, online access on July 3, 2020
8 https://lawendowepole.pl/, online access on July 3, 2020.
9 A destination management company is equated with a national, regional or local tourism organization authorized to manage the development of tourism and responsible for driving, coordinating, stimulating, and monitoring the development of tourism and marketing in the destination territory (Pike & Page, 2014).

10 LARP (live action role-playing), a specific kind of role-play gaming, is an activity between a game and a show in which the participants create and live a story together, each in his/her role. A LARP can be set both in a real or fictional world. LARPs can be historical (medieval) or futuristic (fantasy) games. They are played in an open or closed space adjusted to the realities of the world they are based in. The form of LARPs is similar to that of an improvisational theatre where the game master conceives the plot and the participants are actors. What the players say or do is considered to be what their characters say or do. It is important that the players fit within the defined reality while not stepping out of their roles.

11 https://gardenrangers.pl/ogrody-sensoryczne/, online access on July 3, 2020.

12 European Commission. *Rural development 2014–2020* (2017), https://ec.europa.eu/agriculture/rural-rozwój-2014-2020_en,https://ec.europa.eu/agriculture/rural-rozwój-2014-2020_en, online access on July 3, 2020.

13 https://www.zagroda-edukacyjna.pl/, online access on June 9, 2020

14 See Thieme and Birkigt (2006).

15 Working Weekends On Organic Farms (WWOOF) started in the early 1970s in the UK when Sue Coppard established the first WWOOF group. https://wwoof.net/, online access on April 22, 2020.

16 https://wwoof.net/, online access on April 22, 2020.

17 https://kreatywnaromantyczka.blogspot.com/2011/06/wwoof.html, online access on June 9, 2020.

18 Gap year is a "time (not necessarily one year) planned to be spent on traveling abroad (before, during or after studies), often to exotic countries, to gain experience in a new, different environment, to work as a volunteer and have the trip of a lifetime." www.perspektywy.pl, online access on January 2, 2014.

19 The way young European aristocrats traveled to enhance their knowledge of the world and culture and expand their intellectual horizons. It was particularly popular in England in the second half of the 17th century and in the 18th century (Gralak, 2012).

20 https://enrd.ec.europa.eu/sites/enrd/files/publi-enrd-rr-22-2016-pl.pdf, online access on April 22, 2020.

21 https://enrd.ec.europa.eu/sites/enrd/files/publi-enrd-rr-22-2016-pl.pdf, online access on April 22, 2020.

22 Culinary trails result from the need to emphasize the uniqueness of local cuisine, extraordinary traditions, and particularities of food production. Importance is also attached to authentic production techniques. The remains of former workshops or equipment, such as mills, dryers, pressing facilities, and smokehouses, stimulate imagination and inspire people to revive the old cuisine (Woźniak & Batyk, 2017).

23 http://www.gospodarstwa-opiekuncze.pl/, online access on June 18, 2020.

24 http://www.opieka.kpodr.pl/pl/front/o-projekcie-opieka-z-zagrodzie/, online access on April 27, 2020.

25 The Digital Economy and Society Index (DESI) is a composite index that summarizes relevant indicators on Europe's digital performance and tracks the evolution of EU Member States in digital competitiveness.

26 https://ec.europa.eu/digital-single-market/en/use-internet, online access on July 2, 2020.

27 According to Merski (2002, p. 7), "ITUR is a set of actions that aid the tourists in efficiently moving across time and space, and help making optimal use of cognitive and relaxing tourism assets and of tourism and related services."

28 https://stat.gov.pl/obszary-tematyczne/kultura-turystyka-sport/turystyka/, online access on July 2, 2020.

29 Established in 1996, the "Hospitable Farms" Polish Federation of Rural Tourism is a countrywide umbrella organization for 33 local and regional associations of rural tourism facilities. Its main objective is to take measures to promote and develop

Polish rural tourism. The Federation's key missions include the classification of the Rural Accommodation Directory. This is a voluntary rating scheme for rural tourism facilities supervised by licensed inspectors. The purpose of the classification is to recommend and promote rural tourism facilities in Poland and beyond and improve the quality of services they deliver; http://pftw.pl/o_federacji, online access on July 2, 2020.

30 https://tripadvisor.mediaroom.com/pl-about-us, online access on June 12, 2020.

31 https://tripadvisor.mediaroom.com/pl-about-us, online access on June 10, 2020.

32 https://www.booking.com/content/about.pl.html?label=gen173nr-, online access on July 1, 2020.

33 https://www.booking.com/content/about.pl.html?label=gen173nr-, online access on June 16, 2020.

34 https://www.booking.com/farm-holidays/index.pl.html?label=gen173nr-1FCAEogg I46AdIHlgEaLYBiAEBmAEeuAEXyAEM2AEB6AEB-AECiAIBqAIDuAL5uvf 3BcACAdICJGFiMWU1ZGJiLTdlNzEtNDdiMC04YzFiLWY1NWQ4MThiOD c4YdgCBeACAQ;sid=d112f5b38b1d801347254167c460dc1e;from_booking_home_ promotion=1&, accessed on July 2, 2020.

35 https://www.airbnb.pl/d/onlinehost, online access on July 2, 2020.

36 These are people born after 1995, referred to as "connect, communicate, change" in the literature. Today, that expression has become widespread and is used in the literature equally often as *generation Z* and *iGen* (Aubé & Lamy, n.d.).

37 See Herman (2010).

38 http://zlotyjar.pl/, online access on July 1, 2020.

39 http://siolo-budy.pl/, online access on June 30, 2020.

40 https://digitevent-images.s3.amazonaws.com/5c0e6198801d2065233ff996-registra tionfiletexteditor-1551115459927-smart-villages-briefing-note.pdf, online access on April 28, 2020.

41 https://enrd.ec.europa.eu/sites/enrd/files/enrd_publications/publi-enrd-rr-26-2018-pl. pdf, online access on April 29, 2020.

42 The ENRD Smart Villages Thematic Group was established in 2017 as a major part of "EU action for Smart Villages." It was the focal point for exchanging experiences between European smart village initiatives and projects and for pooling relevant good practices. Furthermore, its objective was to accelerate changes to and improvements in the process of implementing the rural development policy. The basic mission of the ENRD research group was to analyze the progress in rural digitization. In doing so, the group found that many rural areas exhibit two trends that reinforce each other: the first is the absence of jobs and of sustainable economic activity; the second is the insufficient and narrowing scope of services. The smart villages thematic group largely focused its efforts on presenting social and digital innovation in rural services; https://www.smartrural21.eu/, online access on April 28, 2020.

43 http://www.pilotproject-smartvillages.eu/, online access on April 28, 2020.

44 https://www.smartrural21.eu/, online access on April 28, 2020.

45 https://enrd.ec.europa.eu/smart-and-competitive-rural-areas/smart-villages/sma rt-villages-portal/eu-policy-initiatives-strategic-approaches_en, online access on April 29, 2020.

46 http://smart.irwirpan.waw.pl/, online access on June 18, 2020.

47 http://szubin.pl/inwestycja/centrum-astronomiczno-kulturalno-dydaktyczne go-w-niedzwiadach.html, online access on June 17, 2020.

48 The two institutions are Kaczawskie Association and the Voivodeship Fund for Environmental Protection and Water Management in Wrocław, https://www. sudeckazagroda.pl/o-zagrodzie/, online access on June 13, 2020.

49 Kaczawskie is a mountain range located in southwest Poland in Lower Silesia, in the northwest part of Sudetes (West Sudetes).

50 https://www.sudeckazagroda.pl/o-zagrodzie/, online access on April 29, 2020.
51 An education center based in Lower Silesia, the Sudety Educational Farm is run by non-government organizations, including the Kaczawskie Association and the Voivodeship Fund for Environmental Protection and Water Management in Wrocław. It is a modern, interactive education center mostly focused on geo-education and regional education. It is a place were knowledge is transferred in an interactive, dynamic way based on models, digital visualizations, and experiments carried out by the participants themselves. https://www.sudeckazagroda.pl/o-za grodzie/, online access on June 13, 2020.
52 Skierbieszów is a rural commune in the Lubelskie voivodeship, Zamość district. It is located southeast of Lublin, north of Zamość and south of Biała Podlaska; http://www.solary-skierbieszow.eu/o-gminie/, online access on June 19, 2020.
53 The total value of the project is PLN 4,826,644.38; the co-financing from the UE budget amounts to PLN 3,879,201.44, http://www.solary-skierbieszow.eu/o-projek cie/, online access on June 18, 2020.
54 "Internet village: remote education in rural areas" is a project co-financed by the European Union under the European Social Fund and with state budget funds allocated to the Sectoral Operating Program for the Development of Human Resources as part of Priority 2: Development of a knowledge society, Measure 2.1: Enhancing access to education: promoting lifelong learning, Scheme a): Bridging the educational gap between urban and rural areas. The project was implemented by the Foundation of the Regional Agency for the Promotion of Employment. The budget of PLN 49,356,320.00 was disbursed during 2007–2008; (Kamiński & Leśniak, 2019, p. 32).

References

Abel, P. J., Buff, C. L., & Burr, S. A. (2016). Social media and fear of missing out: Scale development and assessment. *International Journal of Economics and Business Research*, 14, 34–44.
Airbnb (2020). Retrieved July 2, 2020, from https://www.airbnb.pl/d/onlinehost.
Albino, V., Berardi, U., & Dangelico, R. M. (2015). Smart cities: Definitions, dimensions, performance and initiatives. *Journal of Urban Technology*, 1 (22), 3–21.
Alejziak, W. (2011). Globalna polityka turystyczna, *Folia Turistica*, 25 (2), 343–393.
Alexiou, M.-V. (2020). Experience economy and co-creation in a cultural heritage festival: consumers' views. *Journal of Heritage Tourism*, 15 (2), 200–216.
Allan, M. (2015). Geotourism: an opportunity to enhance geoethics and boost geoheritage appreciation. *Geological Society, London, Special Publications*, 419, 25–29.
Andereck, K. L., & Nyaupane, G. P. (2011). Exploring the nature of tourism and quality of life perceptions among residents. *Journal of Travel Research*, 50 (3), 248–260.
Andersson, T. (2007). The tourist in the experience economy. *Journal Scandinavian Journal of Hospitality and Tourism*, 7 (1), 46–58.
Angelkova, T. Koteski, C., & Jakovlev, Z. (2012). Sustainability and competitiveness of tourism. *Procedia—Social and Behavioral Sciences*, 44, 221–227.
Aramberri, J. (2010). *Współczesna turystyka masowa*. Emerald, London.
Aşan, K., & Emeksiz, M. (2018). Outdoor recreation participants' motivations, experiences and vacation activity preferences. *Journal of Vacation Marketing*, 24 (1), 3–15.
Aubé, P., & Lamy, C. (n.d.). *Génération C. Les technologies de l'information chez les québécois de 12 à 24 ans*. Retrieved May 21, 2020, from https://cefrio.qc.ca/media/1722/generation-c-rapport-synthese.pdf.

Ayaz, N., & Şamata, N. (2017). Integration of congress tourism to the cultural tourism destinations as an economic product. *Journal of Tourism and Hospitality Management*, 5 (1), 53–61.

Bajdalska, A., & Knefel, M. (2018). Wellness w turystyce aktywnej. *Komitet Przestrzennego Zagospodarowania Kraju Polskiej Akademii Nauk*, 269, 204–215.

Bajer, M. (2012). Technologie informacyjno-komunikacyjne jako narzędzie ograniczania dysparytetów społeczno-ekonomicznych wsi w Unii Europejskiej. *Wieś i Rolnictwo*, 154 (1), 65–84.

Bakas, F. E., Duxbury, N., & Vinagre de Castro, T. (2019). Creative tourism: catalysing artisan entrepreneur networks in rural Portugal. *International Journal of Entrepreneurial Behaviour and Research*, 2 (4), 731–752.

Balińska, A. (2019). Wiejska turystyka kulturowa w kontekście teorii rozwoju endogenicznego. *Zagadnienia Doradztwa Rolniczego*, 3 (19), 24–34.

Balińska, A. & Wojcieszak-Zbierska, M. (2020). Tourism activity of Polish seniors. *The Małopolska School of Economics in Tarnów Research Papers Collection*, 46 (2), 107–118. doi:10.25944/znmwse.2020.02.107118.

Ban, J., & Ramsaran, R. R. (2016). An exploratory examination of service quality attributes in the ecotourism industry. *Journal of Travel and Tourism Marketing* 34 (1), 132–148.

Bansal, S. P. & Kumar, J. 2011. Ecotourism for community development: a stakeholder's perspective in Great Himalayan National Park. *International Journal of Social Ecology and Sustainable Development*, 2 (2), 31–40.

Barbu, I. (2013). Podejście do Concept turystyki wiejskiej. Badania naukowe. *Zarządzanie Rolne*, 4, 125–128.

Bell, D., Jayne, M. (2010). The creative countryside: policy and practice in the UK rural cultural economy. *Journal of Rural Studies*, 26 (3), 209–218.

Belletti, G. (2010). Ruralità e Turismo. *Agriregionieuropa*, 6 (20).

Benckendorff, P., & Zehrer, A. (2013). A network analysis of tourism research. *Annals of Tourism Research*, 43, 121–149.

Bernard, L. (2012). Rural tourism: an overview. In J. Tazim & M. Robinson (eds.), *The SAGE Handbook of Tourism Studies* (p. 12). SAGE Publications.

Bichler, R., & Schmiderer, H. (2011). Theorizing tourism: bridging the virtual-real-dichotomy?. In I. Ateljevic, A. Pritchard, & N. Morgan (eds.), *The Criticial Turn in Tourism Studies: Promoting an Academy of Hope* (pp. 332–337). Oxford, Elsevier.

Bielińska, A., Bucholz, M., Gajos, A., Lesiewicz, J., Kamiński, R., Kimber-Kubiak, A., Manintveld, K., Piesik, I., van Middelkoop Snoeij, C., & Toby, A. (2014). *Gospodarstwa opiekuńcze w Borach Tucholskich a doświadczenia holenderskie*. Kujawsko-Pomorski Ośrodek Doradztwa Rolniczego, Minikowo.

Bogusz, M., & Kmita-Dziasek, E. (2015). Zagrody edukacyjne jako przykład innowacyjnej przedsiębiorczości na obszarach wiejskich. In W. Kamińska (ed.), *Innowacyjność w turystyce wiejskiej a nowe możliwości zatrudnienia na obszarach wiejskich* (pp. 155–166). Warszawa.

Bogusz, M., & Wojcieszak, M. (2018). Zagrody edukacyjne jako przykład markowego produktu turystyki wiejskiej, *Intercathedra*, 37 (4), 329–334.

Bombach, C., Stohler, R., & Wydler, H. (2015). Farming families as foster families: the findings of an exploratory study on care farming in Switzerland. *International Journal of Child, Youth and Family Studies*, 6 (3), 440–457.

Bostrom, M., & Klintman, M. (2011). *Eco-standards, Product Labeling and Green Consumerism*. Palgrave Macmillan, Basingstoke.

Boztug, Y., Babakhani, N., Laesser, C., & Dolnicar, S. (2015). The hybrid tourist. *Annals of Tourism Research*, 54, 190–203.

Brandth, B., & Haugen, M. S. (2010). Farm diversification into tourism – Implications for social identity, *Journal of Rural Studies*, 27 (1), 35–44.

Brandth, B., Haugen, M. S., & Kramvik, B. (2010). The future can only be imagined. Inno-vations in farm-tourism from a phenomenological perspective. *The Open Social Science Journal*, 3, 51–59.

Bravi, M., & Gasca, E. (2014). Preferences evaluation with a choice experiment on cultural heritage tourism. *Journal of Hospitality Marketing & Management*, 23, 406–423.

Brych, Ł., Grobel, P., & Skowron, Ł. (2013). Więźniowie technologii – wpływ nowych rozwiązań IT na charakter komunikacji interpersonalnej. *Handel Wewnętrzny*, 4, 82–89.

Buckley, R., Gretzel, U., Scott, D., Weaver, D., & Becken, S. (2015). Tourism megatrends. *Journal Tourism Recreation Research*, 40, 1.

Buckley, R. C. (2011). Turystyka i środowisko. *Annual Review Środowiska i Zasobów*, 36, 39–41.

Buglass, S. L., Binder, J. F., Betts, L. R., & Underwood, J. D. (2017). Motivators of online vulnerability: The impact of social network sites use and FOMO. *Computers in Human Behavior*, 66, 248–255.

Buhalis, D., & Darcy, S. (2011). *Accessible tourism: Concepts and issues*. Bristol: Channel View Publications.

Butler, R. (2014). Rural recreation and tourism. *The Geography of Rural Change*, 211–232.

Butler, R. W. (2012). Geografia turystyki lub geografia turystyki: Gdzie my, do diabła, jesteśmy? In W. J. Wilson (ed.), *The Routledge Handbook of Geographies of Tourism* (pp. 26–34). London: Routledge.

Butterlope Farm (2020). Retrieved March 15, 2020, from http://www.butterlopefarm. co.uk/.

Çalışkan, V. (2014). Examining cultural tourism attractions for foreign visitors: The case of camel wrestling in Selçuk (Ephesus). *Turizam*, 14 (1), 22–40.

Campos, A. C., Mendes, J., Oom do Valle, P., & Scott, N. (2018). Co-creation of tourist experiences: A literature review. *Current Issues in Tourism*, 21 (4), 369–400.

Capriello, A., Mason, P. R., Davis, B., & Crotts, J. C. (2013). Farm tourism experi-ences in travel reviews: A cross-comparison of three alternative methods for data analysis. *Journal of Business Research*, 66 (6), 778–785.

Centrum Doradztwa Rolniczego (CDR) (2019). *Inteligentna Wieś*, Warszawa.

Chaminuka, P., Groeneveld, R. A., Selomane, A. O., & van Ierland, E. C. (2012). Tourist preferences for ecotourism in rural communities adjacent to Kruger National Park: A choice experiment approach. *Tourism Management*, 33 (1), 168–176.

Chang, Y. W., Chern, J. S. (2016). Ups and downs of space tourism development in 60 years from moon register to spaceshiptwo CRASH, *Acta Astronautica*, 127, 533–541.

Chen, C. F., & Chen, F. S. (2010). Experience quality, perceived value, satisfaction and behavioral intentions for heritage tourists. *Tourism Management*, 31 (1), 29–35.

Chmielewska, B. (2018). Gospodarstwa opiekuńcze jako forma pozyskiwania dochodów przez małe gospodarstwa rolne w Polsce, *Problemy drobnych gospodarstw rolnych*, 3, 21–38.

Chmielewski, M., Daab, M., Miśkowiec, A., Rogalska, E., & Sochacka, M. (2017). *Rozwój rolnictwa społecznego w Europie na przykładzie gospodarstw opiekuńczych w wybranych krajach europejskich.* PCG Polska, Ministerstwo Rolnictwa i Rozwoju Wsi, Warszawa.

Chrysolite, B. P. (2014). Rural tourism a tool for rural revitalization. In R. K. Miryala & J. N. Gade (eds.), *Responsible tourism & human accountability for sustainable business* (pp. 1–6). Hyderabad: Zenon Academic Publishing.

Ciani, A. (2012). The Rural Tourism and Agritourism: a new opportunity for agriculture and rural areas (Between Pluria-activity, Multifunctionality, Sustainable Development Strategy and Green Economy Growth). In *Atas do VIII-CITURDES– Congreso Internacional de Turismo y Desenvolvimiento Sustentavel "Turismo rural em tempos de novas ruralidades".* Chaves: UTAD.

Cigale, D., Lampič, B., & Potočnik-Slavič, I. (2013). Interrelations between tourism offer and tourism demand in the case of farm tourism in Slovenia. *European Countryside,* 5 (4), 339–355.

Cocchia, A. (2014). Smart and digital city: A systematic literature review. In R. P. Dameri & C. Rosenthal-Sabroux (eds.), *Smart City: How to Create Public and Economic Value with High Technology in Urban Space* (pp. 13–43). Cham: Springer International Publishing.

Cohen, E. (1979). A phenomenology of tourist experiences. *Sociology,* 12, 179–202.

Cohen, E. (2010). Tourism, leisure and authenticity. *Tourism Recreation Research,* 35 (1), 67–73.

Coleman, D., Georgiadou, Y., & Labonte, J. (2009). Volunteered geographic information: The nature and motivation of producers. *International Journal of Spatial Data Infrastructures Research,* 4 (1), 332–358.

Collins, M., & Weiss, M. (2015). The role of provenance in luxury textile brands. *International Journal of Retail & Distribution Management,* 43 (10/11), 1030–1050.

Cooper, C., & Hall, M. (2008). *Contemporary Tourism: An International Approach.* London, Butterworth Heinemann Elsevier.

Crouch, G. I. (2010). Destination competitiveness: An analysis of determinant attributes. *Journal of Travel Research,* 20 (10), 1–19.

Czapiewski, K. (2010). *Koncepcja wiejskich obszarów sukcesu społeczno-gospodarczego i ich rozpoznanie w województwie mazowieckim.* Studia Obszarów Wiejskich. Komisja Obszarów Wiejskich PTG. Wydawnictwo IGiPZ. PAN, 22.

Czarnecki, A. (2011). O niewątpliwych pożytkach rozwoju wielofunkcyjnego. In M. Halamska (ed.), *Wieś jako przedmiot badań naukowych* (p. 44). EUROREG, Wyd. Naukowe Scholar, Warszawa.

Czupich, M., Kola-Bezka, M., & Ignasiak-Szulc, A. (2016). Czynniki i bariery wdrażania koncepcji smart city w Polsce, *Studia Ekonomiczne. Zeszyty Naukowe Uniwersytetu Ekonomicznego w Katowicach,* 276, 223–235.

Czyżewski, A., & Staniszewski, J. (2018). Sustainable intensification of agriculture as the composition of economic productivity and environmental pressure measures, *Problemy Rolnictwa Światowego,* 18 (3), 80–90.

Da Rosa Pires, A., Pertoldi, M., Edwards, J., & Hegyis, F. B. (2014). *Smart Specialisation and Innovation in Rural Areas.* Policy Brief Series, 9.

Dąbrowski, A., Stępnik, K., & Król, J. (2017). *Gospodarstwa opiekuńcze – budowanie sieci współpracy* [*Care farms: building a collaborative network*]. Research report, Brwinów Agricultural Consultancy Center, Branch Office in Krakow.

Dall☒Aglio, S. (2011). *Slow Tourism Project. European Cross-Border Programme Italia–Slovenia 2007–2013.* Bled.

Dębski, M. (2016). *Dedicated Use of Cell Phones.* Report. University of Gdansk, Gdansk.

Dej, M., Janas, K., Wolski, O. (2014). *Współpraca miejsko-wiejska w Polsce Uwarunkowania i potencjał.* Kraków, Instytut Rozwoju Miast.

Delener, N. (2010). Current trends in the global tourism industry: evidence from the United States. *Revista de Administração Pública,* 44 (5), 1125–1137.

Deller, S. C., Tsai, T. H. S., Marcouiller, D. W., & English, D. B. (2001). The role of amenities and quality of life in rural economic growth. *American Journal of Agricultural Economics,* 83 (2), 352–365.

Dessein, J., & Bock, B. (2010). *The Economics of Green Care in Agriculture.* Loughborough University Press.

Dhiman, M. C. (2012). Employers' perceptions about tourism management employability skills. *Anatolia – An International Journal of Tourism and Hospitality Research,* 23 (3), 359–372.

Di Iacovo, F. (2014). Agriculture and social sustainability. In *Sustainability of the Agri-Food System: Strategies and Performances.* Universitas Studiorum: Mantova, Italy.

Di Iacovo, F. & O'Connor, D. (2009). *Supporting Policies for Social Farming in Europe: Progressing Multifunctionality in Responsive Rural Areas.* Florence, Arsia, Regione Toscana.

Di Iacovo, F., Moruzzo, R., Rossignoli, C., & Scarpellini, P. (2014). Transition management and social innovation in rural areas: lessons from social farming, *Journal of Agriculture Education and Extension,* 2 (20), 327–347.

Di Iacovo, F., Moruzzo, R., Rossignoli, C. M., & Scarpellini, P. (2016). Measuring the effects of transdisciplinary research: The case of a social farming project, *Futures,* 75, 24–35.

Dickinson, J., & Lumsdon, L. (2010). *Slow Travel and Tourism.* Earthscan, London.

Dickinson, J. E., Hibbert, J. F., & Filimonau, V. (2016). Mobile technology and the tourist experience: (Dis)connection at the campsite. *Tourism Management,* 57, 193–201.

Duda, T. (2016). New forms of cultural tourism as a potential of branded tourism product development in a small town (based on the example of Łobez – Western Pomerania), *Tourism Role in the Regional Economy,* 7, 111–128.

Dudek, M., Karwat-Woźniak, B., & Wrzochalska, A. (2016). *Wybrane determinanty polaryzacji społecznej oraz stabilności ekonomicznej na obszarach wiejskich i w rolnictwie.* Warszawa: IERiGŻ-PIB.

Dudkiewicz, M., Marcinek, B., Tkaczyk, A. (2014). Idea ogrodu sensorycznego w koncepcji zagospodarowania atrium przy szpitalu klinicznym nr 4 w Lublinie. *Architectura,* 13 (3), 71–77.

Dybka, S. (2013). Znaczenie czasu jak elementu wyboru obiektu handlowego przez klienta. *Handel Wewnętrzny,* 3, 187–194.

Dykas, P., & Tokarski, T. (2013). Podażowe czynniki wzrostu gospodarczego – podstawowe modele teoretyczne. *Acta Universitatis Lodziensis Folia Oeconomica,* 294, 9–42.

Dziechciarz, T. (2011). Wykorzystanie witryn internetowych i poczty elektronicznej w marketingu agroturystyki na przykładzie województwa lubelskiego. *Nierówności Społeczne a Wzrost Gospodarczy,* 23, 30–39.

EC (European Commission) (2017). Retrieved September 21, 2019, from https://ec.europa.eu/agriculture/sites/agriculture/files/rural-development-2014-2020/looking-ahead/rur-dev-small-villages_en.pdf.

Eckerling, M. (1996). Guidelines for designing healing gardens. Journal of Therapeutic Horticulture, 8, 21–25.

Edelman, M. (2018). 8 trends that will shape travel in 2018. Retrieved May 22, 2020, from http://cms.edelman.com/sites/default/files/201804/Edelman_Travel_Trends_ Shaping_Travel_2018.pdf.

Elhai, J. D., Levine, J. C., Alghraibeh, A. M., Alafnan, A. A., Aldraiweesh, A. A., & Hall, B. J. (2018). Fear of missing out: Testing relationships with negative affectivity, online social engagement, and problematic smartphone use. *Computers in Human Behavior*, 89, 289–298.

Elings, M. (2012). *Effect of Care Farms. Scientific Research on the Benefits of Care Farms for Clients*. Plant Research International, Wageningen.

Elings, M., & Hassink, J. (2006). Farming for health in The Netherlands. In J. Hassink & M. van Dijk (eds.), *European Network for Rural Development. Overview of Social Farming and Rural Development Policy in Selected EU Member States* (pp. 163–179), Springer, Dordrecht.

Energia słoneczna w Gminie Skierbieszów (2020). Retrieved June 19, 2020, from http://www.solary-skierbieszow.eu/o-gminie/.

Engen, M., & Mehmetoglu, M. (2011). Pine and Gilmore's concept of experience economy and its dimensions: an empirical examination in tourism. *Journal of Quality Assurance in Hospitality & Tourism*, 12 (4), 237–255.

EU (2018). Smart villages revitalizing rural services, *EU Rural Review*, 26.

European Network for Rural Development (2010). *Overview of Social Farming and Rural Development Policy in Selected EU Member States*. NRN Joint Thematic Initiative on Social Farming.

EUROSTAT (2019). Retrieved April 5, 2020, from https://ec.europa.eu/eurostat/web/ tourism/data/database,https://ec.europa.eu/eurostat/web/tourism/publications.

FAO. *Social Farming (also called care farming): an innovative approach for promoting women's economic empowerment, decent rural employment and social inclusion. What works in developing countries?*Proceedings of Global Forum on Food Security and Nutrition, Collection of contributions received, Discussion No. 100 from 15 April to 16 May 2014. Retrieved January 20, 2020, from www.fao.org/fsnforum/ forum/discussions/carefarming.

Ferwerda-van Zonneveld, R. T., Oosting, S. J., & Kijlstra, A. (2012). Care farms as a short-break service for children with Autism spectrum disorders. NJAS Wagening. *Journal of Life Science*, 59, 35–40.

Fiedora, B., & Kociszewski, K. (2010). *Ekonomia rozwoju*. Wydawnictwo UE we Wrocławiu.

Folinas, S., & Metaxas, T. (2020). Tourism: The great patient of coronavirus COVID-2019. *Munich Personal RePEec Archive*, 99666 (17), 1–9.

Foresight Factory (2017). Luxury sector trends. Retrieved May 20, 2020, from www. foresightfactory.co/ffonline/.

FORSAL (2020). Retrieved July 3, 2020, from https://forsal.pl/galeria/1210411,najba rdziej-luksusowe-miejsca-w-polsce-i-na-swiecie-zdaniem-polakow.html.

Frochot, I., & Batat, W. (2013). *Marketing and Designing the Tourist Experience*. Oxford: Goodfellows.

Fullagar, S., Markwell, K., & Wilson, E. (2012). *Slow Tourism – Experiences and Mobilities*, Channel View Publications, Bristol-Buffalo-Toronto.

Galdeano-Gómez, E., Aznar-Sánchez, J. A., & Pérez-Mesa, J. C. (2011). The complexity of theories on rural development in Europe: An analysis of the paradigmatic case of Almería (South-East Spain). *Sociologia Ruralis*, 51 (1), 54–78.

Galvagno, M., & Dalli, D. (2014). Theory of value co-creation: A systematic literature review. *Managing Service Quality*, 24 (6), 643–683.

Galvagno, M., & Giaccone, S. C. (2019). Mapping creative tourism research: Reviewing the Field and outlining future directions. *Journal of Hospitality and Tourism Research*, 43 (8), 1256–1280.

Garnett, T.*et al.* (2013). Sustainable intensification in agriculture: premises and policies. *Science*, 341, 33–34.

Gartner, W. C. (2004). Rural tourism development in the USA. *International Journal of Tourism Research*, 6, 151–164.

González, C. G., Perpinyà, A. B., Pujol, A. F.T. I., Martín, A. V., & Belmonte, N. V. (2014). La Agricultura Social en Catalunya: Innovación Social y Dinamización Agroecológica Para la Ocupación de Personas en Riesgo de Exclusión. Revista de Estudios sobre Despoblación y Desarrollo Rural. *Ager*, 14, 65–97.

Gralak, K. (2012). Slow tourism – nowy trend w rozwoju współczesnej turystyki [Slow tourism: a new development trend in today's tourism]. In I. Ozimek (ed.), *Współczesna turystyka i rekreacja – nowe wyzwania i trendy* [*Modern tourism and recreation: new challenges and trends*] (p. 244). Publishing House of the Warsaw University of Life Sciences, Warsaw.

Gralak, K. (2016). Witryna internetowa jako narzędzie promocji i dystrybucji oferty gospodarstw agroturystycznych. *Zeszyty Naukowe Szkoły Głównej Gospodarstwa Wiejskiego Ekonomika i Organizacja Gospodarki Żywnościowe*, 115, 171–182.

Grigaliūnaitᴇ, V., Pilelienᴇ, L., & Bakanauskas, A. P. (2015). Assessment of the importance of benefits provided by rural tourism homesteads in Lithuania. *Economic Science for Rural Development*, 39, 116–123.

GUS (2020). Retrieved July 2, 2020, from https://stat.gov.pl/obszary-tematyczne/kul tura-turystyka-sport/turystyka/.

Guzal-Dec, D., Siedlecka, A., & Zwolińska-Ligaj, M. (2015). *Ekologiczne uwarunkowania i czynniki rozwoju funkcji gospodarczych na obszarach przyrodniczo cennych województwa lubelskiego*. Biała Podlaska.

Halamska, M. (2011). Wiejskość jako kategoria socjologiczna. *Wieś i Rolnictwo*, 1 (150), 37–54.

Halfacree, K. (2009). Rurality and post-rurality. In R. Kitchin & N. Thrift (eds.), *International Encyclopedia of Human Geography* (p. 449). Elsevier, Amsterdam.

Hall, D., Roberts, L., & Mitchell, M. (2017). New directions in rural tourism: Local impacts and global trends, *New Directions in Rural Tourism*, 1, 225–234.

Han, Ch. & Imran, R. (2018). Cultural tourism: An analysis of engagement, cultural contact, memorable tourism experience and destination loyalty. *Tourism Management Perspectives*, 26, 153–163.

Hardey, M. (2011). Generation, C. content, creation, connections and choice. *International Journal of Market Research*, 53 (6), 749–751.

Hassink, J., Hulsink, W., & Grin, J. (2014). Farming with care: The evolution of care farming in the Netherlands. NJAS Wagening. *Journal of Life Science*, 68, 1–11.

Hennigs, N., Wiedman, K.-P., Klarmann, C., & Behrens, S. (2015). The complexity of value in the luxury industry: from consumers individual value perception to luxury consumption. *International Journal of Retail & Distribution Management*, 43 (10/11), 922–939.

Herman, D. (2010). The Fear of Missing Out (FOMO). Retrieved June 13, 2020, from http://www.danherman.com/ The-Fear-of-Missing-Out-(FOMO)-by-Dan-Herman.html.

Hine, R., Peacock, J., & Pretty, J. (2008). *Care farming in the UK: Evidence and opportunities. Report for the National Care Farming Initiative.* University of Essex.

Hoang, D. M. T. (2015). Acting together: How rural tourism can promote sustainable human development? *Journal of Business and Economics*, 6 (3), 607–612.

Horlings, L. G., & Marsden, T. K. (2014). Exploring the 'New Rural Paradigm' in Europe: Eco-economic strategies as a counterforce to the global competitiveness agenda. *European Urban and Regional Studies*, 21 (1), 4–20.

Ilbery, B., & Saxena, G. (2011). Integrated rural tourism in the English-Welsh cross-border Region: An analysis of strategic. *Administrative and Personal Challenges. Regional Studies*, 45 (8), 1139–1155.

Janiszewska, M. (2017). Increase of senior citizens tourism as a consequence of demographic aging changes. *Research Papers of Wrocław University of Economics*, 473, 257–264.

Jepson, D., & Sharpley, R. (2015). More than sense of place? Exploring the emotional dimension of rural tourism experiences. *Journal of Sustainable Tourism*, 23 (8–9), 1157–1178.

Jovanović, S., & Ilić, I. (2016). Infrastructure as important determinant of tourism development in the countries of Southeast Europe. *Ecoforum Journal*, 5 (1), 288–294.

Kalecińska, J. (2013). *Nowe technologie w branży turystycznej*, AWF, Warszawa, Retrieved May 12, 2020, from http://ecorys.pl/zalaczniki/publikacje/75/ECORYS_nt_TURYSTYKA_lores.pdf.

Kamiński, R., & Leśniak, L. (2019). *Inteligentna wieś, Forum Aktywizacji Obszarów Wiejskich.* Centrum Doradztwa Rolniczego, Warszawa.

Karnafel-Wyka, E. (2011). Turystyka wiejska szansą rozwoju obszarów wiejskich. In C. Jastrzębski (ed.), *Turystyka wiejska na drodze do komercjalizacji* (pp. 19–21). Kielce.

Karwat-Woźniak, B. (2013). Zmiany w społeczno-ekonomicznych uwarunkowaniach rozwojowych rolnictwa. *Journal of Agribusiness and Rural Development* 2, 121–131.

Kastenholz, E., Carneiro, M. J., & Marques, C. (2012). Marketing the rural tourism experience. In R. H. Tsiotsou & R. E. Goldsmith (eds.), *Strategic Marketing in Tourism Services* (pp. 247–264). Bingley: Emerald.

Kastenholz, E., Carneiro, M. J., Marques, C. P., & Loureiro, S. M. C. (2018). The dimensions of rural tourism experience: impacts on arousal, memory, and satisfaction. *Journal of Travel and Tourism Marketing*, 35 (2), 189–201.

Keane, M. (2013). Turystyka wiejska i rozwój obszarów wiejskich. In W. H. Briassoulis & J. Van der Straaten (eds.), *Turystyka i środowisko Regionalne, gospodarcze, Kwestie kulturowe i polityczne*, 2nd ed. (pp. 106–123). Berlin: Springer Science & Business Media.

Kemp, S. (2016). *Digital in 2016 Report: We Are Social's. Compendium of Global Digital, Social and Mobile Data, Trends and Statistics.*

Kim, M. J., Lee, C.-K., & Preis, M. W. (2020). The impact of innovation and gratification on authentic experience, subjective well-being, and behavioral intention in tourism virtual reality: The moderating role of technology readiness. *Telematics and Informatics*, 49 (101349).

Klimek, K. (2010). *Turystyka jako czynnik społeczno-gospodarczego rozwoju Szwajcarii.* PTE, Kraków.

Kłoczko-Gajewska, A. (2013). General characteristics of thematic villages in Poland. *Visegrad Journal on Bioeconomy and Sustainable Development*, 2, 60–63.

Kłodziński, M. (2008). Wielofunkcyjny rozwój obszarów wiejskich w Polsce. In M. Drygas & A. Rosner (eds.), *Polska wieś i rolnictwo w Unii Europejskiej. Dylematy i kierunki przemian* (p. 24), IRWIR PAN, Warszawa.

Knight, K. (2011). Responsive web design: What it is and how to use it. *Smashing Magazine*, 12, 234–262.

Kogiso, K. (2015). The success of health tourism in Thailand: A big demand for asian healing arts. *Sports Management and Sports Humanities*, 6, 149–160.

Kołodziejczak, A. (2010). *Modele rolnictwa a zróżnicowanie przestrzenne sposobów gospodarowania w rolnictwie polskim.* Wydawnictwo naukowe UAM, Poznań.

Komninos, N. (2011). Intelligent cities: Variable geometries of spatial intelligence. *Intelligent Buildings International*, 3 (3), 172–188.

Kosmaczewska, J. (2009). Tworzenie markowych produktów turystyki wiejskiej w oparciu o wybrane dyscypliny plastyki ludowej. In P. Palich (ed.), *Marka Wiejskiego Produktu Turystycznego* (pp. 203–208). Gdynia.

Kosmaczewska, J. (2010). Witryna internetowa jako narzędzie kreowania konkurencyjności w agroturystyce. *Acta Scientiarum Polonorum, Oeconomia* 9 (4), 225–232.

Krievina, A., Leimane, I., & Melece, L. (2015). Role of local action groups in addressing regional development and social problems in Latvia. *Research for Rural Development*, 2, 146–153.

Król, K. (2015). Ocena dostępności ofert turystycznych małych gospodarstw rolnych w Internecie. *Problemy Drobnych Gospodarstw Rolnych. Problems of Small Agricultural Holdings*, 4, 5–23.

Król, K. (2016). Globalne zmiany technologiczne i ich wpływ na promocję agroturystyki w Internecie. *Roczniki Naukowe Ekonomii Rolnictwa i Rozwoju Obszarów Wiejskich*, 103 (3), 84–100.

Król, J. (2017). *Idea rolnictwa społecznego, w tym gospodarstw opiekuńczych na świecie* [*The social farming concept, including care farms, around the world*]. CDR w Brwinowie Oddział w Krakowie, Kraków.

Król, J. (2019). *Koncepcja i przykłady gospodarstw opiekuńczych w Polsce i za granicą* [*Concept and examples of care farms in Poland and other countries*]. CDR w Brwinowie Oddział w Krakowie, Kraków.

Król, K., Bedla, D. (2016). Geoinformation in sales of tourist product. *Marketing and Market*, 3, 20–28.

Król, J., & Stępnik, K. (2018). *Gospodarstwo opiekuńcze w rozwoju obszarów wiejskich – wyjazd studyjny.* CDR w Brwinowie Oddział w Krakowie, Kraków. Król, K., & Wojewodzic, T. (2006). Strona internetowa źródłem przewagi konkurencyjnej gospodarstwa agroturystycznego. *Wieś i Doradztwo*, 1–2 (45–46),59–62. Kruczek, Z. (2014). Nowe trendy i innowacje w sektorze atrakcji turystycznych. In Z. Kruczek & W. Banasik (eds.), *Dynamika przemian rynku turystycznego* (pp. 140–142). Wyższa Szkoła Turystyki i Języków Obcych, Warszawa.

Krzepicka, A. (2015). Technologie mobilne a zachowania konsumentów. *Handel wewnętrzny*, 6 (359), 82–89.

Krzyżanowska, K. (2014). Innowacyjność w turystyce wiejskiej – teoria i praktyka. In K. Nuszkiewicz & M. Roman (eds.), *Innowacje w rozwoju turystyki* (p. 34). Go100ądkowo.

Krzyżanowska, K., & Wojtkowski, R. (2012). Rola Internetu w promocji usług agroturystycznych. *Studia Ekonomiczne i Regionalne* 5 (1), 48.

Kuna, J., & Rzuciło, A. (2015). Jak zobaczyć informację, czyli różnorodne funkcje mapy w procesie wymiany informacji. *Folia Bibliologica*, 57, 87–97.

Kurczewski, R. (2016). Ochrona przyrody i turystyka wiejska (Nature protection and rural tourism). In S. Graja-Zwolińska, A. Spychała & K. Kasprzak (eds.), *Turystyka wiejska. Zagadnienia przyrodnicze i kulturowe* [*Rural tourism. Natural and cultural issues*] (pp. 9–16). Poznań.

Kusnadi, F., & Putra, K. (2019). Emerging travel trends: Joy of Missing Out (JOMO) vs iconic landmarks. *Journal Pariwisata Terapan*, 3 (1), 17–33.

Kuss, D. J., Kanjo, E., Rumsey-Crook, M., Kibowski, F., Wang, G. Y., & Sumich, A. (2018). Problematic mobile phone use and addiction across generations: The roles of psychopathological symptoms and smartphone use. *Journal of Technology in Behavioral Science*, 3, 141–149.

Lacher, R. G., Oh Chi-Ok, Jodice, L. W., & Norman, C. C. (2013). The role of heritage and cultural elements in coastal tourism destination preferences. A choice modeling-based analysis. *Journal of Travel Research*, 52 (4), 534–546.

Lanfranchi, M., Giannetto, C., Abbate, T., & Dimitrova, V. (2015). Agriculture and the social farm: expression of the multifunctional model of agriculture as a solution to the economic crisis in rural areas. *Bulgarian Journal of Agricultural Science*, 21, 711–718.

LaPlante, D. A., Nelson, S. E., & Gray, H. M. (2014). Breadth and depth involvement: Understanding Internet gambling involvement and its relationship to gambling problems. *Psychology of Addictive Behaviors*, 28, 396.

LaSalle, D., & Britton, T. A. (2003). *Priceless. Turning Ordinary Products into Extraordinary Experience.* Harvard Business School Press, Boston, MA.

Latkowska, M., & Miernik, M. (2012). Ogrody terapeutyczne miejsca czynnej i bierniej „zielonej terapii". *Czasopismo techniczne*, 30 (109), 243–251.

Law, R., Shanshan, Qi., & Dimitrios, B. (2010). Progress in tourism management: A review of website evaluation in tourism research. *Tourism Management*, 31 (3), 297–313.

Lawendowe Pole (2020). Retrieved July 3, 2020, from https://lawendowepole.pl/.

Leck, C., Evans, N., & Upton, D. (2014). Agriculture—Who cares? An investigation of "care farming" in the UK. *Journal of Rural Studies*, 34, 313–325.

Lee, J. S., Lee, C. K., & Choi, Y. J. (2011). Examining the role of emotional and functional values in festival evaluation. *Journal of Travel Research*, 50 (6), 685–696.

Lee, K.-H., Packer, J. & Scott, N. (2015). Travel lifestyle preferences and destination activity choices of Slow Food members and non-members. *Tourism Management*, 46, 1–10. Lestari Devita, M., Hardianto, D., & Hidayanto Achmad, N. (2014). Analysis of user experience quality on responsive web design from its informative perspective. *International Journal of Software Engineering and its Applications. Science and Engineering Research Support Society*, 8 (5), 53–62.

Li, J., Pearce, P. L., & Low, D. (2018). Media representation of digital – free tourism: A critical discourse analysis. *Tourism Management*, 69, 317–329.

Lombardi, P., Giordano, S., Farouh, H., & Yousef, W. (2012). Modelling the smart city performance. *Innovation – The European Journal of Social Science Research*, 25 (2), 137–149.

Lowry, L. & Misoon, L. (2016). CittaSlow, slow cities, slow food: searching for a model for the development of slow tourism. *Travel and Tourism Research Association: Advancing Tourism Research Globally*, 40.

Łukasiewicz, K. (2017). Standard jakości w turystyce wiejskiej i agroturystyce w Polsce i na Ukrainie w opinii potencjalnych klientów [Quality standard in rural tourism and agritourism in Poland and Ukraine in the opinion of potential customers].

Roczniki Naukowe Stowarzyszenia Ekonomistów Rolnictwa i Agrobiznesu, 29 (6), 147–152.

MacCannell, D. (2002). *Turysta. Nowa teoria klasy próżniaczej*, Wydawnictwo Muza, Warszawa.

Machnik, A. (2010). Wybrane modele ekoturystyki na obszarach przyrodniczo cennych w Polsce. *Geoturystyka*, 3–4 (22–23), 19–26.

MacLaren, A., O'Gorman, K., Stringfellow, L. & Maclean, M. (2013). Conceptualizing taste: food, culture and celebrities, *Tourism Management*, 37 (1), 77–85, doi:10.1016/j.tourman.2012.12.016.

Majewski, J. (2015). Innowacyjność wiejskich produktów turystycznych. In W. Kamińska (ed.), *Innowacyjność w turystyce wiejskiej a nowe możliwości zatrudnienia na obszarach wiejskich* (pp. 9–19). Warszawa.

Manintveld, K. (2014). Gospodarstwa opiekuńcze w Holandii [Dutch care farms]. In: *Gospodarstwa opiekuńcze w Borach Tucholskich a doświadczenia holenderskie [Care farms in Tuchola Forest vs. the Dutch experience]*. Kujawsko Pomorski Ośrodek Doradztwa Rolniczego, Minikowo.

Marciszewska, B. (2010). *Produkt turystyczny a ekonomia doświadczeń*. Warszawa: C. H. Beck.

Marcotte, E. (2010). Responsive web design. *A List Apart*, no. 4, New York. Retrieved April 24, 2020, from http://alistapart.com/article/ responsive-web-design

Marek, R. (2015). *Determinanty rozwoju polskich biur podróży w Internecie*, Wydawnictwo Promotor, Warszawa.

Markeson, B., Deller, S. (2012). Growth of rural US non-farm proprietors with a focus on amenities. *Review of Urban and Regional Development Studies*, 24 (3), 83–105.

Martinez, J. M. G., Martin, J. M. M., Fernandez, J. A. S., & Mogorron-Guerrerod, H. (2019). An analysis of the stability of rural tourism as a desired condition for sustainable tourism. *Journal of Business Research*, 100, 165–174.

Mastronardi, L., Giaccio, V., Giannelli, A., & Scardera, A. (2015). Is agritourism eco-friendly? A comparison between agritourisms and other farms in Italy using farm accountancy data network dataset. *SpringerPlus*, 4 (1), 1–12.

Mathew, V. (2009). Sustainable tourism: A case of destination competitiveness in South Asia. *South Asian Journal of Tourism and Heritage*, 2 (1), 83–89.

Matuszczak, A. (2013). *Zróżnicowanie rozwoju rolnictwa w regionach Unii Europejskiej w aspekcie jego zrównoważenia*. PWN, Warszawa.

Matysiak, I., & Michalska, S. (2016). Social farming: A new model of dealing with ageing in rural areas in Poland?. *Sociologia e Politiche Sociali*, 19 (3), 65–82.

McCann, P., & Ortega-Argilés, R. (2013). Smart specialization, regional growth and applications to European Union cohesion policy. *Regional Studies*, 49 (8), 1291–1302.

McGladdery, C.A., & Lubbe, B. A. (2017). Rethinking educational tourism: proposing a new model and future directions. *Tourism Review*, 72 (3), 319–329.

McGranahan, D. A., Wojan, T. R., & Lambert, D. M. (2011). The rural growth trifecta: outdoor amenities, creative class and entrepreneurial context. *Journal of Economic Geography*, 11 (3), 529.

Mempel-Śnieżyk, A. (2013). Koncepcje rozwoju regionalnego ze szczególnym uwzględnieniem klastrów i inteligentnych specjalizacji. *Biblioteka Regionalisty*, 13, 1107–1127.

Merski, J. (2002). *Informacja turystyczna w Polsce*. WSE, Warszawa.

Mikos von Rohrscheidt, A. (2011). Sylabus miejsc, czyli jak atrakcyjnie pokazać miasto współczesnemu turyście kulturowemu. In Z. Kruczek (ed.), *Piloci i przewodnicy na styku kultur*. Kraków: Proksenia, 97–120.

Ministerstwo Rozwoju (2020). RetrievedJune 26, 2020, from https://www.gov.pl/web/rozwoj/bonturystyczny.

Mohorovičić, S. (2013). *Implementing responsive web design for enhanced web presence*, 36th International Convention on IEEE Information & Communication Technology Electronics & Microelectronics (MIPRO) (pp. 1206–1210).

Moric, I. (2013). Clusters as a factor of rural tourism competitiveness: Montenegro experiences. *Business Systems Research*, 4 (2), 94–107.

Mroczkowska, D. (2008). Czas wolny jako kategoria społecznie i kulturowo zmienna. Przeobrażenia w czasowej organizacji oraz doświadczaniu czasu wolnego. In W. Muszyński (ed.), *Czas ukoi nas? Jakość życia i czas wolny we współczesnym społeczeństwie* (pp. 89–102). Wydawnictwo Adam Marszałek, Toruń.

Muganda, M., Sirima, A., & Ezra, P. M. (2013). The role of local communities in tourism development: Grassroots perspectives from Tanzania. *Journal of Human Ecology*, 41 (1), 53–66.

Naldi, L., Nilsson, P., Westlund, H., & Wixe, F. (2015). What is smart rural development? *Journal of Rural Studies*, 40, 90–101.

Nam, T., & Pardo, T. A. (2011). *Conceptualizing smart city with dimensions of technology, people and institutions*. Proceedings of the 12th Annual International Conference on Digital Government Research.

Natda, K. (2013). Responsive web design. *Eduvantage, An International Refereed Journal of Business, Accounting, Information Technology & Law*, 1 (1).

Neirotti, P., De Marco, A., Cagliano, A., Mangano, G., & Scorrano, F. (2014). Current trends in Smart Cities initiatives: Some stylised facts. *Cities*, 38, 25–36.

Nicula, V., & Popsa, R. E. (2014). Business tourism market developments. *Procedia Economics and Finance*, 16 (2014), 703–712.

Niedziółka A., Roman, M. (2017). *Agroturystyka jako forma przedsiębiorczości na obszarach wiejskich*. Wydawnictwo SGGW, Warszawa.

Niezgoda, A. (2013). Rola doświadczenia w zachowaniach konsumenta na rynku turystycznym. Koncepcja ekonomii doświadczeń i marketingu doznań, *Folia Turistica*, 28, 91–106.

Niezgoda, A., & Markiewicz, E. (2011). Changes in tourism supply and demand caused by globalization, w: modern word economy. micro – and macroeconomic issues. *Zeszyty Naukowe Uniwersytetu Ekonomicznego w Poznaniu*, 219, 227–242.

Niezgoda, A., & Markiewicz, E. (2014). Slow tourism – idea, uwarunkowania i perspektywy rozwoju. *Rozprawy Naukowe Akademii Wychowania Fizycznego we Wrocławiu*, 46, 82–90.

Ninemeier, J. D., & Perdue, J. (2008). *Discovering Hospitality and Tourism: The World's Greatest Industry*, 2nd ed. Pearson, Upper Saddle River, NJ.

Nowak, S., Ulfik, A., & Herbuś, A. (2012). Zrównoważona turystyka podstawą rozwoju regionalnego i przemian społeczno-gospodarczych. *Logistyka*, 3, 8–87.

Odpoczywaj na wsi (2020). Retrieved July 8, 2020, from https://odpoczywajnawsi.pl/wydawnictwa-turystyczne/.

OECD (2018a). *OECD Tourism Trends and Policies 2018*. OECD Publishing, Paris.

OECD (2018b). *OECD Rural Policy Reviews: Poland 2018*. OECD Publishing, Paris.

OECD (2020a). Key messages: responding to the impact of coronavirus (COVID-19) on the tourism economy. Retrieved April 5, 2020, from https://www.oecd.org/coronavirus/p olicy-responses/tourism-policy-responses-to-the-coronavirus-covid-19-6466aa20/.

OECD (2020b). *Przegląd Polityk Rozwoju Obszarów Wiejskich-Polska (2018)*. Ministerstwo Inwestycji i Rozwoju, Warszawa. Olejniczuk-Merta, A. (2010). Wpływ nowoczesnych technologii na zachowania konsumentów, *Problemy Zarządzania, Finansów i Marketingu*, 15, 137–145.

Ooi, C.-S. (2003). *Crafting tourism experiences: Managing the attention product*, paper presented at the 12th Nordic Symposium on Tourism and Hospitality Research, Stavanger, 2–5 October. Retrieved July 10, 2016, from http://fama2.us.es:8080/ turismo/turismonet1/economia%20del%20turismo/marketing%20turistico/crafting% 20tourism%20experiences%20attenting%20experience%20product.pdf.

Opinia Europejskiego Komitetu Ekonomiczno-Społecznego „Rolnictwo społeczne: zielone usługi terapeutyczno-opiekuńcze oraz polityka społeczna i zdrowotna". *Official Journal of the European Union*, 2013/C 44/07.

Oulasvirta, A., Rattenbury, T., Ma, L., & Raita, E. (2012). Habits make smartphone use more pervasive. *Personal Ubiquitous Computing*, 16, 105–114.

Panasiuk, A. (2011). *Ekonomika turystyki i rekreacji*. Wydawnictwo Naukowe PWN, Warszawa.

Panasiuk, A. (2014). *Rynek turystyczny. Studium strukturalne*. Difin, Warszawa.

Panasiuk, A. (2019). *Rynek turystyczny, Struktura, procesy, tendencje*. Difin, Warszawa.

Pawlak, K., & Poczta, W. (2010). Potencjał polskiego rolnictwa pięć lat po akcesji Polski do UE jako przesłanka jego konkurencyjności. *Wieś i Rolnictwo*, 1, 21–47.

Pawlicz, A. (2012). *E-turystyka. Ekonomiczne problemy implementacji technologii cyfrowych w sektorze turystycznym*. PWN, Warszawa.

Penn, M. (2007). *Microtrends: The Small Forces Behind Tomorrow's Big Changes*. Twelve, New York.

Penn, M., & Fineman, M. (2018). *Microtrends Squared: The New Small Forces Driving Today's Big Disruptions*. Simon & Schuster, New York.

Perechuda, K., & Hołodnik, D. (2012). Nowoczesny model gospodarstwa agroturystycznego oparty na wiedzy. In M. Morawski (ed.), *Zarządzanie wiedzą w turystyce a efektywność gospodarki turystycznej*. Wrocław.

Perenco, J., & Rosa, G. (2011). *Zachowania nabywców*. Uniwersytet Szczeciński, Szczecin.

Peterson, M. (2015). Evaluating mapping APIs. In J. Brus, A. Vondrakova, & V. Vozenilek (eds.), *Modern Trends in Cartography* (pp. 183–197), Springer International.

Pickering, C. (2011). Zmiany w popycie na turystykę ze zmianami klimatu: studium przypadku dotyczące wzorców odwiedzin w sześciu ośrodkach narciarskich w Australii. *Journal of Sustainable Tourism*, 19 (6), 767–781.

Pike, S., & Page, S. J. (2014). Destination marketing organizations and destination marketing. A narrative analysis of the literature. *Tourism Management*, 41, 202–227.

Pine, B. J., & Gilmore, J. H. (1998). Welcome to the experience economy, *Harvard Business Review*, 4.

Pine, B. J., & Gilmore, J. H. (2013). The experience economy: part, present and future. Handbook on the experience economy. In J. Sundbo & F. Sorensen (eds.), *Handbook on the Experience Economy* (pp. 21–44), Edward Elgar Publishing, Cheltenham.

Poczta, J. (2013). Wiejska turystyka kulturowa zgodna z paradygmatem zrównoważonego rozwoju. *Turystyka Kulturowa*, 4, 20–35.

Pretty, J., Toulmin, C., & Williams, S. (2011). Sustainable intensification in African agriculture. *International Journal of Agricultural Sustainability*, 1, 5–24.

Przezbórska-Skobiej, L. (2015). *Uwarunkowania rozwoju turystyki wiejskiej w Polsce. Analiza regionalna, subregionalna, lokalna.* Wydawnictwo Uniwersytetu Przyrodniczego w Poznaniu.

Przybylski, A. K., Murayama, K., DeHaan, C. R., & Gladwell, V. (2013). Motivational, emotional, and behavioural correlates of fear of missing out. *Computers in Human Behavior*, 29, 1841–1848.

Quadri-Felitti, D. & Fiore, A. M. (2015). Experience economy constructs as a framework for understanding wine tourism, *Journal of Vacation Marketing*, 18 (I),3–15.

Ramjit, M. (2015). Sustainable regional development through rural tourism in Jammu and Kashmir. *African Journal of Hospitality, Tourism and Leisure*, 4 (2), 1–16.

Renski, H. C. (2014). The influence of industry mix on regional new firm entry. *Regional Studies*, 48 (8), 1353–1370.

Ritchie, J. R. B., & Crouch, G. I. (2010). A model of destination competitiveness/sustainability: Brazilian perspectives. *Revista de Administracao Publica*, 44 (5), 1049–1066.

Ritchie, J. R. B., Tung, V. W. S., & Ritchie, R. J. B. (2011). Tourism experience management research Emergence, evolution and future directions. *International Journal of Contemporary Hospitality Management*, 23 (4), 419–438. Rizov, M. (2005). Rural development under the European CAP: The role of diversity. *The Social Science Journal*, 42, 621–628.

Roberts, J. A., & David, M. E. (2016). My life has become a major distraction from my cell phone: Partner phubbing and relationship satisfaction among romantic partners. *Computers in Human Behavior*, 54, 134–141.

Roman, M. (2009). Okopska tourism organization as an example of cluster initiative within agritourism. *Infrastructure and Ecology of Rural Areas*, 6, 187–195.

Roman, M. (2015a). Agritourism farms owners' competence in running their economic activities. *Polish Journal of Management Studies* 11 (1), 136–146.

Roman, M. (2015b). Innowacyjność elementem konkurencyjności w turystyce. *Zeszyty Naukowe Małopolskiej Wyższej Szkoły Ekonomicznej w Tarnowie*, 1 (26), 111.

Roman, M. (2018). *Innowacyjność agroturystyki jako czynnik poprawy konkurencyjności turystycznej makroregionu Polski Wschodniej.* Wydawnictwo SGGW.

Roman, M., & Wojcieszak, M. (2018). Znaczenie social farmingu w wybranych krajach Unii Europejskiej jako przykład przedsiębiorczości w turystyce na obszarach wiejskich. *Prace Naukowe Uniwersytetu Ekonomicznego we Wrocławiu*, 535, 161–172.

Roman, M., Roman, M., & Prus, P. (2020). Innovations in agritourism: Evidence from a region in Poland. *Sustainability*, 2 (4858), 1–21.

Rosner, A. (2012). *Zmiany rozkładu przestrzennego zaludnienia obszarów wiejskich. Wiejskie obszary zmniejszające zaludnienie i koncentrujące ludność wiejską.* Wydawnictwo IRWIR PAN, Warszawa.

Rosner, A., & Stanny, M. (2016). *Monitoring rozwoju obszarów wiejskich, Etap* II. Europejski Fundusz Rozwoju Wsi Polskiej. Instytut Rozwoju Wsi i Rolnictwa PAN, Warszawa.

Roxana, D. M. (2012). Considerations about ecotourism and nature-based tourism – realities and perspectives. *International Journal of Academic Research in Economics and Management Sciences*, 1 (5), 215–221.

Rudnicki, L. (2012). *Zachowania konsumentów na rynku.* PWE, Warszawa.

Rudnicki, R. (2010). *Zróżnicowanie przestrzenne wykorzystania funduszy Unii Europejskiej przez gospodarstwa rolne w Polsce.* Bogucki Wyd. Naukowe, Poznań.

Rutka, M. (2017). Cyfryzacja jako szansa na uniknięcie Europy dwóch prędkości - problem wykluczenia cyfrowego w Polsce i województwie pomorskim. *Media - Biznes - Kultura. Dziennikarstwo i komunikacja społeczna* 2, 197–212.

Sadowski, A., & Wojcieszak, M. M. (2019). Geographic differentiation of agritourism activities in Poland vs. cultural and natural attractiveness of destinations at district level. *PLOSE ONE.* https://doi.org/10.1371/journal.pone.0222576.

Santos, M. L., Alencar, D. G., Souza, A. A., & Gândara, J. M. G. (2018). Consumer trends in tourism: the offer in Paraná (Brazil) for experiences seeker. *Investigaciones Turisticas*, 16, 143–164.

Schiff, A., & Becken, S. (2011). Popyt na elastyczność w turystyce w Nowej Zelandii. *Tourism Management*, 32 (3), 564–575.

Scott, H., & Woods, H. C. (2018). Fear of missing out and sleep: Cognitive behavioural factors in adolescents' nighttime social media use. *Journal of Adolescence*, 68, 61–65.

Scott, D., Gössling, S., Hall, C. M. (2012). International tourism and climate change. *Climate Change*, 3 (3), 213–232. Scuderi, A., Timpanaro, G., & Cacciola, S. (2014). Development policies for social farming in the EU2020 strategy, *Quality Access to Success* 15, 7682.

Seken, A., Duissembayev, A., Tleubayeva, A., Akimov, Z., Konurbaeva, Z., & Suieubayeva, S. (2019). Modern potential of rural tourism development in Kazakhstan. *Journal of Environmental Management and Tourism*, 10 (6),1211–1223.

Senni, S. (2013). Uno sguardo europeo sull'agricoltura sociale. Il parere del Comitato Economico e Sociale Europeo. *Agriregionieuropa*, 39.

Seo, Y. & Buchanan-Oliver, M. (2015). Luxury branding: the industry, trends, and future conceptualisations. *Asia Pacific Journal of Marketing and Logistics*, 27 (1), 82–98.

Shu-Yi, Chi. (2019). Some rural attractions which attract tourists. *Asian Journal of Agriculture and Rural Development*, 9 (1), 99–110.

Sikora, J. (2012). *Agroturystyka. Przedsiębiorczość na obszarach wiejskich.* Wydawnictwo C.H. Beck, Warszwa.

Sikora, K. (2010). The application of a clustering in the development of Pomeranian tourist sector – an example of the Vistula Spit. Master thesis, Faculty of Management and Economics, Gdansk University of Technology.

Sikorska-Wolak, I., & Zawadka, J. (2015). Innovative solutions in rural tourism. *Stowarzyszenie Ekonomistów Rolnictwa i Agrobiznesu Roczniki Naukowe*, 18 (4), 207–212.

Sikorska-Wolak, I., Krzyżanowska, K., & Parzonko, A. J. (2014). *Doradztwo w zmieniającej się sytuacji społeczno-ekonomicznej obszarów wiejskich.* SGGW w Warszawie, Warszawa.

Sioło Budy (2020). Retrieved June 30, 2020, from http://siolo-budy.pl/.

Skowronek, I. (2011). Marketing doświadczeń jako wyznacznik wizerunku i wartości przedsiębiorstwa. *Zeszyty Naukowe Uniwersytetu Szczecińskiego*, 46, 209–224.

Smart Rural Areas. Retrieved April 21, 2020, from https://www.smartrural21.eu/.

Smart village. Policy Initiatives & Strategic Approaches. Retrieved April 21, 2020, from https://enrd.ec.europa.eu/smart-and-competitive-rural-areas/smart-villages/smart-villages-portal/eu-policy-initiatives-strategic-approaches_en.

Smart Villages (2019). Retrieved April 28, 2020, from https://digitevent-images.s3.amazonaws.com/5c0e6198801d2065233ff996-registrationfiletexteditor-1551115459927-smart-villages-briefing-note.pdf.

Smart Villages (2020). Retrieved April 28, 2020, from http://www.pilotproject-sma rtvillages.eu/.

SmartRural (2020). Retrieved April 28, 2020, from https://www.smartrural21.eu/.

Solary. Skierbiszow. Retrieved April 21, 2020, from http://www.solary-skierbieszow.eu/galeria/?f=20190917#content.

Souza, L. H., Kastenholz, E., Barbosa, M. L. A., & Carvalho, M. S. S. C. (2019). Tourist experience, perceived authenticity, place attachment and loyalty when staying in a peer-to-peer accommodation. *International Journal of Tourism Cities*, 6 (1), 27–52.

Spencer, D. M., & Nsiah, C. (2013). The economic consequences of community support for tourism: A case study of a heritage fish hatchery. *Tourism Management*, 34, 221–230.

Stanny, M. (2011). Wieś jako przedmiot badań demograficznych. In M. Halamska (ed.), *Wieś jako przedmiot badań naukowych na początku XXI wieku* (pp. 1–16). EUROREG, Wydawnictwo Naukowe Scholar, Warszawa.

Stanny, M. (2013). *Przestrzenne zróżnicowanie rozwoju obszarów wiejskich w Polsce*. IRWIR PAN, Warszawa.

Stanny, M., Śliwowska, Z., & Hoffmann, R. (2016). Miasto - wieś: Dychotomia czy continuum? Rozważania osadzone w trzech kontekstach: socjologicznym, ekono-micznym i geograficznym. *Zeszyty Naukowe Wydziału Nauk Ekonomicznych Poli-techniki Koszalińskiej*, 1 (20), 275–276.

Stasiak, A. (2013). Produkt turystyczny w gospodarce doświadczeń. *Turyzm*, 23 (1), 27–35.

Stasiak, A. (2015). Triada doświadczeń turystycznych i efekt "wow!" podstawą kreo-wania nowoczesnej oferty turystycznej. *Prace Naukowe Uniwersytetu Ekono-micznego we Wrocławiu*, 379, 332–347.

Štefko, R., Kiráľ︎ováb, A., & Mudrík, M. (2014). Strategic marketing communication in pilgrimage tourism. *Procedia – Social and Behavioral Sciences*, 175 (2015), 423–430.

Stoate, C. *et al.* (2001). Ecological impacts of arable intensification in Europe. *Journal of Environmental Management*, 63 (4), 337–365.

Strzelczyk, J. (2007). *Mity, podania i wierzenia dawnych Słowian*. Rebis.

Subocz, E. (2019). Rolnictwo społeczne jako nowa funkcja obszarów wiejskich z per-spektywy europejskiej i polskiej [Social farming as a new function of rural areas from the European and Polish perspective]. *Studia Obszarów Wiejskich*, 54, 69–79.

Szczech-Pietkiewicz, E. (2015). Smart city – próba definicji i pomiaru. *Prace Naukowe Uniwersytetu Ekonomicznego we Wrocławiu [Research Papers of Wrocław Uni-versity of Economics]*, 391, 71–82.

Szubin (2020). Retrieved June 17, 2020, from http://szubin.pl/inwestycja/centrum-astr onomiczno-kulturalno-dydaktycznego-w-niedzwiadach.html.

Tan, S.-K., Kung, S.-F., & Luh, D.-B. (2013). A model of 'creative experience' in creative tourism. *Annals of Tourism Research*, 41, 153–174. doi:10.1016/j.annals.2012.12.002..

Tarssanen, S. & Kylänen, M. (2005). *A Theoretical Model for Producing Experiences – a Touristic Perspective*. Rovaniemi, Finland: Lapland Centre of Expertise for the Experience Industry.

Teräs, J., Dubois, A., Sörvik, J., & Pertoldi, M. (2015). *Implementing Smart Speciali-sation in Sparsely Populated Areas*. S3 Working Papers Series, 10.

The Smart Villages Initiative: Findings 2014–2017. Cambridge: CMEDT – Smart Villages Initiative, Trinity College. Retrieved May 20, 2020, from http://e4sv.org/wp -content/uploads/2017/06/Findings-2014–2017quality.compressed.pdf.

Thieme, M., & Birkigt, K.-U. (2006). Dorfentwicklung mit wirtschaftlichen Leitbildern zur Erhöhung der lokalen Wertschöpfung (Themendörfef). *Infodienst der Sachsischen Landesanstalt für Landwirtschaft*, 2, 15–21.

Tittonell, P. (2014). Ecological intensification of agriculture – sustainable by nature. *Current Opinion in Environmental Sustainability*, 8, 53–61.

Toby, A. (2014). *Modele gospodarstw opiekuńczych według grup docelowych. Gospodarstwa opiekuńcze w Borach Tucholskich a doświadczenia holenderskie.* Wydawnictwo Kujawsko-Pomorskiego Ośrodka Doradztwa Rolniczego w Minikowie, Minikowo.

Tranchenko, L. V. (2015). Tourism as a priority direction for rural economy development. *Actual Problems of Economics* 171 (9), 162–168.

Tripadvisor (2020a). 10 najlepszych atrakcji w Polsce [Top 10 attractions in Poland]. Retrieved July 3, 2020, from https://pl.tripadvisor.com/Attractions-g274723-Activities-Poland.html.

Tripadvisor (2020b). Retrieved June 10, 2020, from https://tripadvisor.mediaroom.com/pl-about-us.

Turþeková, N., Svetlanská, T., Kollár, B., Záhorský, T. (2015). Agri-environmental performance of EU Member states. *Agris on-line Papers in Economics and Informatics*, 7 (4), 199–208.

UNWTO (2019). *International Tourism Highlights*. Retrieved April 5, 2020, from https://www.e-unwto.org/doi/pdf/10.18111/9789284421152.

UNWTO (2020). *Tourism in SIDS – the Challenge of Sustaining Livelihoods in Times of COVID-19.* UNWTO Briefing Note – Tourism and COVID-19, Issue 2. https://www.e-unwto.org/doi/pdf/10.18111/9789284421916.

Urry, J. (2007). *Spojrzenie turysty.* Wydawnictwo Naukowe PWN, Warszawa.

Van Gevelt, T., Holmes, J. (2015). A vision for smart villages. *Smart Villages Briefing*, 5.

Visvizi, A., Lytras, M. D. (2018). Rescaling and refocusing smart cities research: From mega cities to smart villages. *Journal of Science and Technology Policy Management*, 9 (2), 126–133.

Vrsaljko, A, Turalija, A., Grgić, I., & Zrakić, M. (2017). Rolnictwo ekologiczne jako warunku koniecznego dla rozwoju eko – agroturystyki na wyspach – Studium przypadku wyspy Korcula [Ekološka poljoprivreda kao pretpostavka razvoja ekoagroturizma na otocima – Studija slučaja otoka Korčula]. *Journal of Central European Agriculture*, 18, 733–748.

Wang, D., Park, S., & Fesenmaier, D. R. (2012). Rola smartfonów w pośredniczeniu w turystyc e. *Journal of Travel Research*, 51 (4), 371–387.

Weaver, D. (2010). Lokalne etapy turystyki i ich implikacje dla zrównoważonego rozwoju. *Journal of Sustainable Tourism*, 18 (1), 43–60.

Weinberger, M. F., Zavisca, J. R., & Silva, J. M. (2017). Consuming for an imagined future: middle-class consumer lifestyle and exploratory experiences in the transition to adulthood. *Journal of Consumer Research*, 44 (2), 332–360. doi:10.1093/jcr/ucx045.

Wesołowska, M. (2011). Wiejskie obszary peryferyjne – uwarunkowania i czynniki aktywizacj i. *Studia Obszarów Wiejskich Komisja Obszarów Wiejskich PTG, IGiPZ PAN*, 26.

Wieczorkiewicz, A. (2008). *Apetyt turysty. O doświadczeniu świata w podróży.* Wydawnictwo Universitas, Kraków.

Wilkin, J. (2010). *Wielofunkcyjność rolnictwa. Kierunki badań, podstawy metodologiczne i implikacje praktyczne.* IRWIR PAN, Warszawa.

Wilkin, J. (2011). Jak zapewnić rozwój wsi w warunkach zmniejszającej się roli rolnictwa?. In M. Halamska (ed.), *Wieś jako przedmiot badań naukowych*. EUROREG, Wydawnictwo Naukowe Scholar, Warszawa.

Wioska ale internetowa. Retrieved April 21, 2020, from https://dziennikzachodni.pl/wioska-ale-internetowa/ar/78174.

Wojciechowska, J. (2009). *Procesy i uwarunkowania rozwoju agroturystyki w Polsce*. Łódź.

Wojcieszak, M. (2017). *Uwarunkowania rozwoju turystyki przyrodniczej na obszarach metropolitalnych i jego efekty ekonomiczne*. Wydawnictwo Uniwersytetu Przyrodniczego w Poznaniu, Poznań.

Wojcieszak, M., & Wojcieszak, A. (2018). Uwarunkowania funkcjonowania gospodarstw opiekuńczych na terenach wiejskich, *Zagadnienia Doradztwa Rolniczego, Centrum Doradztwa Rolniczego w Brwinowie Oddział w Poznaniu*, 3, 20–30.

Wójcik, M., 2012: *Geografia wsi w Polsce. Studium zmiany podstaw teoretyczno-metodologicznych*. Wydawnictwo Uniwersytetu Łódzkiego, Łódź.

Woods, M. (2009). Rural geography: blurring boundaries and making connections. *Progress in Human Geography*, 33.

World Tourism Organization (UNWTO)2019. Retrieved March 27, 2020, from https://www.e-unwto.org/doi/pdf/10.18111/9789284421152.

World Travel & Tourism Council (WTTC)2019*WORLD Report 2019*. Retrieved March 25, 2020 from https://www.wttc.org/-/media/files/reports/economic-impact-research/regions-2019/world2019.pdf..

Woś, A. (1998). Rolnictwo zrównoważone. In A. Woś (ed.), *Encyklopedia agrobiznesu. Fundacja Innowacja* (p. 735). Warszawa.

WOT (2020). Retrieved June 29, 2020, from https://www.wot.org.pl/2020/06/korona wirus-informacje-dla-branzy-turystycznej/.

Woźniak, M., & Batyk, I. M. (2017). Szlaki kulinarne jako forma konkurencyjności oferty turystycznej. *Zeszyty Naukowe Uczelni Vistula*, 54 (3), 98–111.

WTTC (2019). *Report 2019*. Retrieved April 23, 2020, from https://www.wttc.org/-/media/files/reports/economic-impact-research/regions-2019/world2019.pdf.

Yeoman, I. (2012). *2050 – Tomorrow's Tourism*. Bristol: Channel View Publications.

Yeoman, I. (2016). *The future tourist: fluid and simple identities*. Victoria University of Wellington, 12th TourMIS users' workshop presentation via video conference. Retrieved May 20, 2019, from https://vimeo.com/181103735/ad143522da;www.modul.ac.at/article/view/12th-annual-tourmis-users-workshop-andinternationalseminar-on-co nsumer-trends-tourism/.

Yeoman, I. S., & McMahon-Beattie, U. (2019). The experience economy: micro trends, *Journal of Tourism Futures*, 5 (2), 114–119.

Yeoman, I., McMahon-Beattie, U., Backer, E., Robertson, M., & Smith, K. (2014). *The Future of Events & Festivals*. Routledge, New York.

Zagroda Edukacyjna (2020). RetrievedJune 9, 2020, from https://www.zagroda-eduka cyjna.pl/.

Zavratnik, V., Kos, A., & Stojmenova Duh, E. (2018). Smart villages: Comprehensive review of initiatives and practices. *Sustainability*, 10 (8), 2559–2573.

Zawadka, J. (2010). *Ekonomiczno-społeczne determinanty rozwoju agroturystyki na Lubelszczyźnie (na przykładzie wybranych gmin wiejskich)*. Wydawnictwo SGGW, Warszawa.

Zegar, J. (2012). *Współczesne wyzwania rolnictwa*. Wydawnictwo PWN, Warszawa.

Żemła, M. (2017). The Role of Experience Economy in the Postmodern Interpretation of Market Trends in Tourism and in the Creation of Contemporary Tourism Products. *Studies of the Industrial Geography Commission of the Polish Geographical Society*, 31 (3), 7–16.

Złoty Jar (2020). Retrieved July 1, 2020, from http://zlotyjar.pl/,dostęp 1.07.2020r.

Zrakić, M., Jež Rogelj, M., & Grgić, I. (2017). Organic agricultural production on family farms in Croatia. *Agroecology and Sustainable Food Systems*, 41 (6), 635–649.

Zwolińska-Ligaj, M., Guzal-Dec, D., & Adamowicz, M. (2018). Koncepcja inteligentnego rozwoju lokalnych jednostek terytorialnych na obszarach wiejskich regionu peryferyjnego na przykładzie województwa lubelskiego. *Wieś i Rolnictwo*, 2 (179), 247–279.

Index